T0155888

Lecture Notes in Computer Science 13848

The series Lecture Notes in Computer Science (LNCS), including its subseries Lecture Notes in Artificial Intelligence (LNAI) and Lecture Notes in Bioinformatics (LNBI), has established itself as a medium for the publication of new developments in computer science and information technology research, teaching, and education.

LNCS enjoys close cooperation with the computer science R & D community, the series counts many renowned academics among its volume editors and paper authors, and collaborates with prestigious societies. Its mission is to serve this international community by providing an invaluable service, mainly focused on the publication of conference and workshop proceedings and postproceedings. LNCS commenced publication in 1973.

Yinqiang Zheng · Hacer Yalim Keleş ·
Piotr Koniusz
Editors

Computer Vision – ACCV 2022 Workshops

16th Asian Conference on Computer Vision
Macao, China, December 4–8, 2022
Revised Selected Papers

Editors
Yinqiang Zheng
University of Tokyo
Tokyo, Japan

Hacer Yalim Keleş
Hacettepe University
Ankara, Turkey

Piotr Koniusz
Data61/CSIRO
Canberra, ACT, Australia

ISSN 0302-9743 ISSN 1611-3349 (electronic)
Lecture Notes in Computer Science
ISBN 978-3-031-27065-9 ISBN 978-3-031-27066-6 (eBook)
https://doi.org/10.1007/978-3-031-27066-6

This Springer imprint is published by the registered company Springer Nature Switzerland AG
The registered company address is: Gewerbestrasse 11, 6330 Cham, Switzerland

Preface

Workshops of the 16th Asian Conference on Computer Vision (ACCV) 2022 were held in a hybrid mode in Macau SAR, China during December 4–8, 2022. We present the workshop proceedings from eight successfully held workshops on a variety of selected topics which enjoyed many attendees. The number of accepted workshops (10 in total) was larger than in the previous conferences as ACCV 2022 was held in a hybrid mode to mitigate challenges posed by the COVID-19 pandemic by welcoming virtual and in-person attendees. Indeed, this facilitated numerous paper submissions and the discussion on a variety of timely topics. 25 papers were accepted given 40 submissions in total following the double-blind review process with 3.3 reviewers (on average) per paper. The accepted workshops were of excellent quality.

The list of eight workshops and one grand challenge held in conjunction with ACCV 2022 is as follows:

- Learning with Limited Data for Face Analysis
- Adversarial Machine Learning Towards Advanced Vision Systems
- Multi-view Learning and Its Applications in Computer Vision
- Computer Vision Technology in Electric Power System
- Computer Vision for Medical Computing
- Machine Learning and Computing for Visual Semantic Analysis
- Vision Transformers: Theory and Applications
- Challenges of Fine-Grained Image Analysis
- Deep Learning-Based Small Object Detection from Images and Videos

The workshops primarily covered topics in high-level computer vision, as well as modern applications of computer vision. All workshops and the grand challenge had invited speaker sessions, where leading researchers gave insightful talks. Six workshops accepted papers for these proceedings. Two workshops hosted challenges and presented their results in their sessions, and some even arranged cash awards for the best papers. We thank all the workshop organizers for their efforts to prepare and host these successful events, especially in the challenging environment of hybrid mode hosting. We hope they continue to hold their workshops in future ACCVs. We thank the reviewers for their valuable feedback for paper authors. We also thank the publication chairs for collating the papers presented in the workshops to prepare the workshop proceedings.

January 2023

Yinqiang Zheng
Hacer Yalim Keleş
Piotr Koniusz

Organization

Learning with Limited Data for Face Analysis

Ping Liu	A*STAR, Singapore
Yuewei Lin	Brookhaven National Laboratory, USA
Zibo Meng	OPPO US Research Center, USA
Yawei Luo	Zhejiang University, China
Shangfei Wang	University of Science and Technology of China, China
Joey Tianyi Zhou	A*STAR, Singapore

Adversarial Machine Learning Towards Advanced Vision Systems

Minhui Xue	Data61/CSIRO, Australia
Huaming Chen	University of Sydney, Australia

Multi-view Learning and Its Applications in Computer Vision

Xinwang Liu	National University of Defense Technology, China
Chang Tang	China University of Geosciences, Wuhan, China
Marius Kloft	TU Kaiserslautern, Germany
Xiaojun Chang	University of Technology Sydney, Australia
Tongliang Liu	University of Sydney, Australia

Computer Vision Technology in Electric Power System

Wenqing Zhao	North China Electric Power University, Baoding, China
Yongjie Zhai	North China Electric Power University, Baoding, China
Zhenbing Zhao	North China Electric Power University, Baoding, China

Computer Vision for Medical Computing

Imran Razzak	University of New South Wales, Australia
Xuequan Lu	Deakin University, Australia
Yuejie Zhang	Fudan University, China

Machine Learning and Computing for Visual Semantic Analysis

Xian-Hua Han	Yamaguchi University, Japan
YongQing Sun	NTT, Japan
Rahul Kumar Jain	Ritsumeikan University, Japan

Vision Transformers: Theory and Applications

Fahad Shahbaz Khan	MBZUAI, United Arab Emirates
Salman Khan	MBZUAI, United Arab Emirates
Rao Muhammad Answer	MBZUAI, United Arab Emirates
Hisham Cholakkal	MBZUAI, United Arab Emirates
Muhammad Haris Khan	MBZUAI, United Arab Emirates

Challenges of Fine-Grained Image Analysis

Xiu-Shen Wei	Nanjing University of Science and Technology, China
Peng Wang	University of Wollongong, Australia
Huan Wang	Nanjing University of Science and Technology, China
Zeren Sun	Nanjing University of Science and Technology, China
Yang Shen	Nanjing University of Science and Technology, China

Deep Learning-Based Small Object Detection from Images and Videos

Aref Miri Rekavandi	University of Western Australia, Australia
Lian Xu	University of Western Australia, Australia
Farid Boussaid	University of Western Australia, Australia

Abd-Krim Seghouane	University of Western Australia, Australia
Mohammed Bennamoun	University of Western Australia, Australia

Contents

[illegible] Theory and Applications

Learning with Limited Data for Face Analysis

FAPN: Face Alignment Propagation Network for Face Video Super-Resolution

Sige Bian[1], He Li[1], Feng Yu[2], Jiyuan Liu[1], Song Changjun[1], and Yongming Tang[1(✉)]

[1] Joint International Research Laboratory of Information Display and Visualization, Southeast University, Nanjing 210096, China
{220211638,erie,scj,tym}@seu.edu.cn, he.li@ieee.org

[2] School of Computing, National University of Singapore, Singapore, Singapore
yufeng@nus.edu.sg

Abstract. Face video super-resolution (FVSR) aims to use continuous low resolution (LR) video frames to reconstruct face and recover facial details under the premise of ensuring authenticity. The existing video super-resolution (VSR) technology usually uses inter-frame information to achieve better super-resolution (SR) performance. However, due to the complex temporal dependence between frames, as the number of input frames increases, the information cannot be fully utilized, and even wrong information is introduced, resulting in poor performance. In this work, we propose an alignment propagation network for accumulating facial prior information (FAPN). We design a neighborhood information coupling (NIC) module based on optical flow estimation and alignment, where the current frame, the adjacent frames and the SR results of the previous frame are locally fused. The coupled frames are sent to a unidirectional propagation (UP) structure for propagation. Meanwhile, in the UP structure, the facial prior information is filtered and accumulated in the face super-resolution cell (FSRC), and the high-dimensional hidden state is introduced to propagate effective temporal information between frames along the unidirectional structure. Extensive evaluations and comparisons validate the strengths of our approach, FAPN can accumulate more facial details while ensuring the authenticity of the face. And the experimental results demonstrated that the proposed framework achieves better performance on PSNR (up to 0.31 dB), SSIM (up to 0.15 dB) and face recognition accuracy (up to 1.99%) compared with state-of-the-art methods.

Keywords: Face video super-resolution · Alignment propagation network · Face recognition accuracy

1 Introduction

With the development of artificial intelligence technology, face video super-resolution (FVSR) has been widely used in intelligent transportation, personal

This work is supported by National Natural Science Foundation of China (grant number 62275046).

Y. Zheng et al. (Eds.): ACCV 2022, LNCS 13848, pp. 3–18, 2023.
https://doi.org/10.1007/978-3-031-27066-6_1

identification, public security and other fields [1]. In the field of video surveillance, due to the hardware limitations of video capture equipment, it is difficult to obtain clear frames of target face far away from equipment [2]. FVSR aims to effectively enhance facial details in video [3], improve the accuracy of face recognition, and provide greater utilization of raw data (Fig. 1).

Fig. 1. Visual results of our FAPN on scale factor 4. Six consecutive LR video frames (1st row), HR frames (2st row, generated by our method) and groundtruth (GT) frames (3st row) are shown.

Compared with single image super-resolution (SR), video super-resolution (VSR) can achieve better performance by utilizing inter-frame information. At present, there are two main categories to utilize inter-frame information, alignment-based and non-alignment-based. RBPN [4] integrates the spatial and temporal context of consecutive video frames using a recurrent encoder and decoder module in the multi-projection stage. EDVR [5] uses deformable convolution to complete frame alignment at the feature level in a coarse-to-fine manner. The alignment-based approach is efficient, but it only uses information from adjacent frames and cannot effectively use input information far from the current frame. To effectively utilize more inter-frame information, non-alignment-based approaches [6–8] have also been proposed. Although these recurrent structures can accumulate more information from frames, it will inevitably introduce interference or even erroneous information useless for the current frame SR.

In the aspect of face super-resolution (FSR), rational use of strong constraints of human face will bring abundant prior information to the SR process. In addition, if a reasonable information accumulation mechanism is used to accumulate correct and useful information for the face region in the video, the details of the face region can be effectively enhanced under the premise of ensuring the authenticity of the generated face.

In this work, we propose a face video super-resolution network (FAPN), which combines the advantages of alignment-based and non-alignment-based methods. The current frame, the adjacent frames and the SR results of the previous frame are locally fused, and then sent to a unidirectional propagation (UP) structure for propagation, which is based on optical flow estimation and alignment. In addition, in the propagation structure, we filter and accumulate face information, and introduce hidden state to propagate the effective information forward with UP structure. FAPN outperforms existing state of the arts in PSNR (up to 0.31 dB), SSIM (up to 0.15 dB) and face recognition accuracy (up to 1.99%).

Our main contributions can be summarized as follows: 1) We redesigned the inter-frame alignment structure and the propagation structure so that the coupled inter-frame information could be propagated between frames after neighborhood information coupling (NIC) module, thus improving the accuracy of the model. 2) We propose an effective facial prior information filtering mechanism, retain correct and effective information, realize the accumulation of face information to enrich face details. 3) Combined with the prior information of human face, we introduce the pixel loss of facial features and the loss of high-level feature vectors of face to constrain the network training process.

2 Related Work

2.1 Video Super-Resolution

Alignment-based VSR methods mainly include motion compensation and deformable convolution. For motion compensation, VESPCN [9] consists of an alignment network and a fusion spatiotemporal sub-pixel network, which can effectively utilize temporal redundancy, but with low accuracy of generated images. EDVR [5] achieves frame alignment at the feature level in a coarse-to-fine fashion using deformable convolutions. Alignment-based methods usually utilize image information of adjacent frames but cannot effectively utilize input information far from the current frame.

As for non-alignment-based approaches, DUF [6] generates dynamic upsampling filters and residual images based on the local spatiotemporal neighborhood of each pixel to avoid explicit motion compensation. This method is computationally intensive and will introduce memory burden. RLSP [8] introduces a recurrent structure, where information propagates through hidden states. BasicVSR [10] combines feature-level alignment with the bidirectional propagation mechanism, which has a good performance improvement. However, the bidirectional propagation mechanism needs to acquire all the video sequences before VSR, which is not easy to be applied in real-time VSR.

As for face VSR, it aims to improve the resolution of facial regions. Xin [11] proposes a Motion Adaptive Feedback Unit (MAFC) that filters out unimportant motions such as background motions, preserves facial normal rigid motions and non-rigid motions of facial expressions, and feeds them back to the network. Yu [3] optimized the network parameters through three search strategies of TPE, random search and SMAC, and proposed a lightweight FVSR network HO-FVSR. Different from previous works, we propose an end-to-end FVSR framework (FAPN) to accumulate correct facial information. It is demonstrated that our face information accumulation mechanism facilitates our framework to achieve the state-of-the-art performance.

2.2 Facial Information Extraction and Recognition

Different from general VSR, in terms of FVSR, we mainly focus on the authenticity of the generated face, with the purpose of improving the accuracy of face recognition. In the field of face recognition, high-dimensional feature vectors are usually used to represent facial information. Euclidean distance is calculated for the generated feature vectors. When Euclidean distance becomes smaller, the face similarity increases. Representative methods in this field include Openface [12], Face_recognition [13], Insightface [14] and other networks. The accuracy of Face_recognition and Insightface is 99.38% and 99.74% respectively, which can be used for feature extraction and result evaluation.

3 Network Architecture

3.1 Framework of FAPN

VSR aims to map a LR video sequence $I^{LR} \in \mathbb{R}^{H \times W \times C}$ to a HR video sequence $I^{HR} \in \mathbb{R}^{kH \times kW \times C}$, where k is the upsampling factor, H and W are the height and width, and C is the number of channels. We propose a Face Alignment Propagation Network (FAPN) to filter and accumulate facial information. Our network structure is shown in Fig. 2.

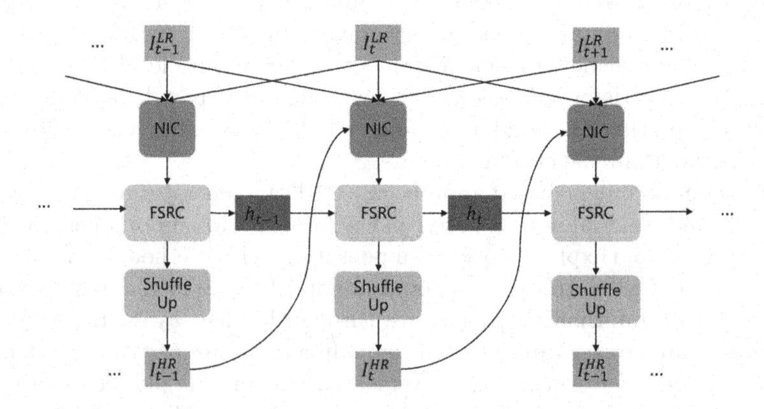

Fig. 2. Overview of the proposed framework FAPN.

Our framework takes consecutive LR frames $I_t^{LR} \in \mathbb{R}^{H \times W \times C}$ $(t = 0, 1, 2 \ldots)$ as input to the network. We design a neighborhood information coupling (NIC) module to couple the information of adjacent frames in the first stage. In addition, in order to utilize previous SR results, feedback is introduced, thus I_{t-1}^{HR} is also sent to the NIC module at time t. In the second stage, the coupled information is sent to a unidirectional propagation (UP) structure. By introducing the hidden state $h_t(t = 0, 1, 2...)$, the UP structure can propagate information from the first frame to the current frame to supplement information. FSRC stands for facial information SR module, which is used to extract facial prior information in the network and generate face video streams with rich details and high authenticity. At time t, the output of FSRC are hidden state h_t and the filtered facial features I_t^{FSR}. I_t^{FSR} is shuffled up to get the final output.

3.2 Neighborhood Information Coupling

Our NIC module aims to add details to the current frame. It is first coupled with the information of adjacent frames and previous SR results, then the coupled information is sent to the propagation structure. The NIC module is mainly composed of optical flow estimation (OFE) module, alignment module and shuffling module (see Fig. 3).

The role of OFE is to predict HR optical flow from LR frame, and the predicted HR optical flow can be used to align the frames in the neighborhood with the current frame, which helps to reconstruct more accurate temporal information. In NIC, we adopt OFRnet [15] for optical flow information estimation in OFE module due to its simplicity and effectiveness:

$$OF_{t-1}^{HR} = OFE(I_{t-1}^{LR},\ I_t^{LR}) \tag{1}$$

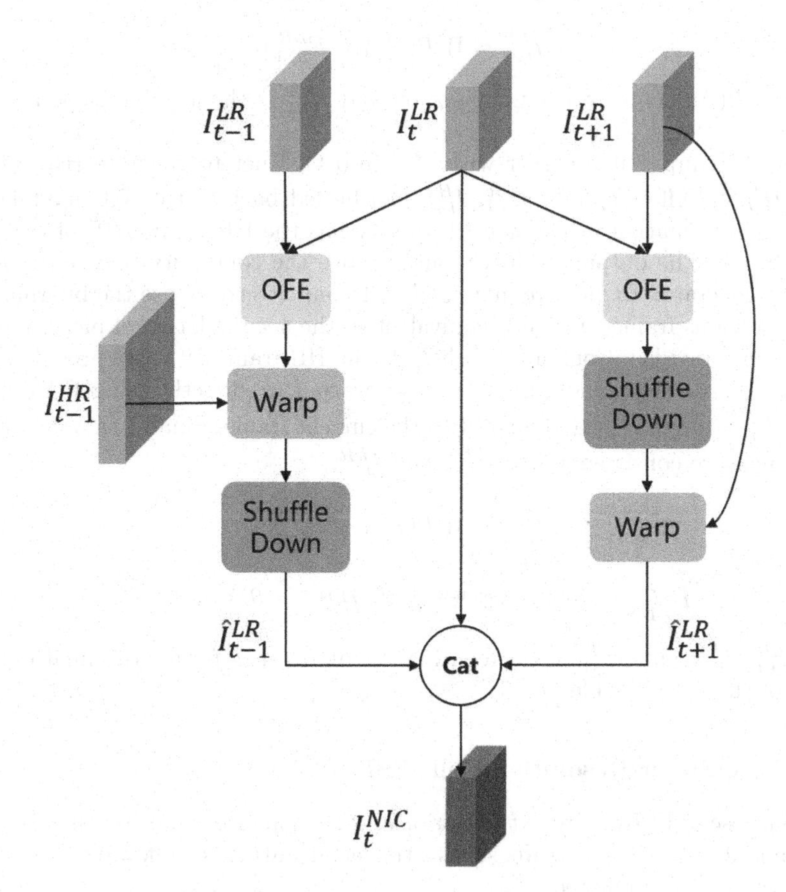

Fig. 3. Architecture of our NIC module.

$$OF_{t+1}^{HR} = OFE(I_{t+1}^{LR},\ I_t^{LR}) \tag{2}$$

where OF_{t-1}^{HR} and OF_{t+1}^{HR} represent the HR optical flow generated by OFE module respectively.

Shuffling is used for space-to-depth conversion, which uses the scale factor k to map LR to HR space. The operation is reversible and can achieve the inverse mapping from HR to LR. In previous work [8,16,17] shuffling has been used to change the spatial resolution of feature.

$$s^{LR} \in \mathbb{R}^{H\times W\times C} \xrightarrow{\text{shuffle up}} s^{HR} \in \mathbb{R}^{kH\times kW\times C/k^2} \tag{3}$$

$$s^{HR} \in \mathbb{R}^{H\times W\times C} \xrightarrow{\text{shuffle down}} s^{LR} \in \mathbb{R}^{H/k\times W/k\times k^2C} \tag{4}$$

For I_{t+1}^{LR}, input it together with I_t^{LR} into OFE module to generate HR optical flow OF_{t+1}^{HR}. After shuffling down, the LR flow cube OF_{t+1}^{LR} is generated.

$$OF_{t+1}^{HR} \in \mathbb{R}^{H\times W\times C} \xrightarrow{\text{shuffle down}} OF_{t+1}^{LR} \in \mathbb{R}^{H/k\times W/k\times k^2C} \tag{5}$$

$$\hat{I}_{t+1}^{LR} = WP(I_{t+1}^{LR}, OF_{t+1}^{LR}) \tag{6}$$

where $WP(\cdot)$ denotes warping operation and $\hat{I}_{t+1}^{LR}$ is the aligned frame at time $t+1$.

For I_{t-1}^{LR}, input it together with I_t^{LR} into OFRnet to generate HR optical flow OF_{t-1}^{HR}. Different from I_{t+1}^{LR}, I_{t-1}^{HR} can be fed back to the NIC module for auxiliary information fusion since it already has the HR result I_{t-1}^{HR} of the previous frame. The output feedback can improve the continuity between frames, and the generated frames are more stable. It can effectively use the information of the previous frames, which is equivalent to the accumulation of picture information in the video. Compared with I_{t-1}^{LR}, the HR frame I_{t-1}^{HR} can provide more information to help generate I_t^{HR}. So we warp I_{t-1}^{HR} directly using HR optical flow OF_{t-1}^{HR} to get $\hat{I}_{t-1}^{HR}$ and align it to the current frame. Finally, shuffling down $\hat{I}_{t-1}^{HR}$ to ensure consistency with $\hat{I}_{t+1}^{LR}$ and $\hat{I}_t^{LR}$.

$$\hat{I}_{t-1}^{HR} = WP(I_{t-1}^{HR}, OF_{t-1}^{HR}) \tag{7}$$

$$\hat{I}_{t-1}^{HR} \in \mathbb{R}^{H\times W\times C} \xrightarrow{\text{shuffle down}} \hat{I}_{t-1}^{LR} \in \mathbb{R}^{H/k\times W/k\times k^2C} \tag{8}$$

For $\hat{I}_{t+1}^{LR}$, $\hat{I}_t^{LR}$ and $\hat{I}_{t-1}^{LR}$, concatenating them together to get the final output I_t^{NIC} of the NIC module.

3.3 Face Super-Resolution Cell (FSRC)

The purpose of FSRC is to extract prior information of face, utilize the previous NIC module and UP structure, so that the facial features can flow over time and be selectively accumulated.

FSRNet [18] proposed the method of combining image features and prior information to carry out FSR. Inspired by FSRNet, we designed the following network structure (see Fig. 4) to extract prior information of face.

The network consists of FSR encoder, FSR decoder and prior information extraction (PIE) module. The input of the network is the video frames coupled with the neighborhood information and the historical hidden state. After the coupled video frame I_t^{NIC} is concatenated with the hidden state h_{t-1}, it is sent to the FSR encoder to extract image features. In another branch, the coupled video frames I_t^{NIC} are sent to the PIE module to extract facial prior information. The extracted prior information is concatenated with the image features generated by the encoder, and then sent to the FSR decoder to generate hidden state h_t and the filtered facial features I_t^{FSR}.

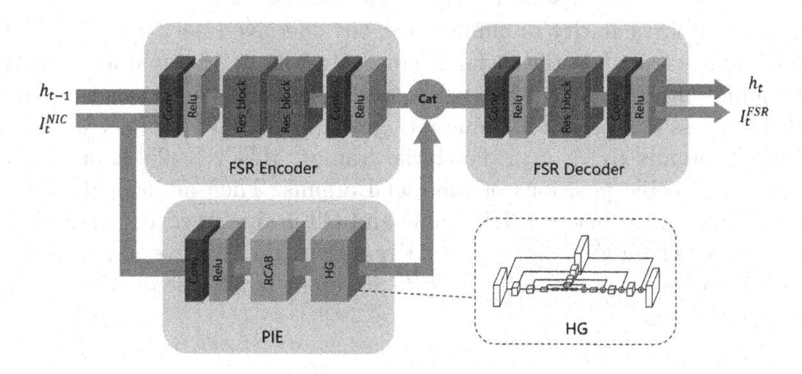

Fig. 4. Architecture of our FSRC. 'Cat' denotes concatenation along the channel dimension, 'RCAB' denotes residual channel attention block. 'HG' denotes HourGlass structure, which uses a skip connection mechanism between symmetrical layers.

Encoder-Decoder Structure. Inspired by the success of ResNet [19] in SR, we use residual blocks for feature extraction, and the network structure is shown in Fig. 4. The FSR encoder consists of convolutional layers, ReLU [20] layers and two residual blocks to extract the features of the coupled video frame I_t^{NIC} and the hidden state. The FSR decoder consists of convolutional layers, ReLU layers and a residual block, which jointly utilizes features and prior information for face image restoration.

Prior Information Extraction (PIE). In many CNN-based methods [21,22], information is treated equally in all channels during feature extraction, which makes the network lack the ability of discriminative learning. RCAN [23] proposes a deep residual channel attention network to obtain better performance. Inspired by RCAN, we add the residual channel attention block (RCAB) to re-weight the distribution of different channels.

In addition, inspired by the success of stacked heatmap regression in human pose estimation [24] and human face image SR [18], we added an HourGlass (HG) structure [24] after RCAB to effectively integrate cross-scale features and preserve spatial information at different scales. The network structure of PIE is shown in Fig. 4, consisting of convolutional layers, ReLU layers, RCAB and HG structure.

3.4 Loss Function

Adding constraints will bring more prior information to the SR process, and can effectively constrain the distribution of solutions, so as to obtain more accurate results. In the process of FVSR, we can also use the features of face to constrain the spatial distribution of solutions in a more precise way.

Due to the particularity of human face, the effective information in the face is mainly concentrated in the facial organs, so adding additional loss functions in the training process can achieve better results. In addition to the mean square error (MSE) loss $\mathcal{L}_{MSE}$ of SR frame and groundtruth (GT) frame, we add the loss of facial organs $\mathcal{L}_{face\_organ}$. For facial frames, MTCNN [25] is used to pre-calibrate the specific positions of the facial organs. Then additional MSE calculation is performed between HR frame and GT frame in corresponding facial organ regions, where $\Phi_i(i = 1,2,3,4)$ represent the left eye, right eye, nose and mouth components respectively, I_t^{SR} represent HR frame of time t and I_t^H represent GT frame.

$$\mathcal{L}_{MSE} = \| I_t^{SR} - I_t^H \|_2^2 \tag{9}$$

$$\mathcal{L}_{face\_organ} = \sum_{i=1}^{4} \left\| \Phi_i(I_t^{SR}) - \Phi_i(I_t^H) \right\|_2^2 \tag{10}$$

In addition to pixel-level differences, we can also add loss of high-level information such as image structure, texture, and style. With reference to Insightface [14], face feature extraction can be performed on face images to generate a vector containing face identity information. The loss function $\mathcal{L}_{face\_vector}$ is constructed according to the Euclidean distance of the feature vector between HR frame and GT frame, where Θ represents using Insightface to extract face features. According to this training method, more accurate recognition results can be obtained after FSR.

$$\mathcal{L}_{face\_vector} = \left\| \Theta(I_t^{SR}) - \Theta(I_t^H) \right\|_2^2 \tag{11}$$

Based on the above analysis, we design three loss terms, the MSE loss of HR frame and GT frame $\mathcal{L}_{MSE}$, and the MSE loss of the pixels in facial features area $\mathcal{L}_{face\_organ}$ and the loss of high-level feature vectors of faces $\mathcal{L}_{face\_vector}$.

$$\mathcal{L} = \mathcal{L}_{MSE} + \lambda_1 \mathcal{L}_{face\_organ} + \lambda_2 \mathcal{L}_{face\_vector} \tag{12}$$

4 Experiments

In this section, we first compare our framework to several existing VSR methods. Then, we further conduct ablation experiments to evaluate our framework.

4.1 Dataset

Due to the lack of recognized FVSR Dataset, we make a Face Video Dataset made by 300 Videos in the Wild (300-VW) and conducted experiments on it. It was first used in ICCV's face recognition contest in 2015 and can be downloaded at ibug.doc.ic.ac.uk. According to requirements, our dataset production process is as follows. First, the original videos are intercepted into sequences of consecutive frames, 32 frames per sequence, and a total of 400 video sequences are generated. MTCNN [25] network is used to select and cut facial area of each frame, and then adjust the size to 160*160 to obtain HR frames. Then we performed downsampling to generate LR frames and obtain final Face Video Dataset. In this work, we only focus on the downsampling factor of 4 since it is the most challenging case.

Finally, We divide 400 generated video sequences into training sets, verification sets and testing sets (see Table 1).

Table 1. Datasets used in FVSR.

Face video dataset	Sequences	Frames
Training	340	10880
Validation	15	480
Testing	45	1440

4.2 Implementation Details

To train our network, we randomly selected 10 consecutive frames from 32 frames. Due to UP structure, the hidden state h_{t-1} and the HR result I_{t-1}^{HR} of the previous frame need to be initialized. Both tensors are initialized with zeros.

We implemented our framework in Pytorch. We set the batch size to 4, the learning rate to 10^{-4}, and λ_1 and λ_2 to 0.05 and 0.01, respectively. All experiments are conducted on a PC with an Nvidia GTX 1080Ti GPU.

4.3 Comparisons to the State-of-the-Art

Quantitative Comparisons. We compare our proposed FAPN with the state-of-the-art VSR methods, including VESPCN [9], FRVSR [16], SOF-VSR [15] and

Table 2. Quantitative comparisons. Best results are shown in boldface.

	SOF-VSR [15]	FRVSR [16]	TecoGan [26]	VESPCN [9]	Ours
PSNR (dB)	30.75	25.25	24.85	30.00	**31.06**
SSIM (dB)	0.917	0.823	0.811	0.907	**0.924**
Face distance	0.201	0.276	0.252	0.228	**0.197**

TecoGan [26]. Quantitative comparison with other state-of-the-art VSR methods is shown in Table 2.

We conducte experiments on 45 test sets and measure PSNR, SSIM and Face distance, which measures the difference between faces. Face_recognition [13] is a concise and powerful face recognition library, tested with the Labeled Faces in the Wild face dataset, with an accuracy rate of 99.38%. We use it to measure the Face distance metric. Face_recognition generates high-dimensional feature vectors for face images, and then calculates the Euclidean distance between corresponding face feature vectors of HR frames and GT frames to quantify the differences between faces. The smaller the Face distance is, the higher the face similarity is.

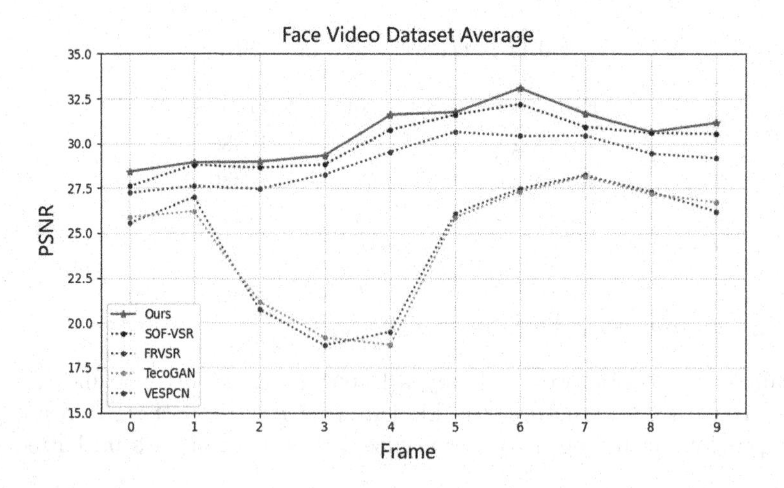

Fig. 5. Average PSNR values for the first 10 frames on the test sets. Average PSNR values of each frame is calculated from the average of the corresponding frames on all test sets

The experimental results demonstrated that the proposed framework FAPN achieves better performance on PSNR, SSIM and Face distance compared with state-of-the-art methods. Specifically, the PSNR and SSIM values achieved by our framework are better than other methods by over 0.31 dB and 0.15 dB.

This is because we use NIC to supplement and fuse information, and the information provided to the UP structure is more abundant and accurate, therefore, more reliable spatial details and temporal consistency can be recovered well.

For Face distance, we achieved an improvement of 1.99%, which indicates that the face generated by our network is closer to real face and has high reliability.

It can be seen that the PSNR and SSIM values of the TecoGan [26] and FRVSR [16] networks are only about 25 dB and 0.8 dB, and the Face distance value exceeds 0.25. This is due to the non-natural generation of SR frames, with severe facial deformation or incorrect information generation. The specific test images will be displayed in the Qualitative comparisons. In addition, we plot the average PSNR for the first 10 frames on the test sets. As can be seen from Fig. 5, our network achieves the best PSNR values for each frame.

Qualitative Comparisons. A qualitative comparison between our method and other SR methods [9,15,16,26] are shown in Fig. 6. There are three face images, each of which is reconstructed from ten consecutive frames. The results obtained by our network are closer to real frames than other methods, especially in facial features.

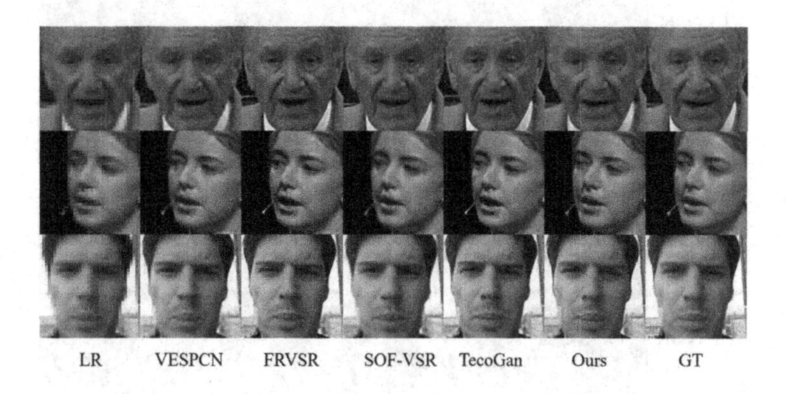

Fig. 6. Qualitative comparisons, where all the examples come from our test sets.

It can be seen that the visual effect represented by TecoGan [26] is better, but their PSNR, SSIM and Face distance metrics are poor. Although they generate richer details, they are generated incorrectly, resulting in a certain degree of distortion of human facial features. For example, the wrinkles on the forehead of the old man in the first row are falsely generated. In addition, the mouth of the little girl generated by TecoGan [26] and FRVSR [16] in the second row is deformed. These additionally generated textures and deformations will make the visual effect better to a certain extent, but will reduce the authenticity of the generated face. From the perspective of the actual application scenario of FVSR, it is expected to improve the accuracy of face recognition through super-resolution. Therefore, it is necessary to include more details as much as possible on the premise of ensuring the authenticity of the generated face. So compared with TecoGan [26] and FRVSR [16], our results are closer to the real situation.

From the 32 frames of the test sets, we selected 6 consecutive frames for testing, and the results are shown in Fig. 7. Expanding horizontally in chronological order, it can be seen that for TecoGan [26] and FRVSR [16], the faces in columns 4 and 5 are severely distorted. This is because when the inter-frame motion is intense to a certain extent, inaccurate motion estimation and compensation not only cannot effectively utilize the inter-frame information, but also introduce error information and seriously interfere with SR results. Our network utilizes NIC module, which can not only effectively utilize the information of previous frame to supplement the information of the current frame, but also ensure the authenticity of the introduced information to avoid unnatural generation.

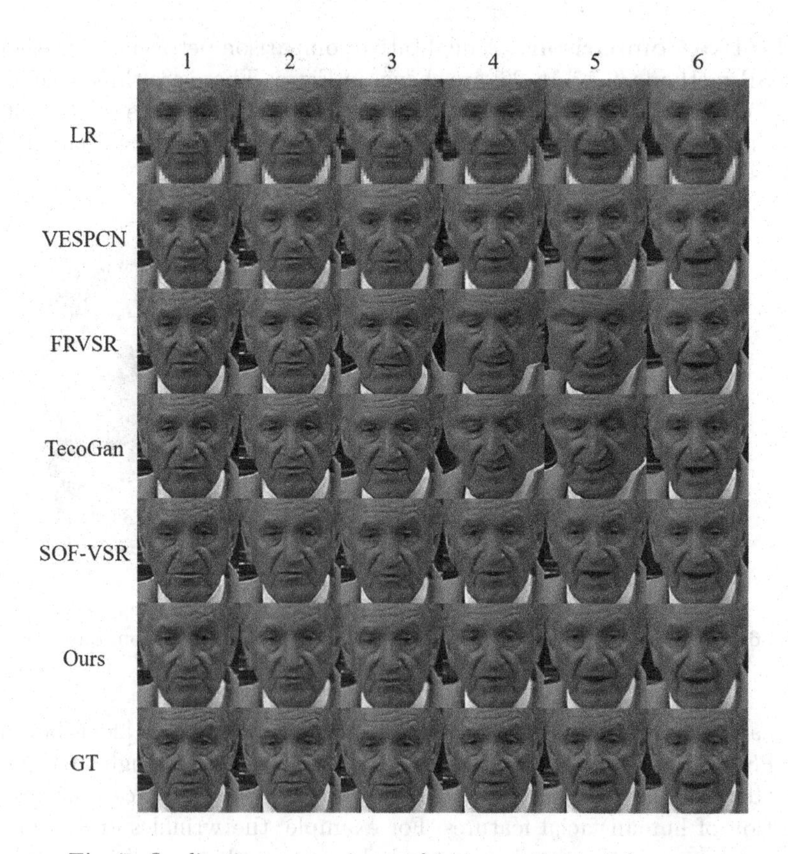

Fig. 7. Qualitative comparisons of 6 consecutive test frames.

Due to UP structure, our network relies on historical input frames. Initially, the information content available is minimal, and a certain number of input frames are required to accumulate information. This phenomenon can be seen in Fig. 7. In the first two frames, the texture on the forehead of the old man, and the eyes have a certain degree of distortion, and the facial structure of the human face cannot be completely reconstructed. But with more information, the textures on the facial features and forehead gradually approach the real face.

In addition, compared with SOF-VSR [15] and VESPCN [9], we slightly improve visual effects and indicators, introducing richer details on the premise of ensuring authenticity.

4.4 Ablation Study

Effects of NIC and FSR. We conducted 3 experiments to evaluate the effects of NIC and FSR, respectively. Specifically, we remove the NIC and FSRC from our network, the remaining parts constitute the first network, named 'BasicNet v1'. The second network, named 'BasicNet v2', has the same structure as FAPN except for NIC module. In this part, we study the effects of different networks. For fairly comparison, we train all those models with other same implementation details.

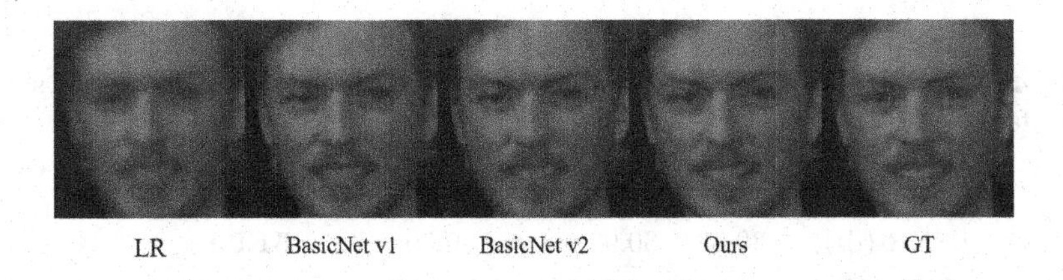

Fig. 8. Qualitative results of ablation study.

For 'BasicNet v1', as Fig. 8 shows, the generated face is blurry, and the left eye is distorted relative to GT. For 'BasicNet v2', the face is clear, but there exist distortions and false generation. This is because 'BasicNet v2' introduced FSRC compared to 'BasicNet v1', which can advance the high-level information of the face, but without NIC, it may lead to the accumulation of wrong information. The results of FAPN are closest to GT, achieving correct and sufficient information extraction. The results of the quantitative comparison are shown in Table 3. PSNR, SSIM and Face distance of 'BasicNet v2' are the worst, which indicates that we should not only focus on visual effects, but should take authenticity as an important evaluation factor. Compared to 'BasicNet v1', our final network has 0.68 dB improvement in PSNR, 0.004 dB improvement in SSIM, and 0.005 improvement in Face distance.

Effects of Loss Function. In this section, we adopt 4 training methods, and evaluated the results respectively for our PAFN network. Where 'MSE' stands for MSE loss. 'MSE+GAN' refers to adding an adversarial network on the basis of MSE, and forms an adversarial loss [27] with the help of the generator and

Table 3. Quantitative results of ablation study. Best results are shown in boldface.

	BasicNet v1	BasicNet v2	FAPN
PSNR (dB)	29.58	29.37	**30.26**
SSIM (dB)	0.918	0.823	**0.922**
Face distance	0.197	0.206	**0.192**

the discriminator. 'MSE+VGG' refers to using the VGG network for high-level feature extraction of the face based on the MSE and drawing on the powerful feature extraction characteristics of the VGGNet [28]. The loss function includes not only the MSE loss, but also the mean squared loss of each layer of feature maps of the VGG feature extraction module, in order to obtain the similarity between high-level features of the image. 'MSE+FACE' refers to the loss function we constructed in Sect. 3.5. We omit the different loss weight adjustment steps for each method, and the results of each method under the optimal parameters are shown in Table 4.

Table 4. Ablation study on the effects of different loss terms. Best results are shown in boldface.

	MSE	MSE+GAN	MSE+VGG	MSE+FACE
PSNR (dB)	30.88	30.92	30.50	**31.00**
SSIM (dB)	0.912	**0.913**	0.902	**0.913**
Face distance	0.215	0.206	0.214	**0.203**

It can be seen that 'MSE+FACE' achieves the best performance on PSNR (up to 0.08 dB), SSIM (same as 'MSE+GAN') and face recognition accuracy (up to 1.46%), because it not only emphasizes the facial features, but also uses the Insightface [14] network to extract the high-level features of the face.

5 Conclusion

In this paper, we propose an end-to-end face alignment propagation network (FAPN) for face video super-resolution. Our NIC module first fuses adjacent frames and previous HR frame. UP structure and FSRC are then performed to accumulate correct facial prior information. Extensive experiments have demonstrated that our FAPN can recover facial details and improve the accuracy of face recognition on the premise of ensuring the authenticity of the generated face. Comparison to existing video SR methods has shown that our framework achieves the state-of-the-art performance on PSNR (up to 0.31 dB), SSIM (up to 0.15 dB) and face recognition accuracy (up to 1.99%).

References

1. Wang, M., Deng, W.: Deep face recognition: a survey. Neurocomputing **429**, 215–244 (2021)
2. Farooq, M., Dailey, M., Mahmood, A., Moonrinta, J., Ekpanyapong, M.: Human face super-resolution on poor quality surveillance video footage. Neural Comput. Appl. **33**, 13505–13523 (2021)
3. Yu, F., Li, H., Bian, S., Tang, Y.: An efficient network design for face video super-resolution. In: 2021 IEEE/CVF International Conference on Computer Vision Workshops (ICCVW), pp. 1513–1520 (2021)
4. Haris, M., Shakhnarovich, G., Ukita, N.: Recurrent back-projection network for video super-resolution. In: 2019 IEEE/CVF Conference on Computer Vision and Pattern Recognition (CVPR), pp. 3892–3901 (2019)
5. Wang, X., Chan, K.C., Yu, K., Dong, C., Loy, C.C.: EDVR: video restoration with enhanced deformable convolutional networks. In: 2019 IEEE/CVF Conference on Computer Vision and Pattern Recognition Workshops (CVPRW), pp. 1954–1963 (2019)
6. Jo, Y., Oh, S.W., Kang, J., Kim, S.J.: Deep video super-resolution network using dynamic upsampling filters without explicit motion compensation. In: 2018 IEEE/CVF Conference on Computer Vision and Pattern Recognition, pp. 3224–3232 (2018)
7. Isobe, T., Jia, X., Gu, S., Li, S., Wang, S., Tian, Q.: Video super-resolution with recurrent structure-detail network. In: Vedaldi, A., Bischof, H., Brox, T., Frahm, J.-M. (eds.) ECCV 2020. LNCS, vol. 12357, pp. 645–660. Springer, Cham (2020). https://doi.org/10.1007/978-3-030-58610-2_38
8. Fuoli, D., Gu, S., Timofte, R.: Efficient video super-resolution through recurrent latent space propagation. In: 2019 IEEE/CVF International Conference on Computer Vision Workshop (ICCVW), pp. 3476–3485 (2019)
9. Caballero, J., et al.: Real-time video super-resolution with spatio-temporal networks and motion compensation. In: 2017 IEEE Conference on Computer Vision and Pattern Recognition (CVPR), pp. 2848–2857 (2017)
10. Chan, K.C., Wang, X., Yu, K., Dong, C., Loy, C.C.: BasicVSR: the search for essential components in video super-resolution and beyond. In: 2021 IEEE/CVF Conference on Computer Vision and Pattern Recognition (CVPR), pp. 4945–4954 (2021)
11. Xin, J., Wang, N., Li, J., Gao, X., Li, Z.: Video face super-resolution with motion-adaptive feedback cell. In: Proceedings of the AAAI Conference on Artificial Intelligence, vol. 34, pp. 12468–12475 (2020)
12. Amos, B., Ludwiczuk, B., Satyanarayanan, M.: Openface: a general-purpose face recognition library with mobile applications (2016)
13. Parkhi, O.M., Vedaldi, A., Zisserman, A.: Deep face recognition. In: British Machine Vision Conference (2015)
14. Deng, J., Guo, J., Xue, N., Zafeiriou, S.: Arcface: additive angular margin loss for deep face recognition. In: 2019 IEEE/CVF Conference on Computer Vision and Pattern Recognition (CVPR), pp. 4685–4694 (2019)
15. Wang, L., Guo, Y., Lin, Z., Deng, X., An, W.: Learning for video super-resolution through HR optical flow estimation. In: Jawahar, C.V., Li, H., Mori, G., Schindler, K. (eds.) ACCV 2018. LNCS, vol. 11361, pp. 514–529. Springer, Cham (2019). https://doi.org/10.1007/978-3-030-20887-5_32

16. Sajjadi, M.S.M., Vemulapalli, R., Brown, M.: Frame-recurrent video super-resolution. In: 2018 IEEE/CVF Conference on Computer Vision and Pattern Recognition, pp. 6626–6634 (2018)
17. Shi, W., et al.: Real-time single image and video super-resolution using an efficient sub-pixel convolutional neural network. In: 2016 IEEE Conference on Computer Vision and Pattern Recognition (CVPR), pp. 1874–1883 (2016)
18. Chen, Y., Tai, Y., Liu, X., Shen, C., Yang, J.: FSRNet: end-to-end learning face super-resolution with facial priors. In: 2018 IEEE/CVF Conference on Computer Vision and Pattern Recognition, pp. 2492–2501 (2018)
19. He, K., Zhang, X., Ren, S., Sun, J.: Deep residual learning for image recognition. In: 2016 IEEE Conference on Computer Vision and Pattern Recognition (CVPR), pp. 770–778 (2016)
20. Nair, V., Hinton, G.E.: Rectified linear units improve restricted Boltzmann machines vinod nair. In: International Conference on International Conference on Machine Learning (2010)
21. Dong, C., Loy, C.C., Tang, X.: Accelerating the super-resolution convolutional neural network. In: Leibe, B., Matas, J., Sebe, N., Welling, M. (eds.) ECCV 2016. LNCS, vol. 9906, pp. 391–407. Springer, Cham (2016). https://doi.org/10.1007/978-3-319-46475-6_25
22. Lim, B., Son, S., Kim, H., Nah, S., Lee, K.M.: Enhanced deep residual networks for single image super-resolution. In: 2017 IEEE Conference on Computer Vision and Pattern Recognition Workshops (CVPRW), pp. 1132–1140 (2017)
23. Basak, H., Kundu, R., Agarwal, A., Giri, S.: Single image super-resolution using residual channel attention network. In: 2020 IEEE 15th International Conference on Industrial and Information Systems (ICIIS), pp. 219–224 (2020)
24. Newell, A., Yang, K., Deng, J.: Stacked hourglass networks for human pose estimation. In: Leibe, B., Matas, J., Sebe, N., Welling, M. (eds.) ECCV 2016. LNCS, vol. 9912, pp. 483–499. Springer, Cham (2016). https://doi.org/10.1007/978-3-319-46484-8_29
25. Zhang, L., Wang, H., Chen, Z.: A multi-task cascaded algorithm with optimized convolution neural network for face detection. In: 2021 Asia-Pacific Conference on Communications Technology and Computer Science (ACCTCS), pp. 242–245 (2021)
26. Chu, M., Xie, Y., Mayer, J., Leal-Taixé, L., Thuerey, N.: Learning temporal coherence via self-supervision for GAN-based video generation. ACM Trans. Graph. (TOG) **39** (2020)
27. Goodfellow, I.J., et al.: Generative adversarial nets. In: Proceedings of the 27th International Conference on Neural Information Processing Systems - Volume 2, NIPS 2014, pp. 2672–2680. MIT Press, Cambridge (2014)
28. Simonyan, K., Zisserman, A.: Very deep convolutional networks for large-scale image recognition. CoRR abs/1409.1556 (2015)

Micro-expression Recognition Using a Shallow ConvLSTM-Based Network

Saurav Shukla[1], Prabodh Kant Rai[1], and Tanmay T. Verlekar[1,2](✉)

[1] Department of CSIS, BITS Pilani, K.K Birla Goa Campus, Goa 403726, India
{f20180653,f20180748,tanmayv}@goa.bits-pilani.ac.in

[2] APPCAIR, BITS Pilani, K.K Birla Goa Campus, Goa 403726, India

Abstract. Micro-expressions reflect people's genuine emotions, making their recognition of great interest to the research community. Most state-of-the-art methods focus on the use of spatial features to perform micro-expression recognition. Thus, they fail to capture the spatiotemporal information available in a video sequence.

This paper proposes a shallow convolutional long short-term memory (ConvLSTM) based network to perform micro-expression recognition. The convolutional and recurrent structures within the ConvLSTM allow the network to effectively capture the spatiotemporal information available in a video sequence. To highlight its effectiveness, the proposed ConvLSTM-based network is evaluated on the SAMM dataset. It is trained to perform micro-expression recognition across three (positive, negative, and surprise) and five (happiness, other, anger, contempt, and surprise) classes. When compared with the state-of-the-art, the results report a significant improvement in accuracy and the F1 score. The proposal is also robust against the unbalanced class sizes of the SAMM dataset.

Keywords: Micro-expression recognition · Shallow neural networks · convolutional LSTM

1 Introduction

Micro-expressions are twitches on people's faces that reveal their genuine emotions. They are spontaneous reactions to an external stimulus that people consciously suppress, resulting in voluntary and involuntary emotional responses co-occurring and conflicting with one another [1]. The fleeting and subtle nature of the micro-expressions make their detection and recognition a challenging task.

Psychologists believe that 55% of people's emotional states can be conveyed through facial expressions [2]. However, most macro-expressions can be forced, making them unreliable. Micro-expressions being difficult to suppress, hide, disguise, or conceal, make them the preferred choice in understanding people's emotional states. Thus, successful micro-expression recognition has several applications, ranging from criminal investigation to psychological and medical treatment.

Y. Zheng et al. (Eds.): ACCV 2022, LNCS 13848, pp. 19–30, 2023.
https://doi.org/10.1007/978-3-031-27066-6_2

Traditionally, professionals performed micro- expressions recognition through visual inspection. The technique was highly ineffective even after obtaining sufficient training, as most micro-expressions last less than 0.33 s [3]. With the development of high frame rate cameras, computer vision and machine learning, researchers have found some success detecting and recognising micro-expressions.

This paper proposes a novel strategy for recognising micro-expressions using ConvLSTM-based network. The convolutional structures in both the input-to-state and state-to-state transition of ConvLSTM allow the network to capture complex information, which is not possible with just convolutional or recurrent neural networks (CNN, RNN). The structure, in turn, enables the construction of a shallow network for micro-expression recognition.

2 Literature Review

The biggest challenge faced by the researchers in addressing the problem of micro-expression recognition is the availability of data. One of the first publically available datasets was collected in 2009, called the Polikovsky dataset [4]. The dataset contains 42 video sequences from 11 people captured at 200 fps. Apart from the limited number of video sequences, another drawback of the dataset is that the participants posed the six different micro-expressions. The simulation of the micro-expressions was a common problem with most initial datasets. The spontaneous micro-expression corpus (SMIC) series dataset was the first to capture spontaneous micro-expression. The dataset contains 77 sequences of just six people recorded using a 100 fps camera [5]. The micro-expressions were induced by making people watch video clips with emotional stimulation. The captured micro-expressions were labelled as positive, negative and surprise. Finer micro-expressions such as happiness, fear, disgust, and sadness, among others, were made available in the Chinese academy of sciences micro-expressions (CASME) series datasets [6].

The most recent dataset called the spontaneous actions and micro-movements (SAMM) series dataset, has significant improvements over the other existing datasets [7]. It involves the participation of 32 people to capture 159 video sequences comprising of seven spontaneous micro-expression using a 200 fps camera. The participants are widely distributed in terms of age, gender and ethnicity, making it the preferred dataset for evaluating micro-expression recognition techniques. Hence, this paper reports its results using the SAMM dataset.

The initial work on micro-expressions involved detecting the time stamp indicating their beginning (onset), end (offset), and their peak manifestation (apex). They were detected by analysing the difference between consecutive frames using techniques such as local binary patterns (LBP) [8] and histogram of oriented gradients [9]. Recently, local temporal patterns have been used to detect facial movement across frames. The specific signature associated with the micro-expressions identified in those facial movements has led to the most promising results [10].

2.1 Traditional Micro-expression Recognition Techniques

Typically, once the onset and offset of a micro-expression are detected, the frames belonging to that time interval are used to recognise the micro-expression. Initial works on micro-expression recognition were evaluated on posed datasets. Because of the small size of the datasets, several hand-crafted techniques were developed. The most successful among them involved the use of LBP across the three orthogonal planes (LBP-TOP) of a stacked video sequence tensor [11]. The recognition was then performed using a support vector machine (SVM). The LBP-TOP was further improved by reducing the redundancy and sparsity in the feature descriptors. Redundancy was reduced by identifying six neighbourhood points associated with the three intersecting lines of the orthogonal plane, where the LBP could be applied [12]. The sparsity problem was addressed using components of each orthogonal plane, such as magnitude, sign and orientation, to generate compact feature descriptors [13]. Apart from LBP, optical flow was used to generate dynamic feature maps that captured the movements observed on the face during the manifestation of a micro-expression [14]. SVM then used these dynamic feature maps to perform micro-expression recognition. Other operations were also performed over the optical flow to obtain better feature descriptors, such as computing its derivative [15]. In contrast to global operations, local operations, such as assigning weights to specific regions of the dynamic feature maps, contributed to an increase in the micro-expression recognition accuracy [16].

2.2 Micro-expression Recognition Using Neural Networks

Recently, CNNs have been extensively used to address image/video-based problems. With larger datasets being captured in recent years, CNNs have also been explored to address the problem of micro-expression recognition. In works such as [17], cropped apex frames are used as inputs to deep CNNs to classify them across six different micro-expression classes. DeepCNNs have also been used to detect facial landmarks, which in turn are used to generate optical flow-based dynamic feature maps [18]. Popular deep CNNs such as ResNet trained on ImageNet, have also been repurposed to perform micro-expression recognition through fine-tuning [19]. These deep CNNs are altered by introducing attention units in residual blocks to capture subtle features associated with micro-expression.

Besides deep CNNs, shallow CNNs have also been explored to obtain lightweight networks that can effectively recognise micro-expressions. The shallow CNNs, in some cases, are truncated versions of deep CNNs such as AlexNet with optical flow-based dynamic feature maps as their input [20,21]. They are used in a dual-stream configuration to capture discriminative features in salient facial areas. In other cases, novel architectures are proposed, such as shallow CNNs with 3D convolutions to capture temporal information across frames [22]. Temporal information can be effectively captured using RNNs. The idea is explored in works such as [23], where spatial features captured using VGG-16 are fed to LSTM to perform micro-expression recognition.

The review suggests that within the state-of-the-art, CNNs are the most effective in performing micro-expression recognition. However, most CNNs cannot capture the spatiotemporal information available in video sequences. A second drawback of the state-of-the-art is the use of deep CNNs through fine-tuning. These networks are usually proposed for significantly more complex problems making them computationally expensive. Also, the features used for micro-expression recognition are usually more subtle than other vision problem that these networks address. Shallow CNNs have shown some promise, but their results have scope for improvement, especially on the SAMM dataset that containd unbalanced micro-expression classes.

This paper presents a shallow neural network that uses convolution in tandem with LSTM. The merging of convolution in LSTM allows the network to capture spatiotemporal information more effectively than other neural networks. Being shallow enables the network to be trained end-to-end on small datasets. Thus we hypothesise that using ConvLSTM in our proposed micro-expression recognition network allows it to perform better than other shallow CNNs, and deep CNN+LSTM networks.

3 Proposal

Inspired by the shallow networks for micro-expression recognition discussed in Sect. 2.2, we propose a ConvLSTM-based network to effectively capture the spatiotemporal information available in video sequences. The entire architecture of the proposal is presented in Fig. 1. Its details are discussed in the following three sub-sections:

- Pre-processing: To prepare the video sequences for processing by the network;
- ConvLSTM: To capture spatiotemporal information available in the video sequence;
- Dense layers: To perform the final classification.

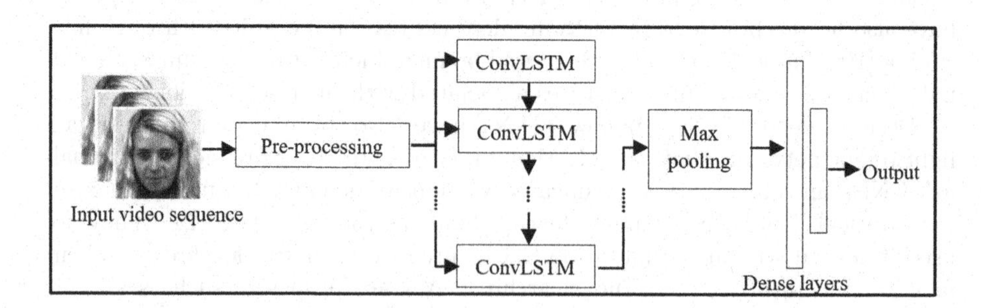

Fig. 1. The architecture of the proposed ConvLSTM based network for micro-expression recognition.

3.1 Pre-processing

Since each micro-expression lasts for a variable length, and the micro-expressions themselves are captured using a high fps camera, we first downsample the video sequence to t frames. These t frames include the onset, offset and the apex frames. The remaining frames are selected by uniformly sampling between onset and offset frames. Thus, given a video sequence, downsampling assures that the temporal redundancy across consecutive frames is reduced while still maintaining the facial movement information across frames. The frames are then resized to $mtimesn$ and stacked together to create a $mtimesntimest$ dimensional input to the ConvLSTM-based network. The values m, n, and t are set according to the resolution and fps of the camera used. In the case of the SAMM dataset, the values m, n, and t are set to 90, 90, and 9, respectively.

3.2 ConvLSTM

The ConvLSTM proposed in [24] improves the capability of LSTM by capturing spatiotemporal correlations. LSTM, which forms the basis for the ConvLSTM, is an RNN structure capable of capturing long-term dependencies in sequential data. It achieves this by introducing a memory cell C_t within the RNN. The information within the memory cell is regulated using four special neural network layers called gates. They interact with each other to decide whether the memory cell will be accessed, written or cleared during the processing of an input X_t. The memory cell C_t maintains the long-term dependencies observed within the input X_t, thus addressing the vanishing gradient problem.

A limitation of the LSTM is that it creates too much redundancy when operating on video sequences. The ConvLSTM deals with this issue by arranging the input as 3D tensors and processing it through convolution. The steps involved are as follows:

Given an input X with dimensions $mtimesntimest$, the gate i_t, decides whether any new information will be added to the memory cell C_t according to Eq. (1). Where H_t is the hidden state, $*$ is the convolution operator and $\circ$ is the Hadamard product.

$$i_t = \sigma\left(W_{xi} * X_t + W_{hi} * H_{t-1} + W_{ci} \circ C_{t-1} + b_i\right) \quad (1)$$

The decision to clear the past content on the memory cell C_{t-1} is made by the gate f_t following Eq. (2). While, the actual update to the memory cell C_t is performed according to Eq. (3).

$$f_t = \sigma\left(W_{xf} * X_t + W_{hf} * H_{t-1} + W_{cf} \circ C_{t-1} + b_f\right) \quad (2)$$

$$C_t = f_t \circ C_{t-1} + i_t \circ \tanh\left(W_{xc} * X_t + W_{hc} * H_{t-1} + b_c\right) \tag{3}$$

The gate o_t decides whether the content of C_t is accessible to the hidden state/output H_t according to Eqs. (4) and (5).

$$o_t = \sigma\left(W_{xo} * X_t + W_{ho} * H_{t-1} + W_{co} \circ C_{t-1} + b_o\right) \tag{4}$$

$$H_t = o_t \circ \tanh\left(C_t\right) \tag{5}$$

Thus, in the context of micro-expression recognition, the ConvLSTM can capture facial movement across frames using the convolution kernels. Since the frames maintain their 2D structure, the network can capture better spatiotemporal correlation.

In this paper, to capture the spatiotemporal correlation across frames, we use a single layer of ConvLSTM. The number of kernels is set to 32, with a kernel size of 3×3 and strides of 1×1. The output from the ConvLSTM is downsampled using max-pooling with a kernal size of 2×2 and stride of 1×1.

3.3 Dense Layers

The output of the max-pooling layer is flattened and passed on to the dense layers to perform recognition. In this proposal, we use two dense layers as the final two layers of the network. The first layer consists of 48 neurons with sigmoid activation function. Neurons in the second layer are set according to the number of micro-expressions to be recognised by the network. Being the final layer, it has a softmax activation function. We also introduce a dropout of 30% between the two layers.

4 Evaluation Result

As discussed in Sect. 2, the state-of-the-art dataset available for evaluating the proposal is the SAMM dataset. The dataset contains a total of 159 video sequences. Each of those sequences belongs to one of the seven micro-expressions or other class - see Fig. 2. However, the total number of sequences belonging to a micro-expression class is not equal. The disparity is quite significant, with the anger class having 57 sequences while the sad class has just 6. To address this issue, the works discussed in Sect. 2 have introduced several evaluation protocols. We adopt a combination of protocols presented in [20,21] and [22] to perform our evaluations.

4.1 Evaluation Protocol

Following the protocol in [20] and [21], we evaluate our proposal using two subsets of the SAMM dataset containing 3 and 5 classes, respectively. The evaluation is carried out using stratified k-fold cross-validation, where the value of k is set to 5. This protocol is an upgrade over the 80–20 split employed in [22]. We ensure that the five groups created by the k-fold cross-validation are mutually exclusive with respect to the participants. Also, the imbalanced class distribution is maintained in each group. The k-fold cross-validation is performed by iteratively selecting one group for validation while the remaining four groups are used for training. Once the validation is completed, the trained weights are discarded, and the process is repeated using the new groups. The performance is reported in terms of the accuracy and the F1 score. The F1 score is the harmonic mean of precision and recall, which is quite insightful when working with an unbalanced dataset.

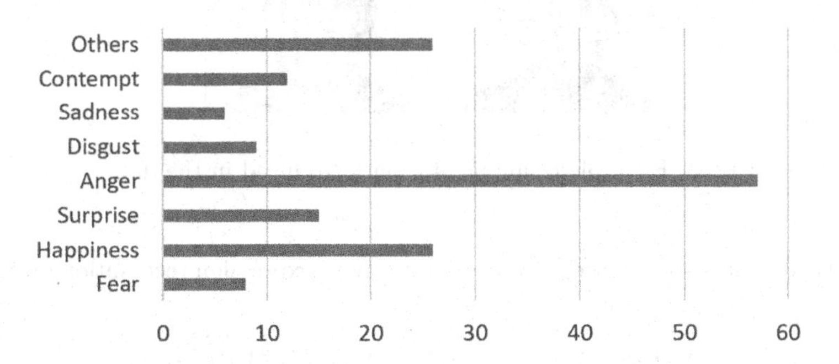

Fig. 2. Statistics of the SAMM dataset.

The problem of data imbalance is also partly addressed using data augmentation. We introduce a horizontal flip, change in brightness and salt and pepper noise in classes with fewer sequences, as illustrated in Fig. 3. The choice of introducing the augmentation is random, which prevents saturating a class with augmented data. We then train the model using the ADAM optimiser with a learning rate of 10^{-4}. The learning rate is set to reduce on plateau with a factor of 0.5. The loss is set to categorical cross-entropy loss, the epochs is set to 50 with early stopping, and the batch size is set to 16.

4.2 Result and Discussion

Table 1 reports the result of the three class micro-expression recognition. The three classes are positive (happiness), negative (anger + contempt + disgust + fear + sadness), and surprise, containing 26, 92 and 15 sequences, respectively. Table 1 also compares the results of the proposal with the state-of-the-art. Similarly, Table 2 reports the results of micro-expression recognition across five classes. The five classes are happiness, other, anger, contempt, surprise, containing 26, 26, 57, 12 and 15 sequences, respectively - see Fig. 2.

The results in Table 1 report an improvement of 10.15% and 24.69% in accuracy and F1 score, respectively, over the state-of-the-art for the three class problem. The improvement in the five class problem is also significant, with a 10.29% and 15.25% increase in the accuracy and F1 score, respectively - see Table 2.

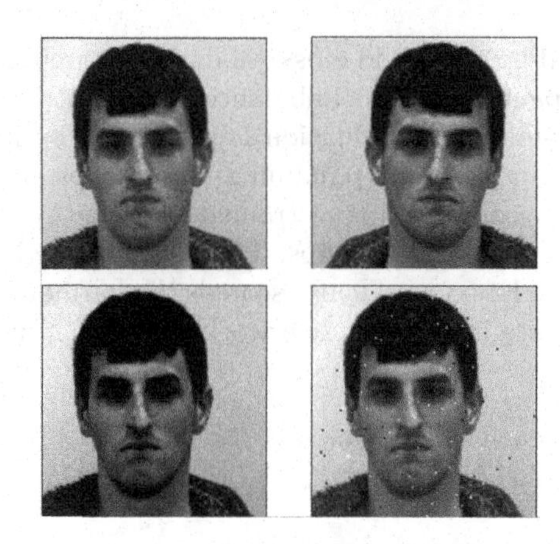

Fig. 3. Example of augmentation introduced in the data.

Table 1. Accuracy and F1 score of three class micro-expression recognition on SAMM dataset.

Method	Accuracy	F1 score
Weighted optical flow [16]	58.30	39.70
Shallow CNN [21]	68.18	54.23
Proposed ConvLSTM	**78.33**	**78.92**

Table 2. Accuracy and F1 score of five class micro-expression recognition on SAMM dataset.

Method	Accuracy	F1 score
LBP-TOP [11]	39.68	35.89
Truncated AlexNet [20]	57.35	46.44
CNN+LSTM [23]	50.00	52.44
Proposed ConvLSTM	**67.64**	**67.69**

The difference in performance can be attributed to the fact that most state-of-the-art methods emphasise capturing spatial information. The features across frames are aggregated using simple concatenation techniques. The lack of good spatiotemporal features leads to poor micro-expression recognition in [11,16,21] and [20]. Some state-of-the-art methods that consider temporal features, such as [23], employ a two-step approach of using CNNs followed by the LSTM, which improves its F1 score. However, our proposal beats the use of CNN+LSTM with an improvement in accuracy and F1 score of 67.64% and 15.25%, respectively. These results support our hypothesis that capturing spatiotemporal information within a video sequence using ConvLSTM is significantly more effective than using shallow CNNs, deep CNN+LSTM or hand-crafted techniques.

Some concerns about the results can arise due to the data imbalance in the dataset. Hence, to obtain better insights, we report the confusion matrix for 5 class micro-expression recognition results in Table 3. The results suggests that all classes contribute approximately evenly to the proposed network's accuracy. In Table 3 surprise class has the highest accuracy of 75.93%, while the lowest accuracy is reported for the happiness class at 60.94%. This result is in sharp contrast to the Truncated AlexNet [20], where the anger class has an accuracy of 90%, while the happiness, contempt, surprise and other classes have an accuracy of 35%, 42%, 33% and 35%, respectively. Here, the accuracy of the state-of-the-art network largely depends on the correct recognition of the class with the highest number of sequences. The reasonably uniform accuracies reported in Table 3 suggest that the proposal handles the data imbalance significantly better than the state-of-the-art. The uniformity in the accuracy is also present in the three class micro-expression recognition problem, where the negative class containing 92 sequences and surprise class containing 15 sequences report an accuracy of 79.41% and 78.85%, respectively- see Table 4.

Table 3. Confusion matrix for the five class micro-expression recognition problem

	Prediction					
	Class	Happiness	Contempt	Anger	Other	Surprise
Ground truth	Happiness	**60.94**	7.81	14.06	4.69	12.50
	Contempt	9.21	**64.48**	13.16	5.26	7.89
	Anger	8.70	2.17	**65.22**	17.39	6.52
	Other	3.34	3.34	10.00	**71.67**	11.67
	Surprise	9.26	5.56	9.26	0.00	**75.93**

Finally, we would like to comment on the other class. While anger and happiness are micro-expressions, other contain sequences that do not belong to any of the seven micro-expressions. Thus, a probable cause for the low accuracy in Table 3 can be the absence of similar features among the other class sequences. Therefore, we conducted another experiment using only the seven

Table 4. Confusion matrix for the three class micro-expression recognition problem

	Prediction			
	Class	Positive	Negative	Surprise
Ground truth	Positive	**76.74**	19.77	3.49
	Negative	13.24	**79.41**	7.36
	Surprise	2.88	18.27	**78.85**

micro-expression available in the SAMM dataset. Following the protocol discussed in Sect. 4.1, we achieved an accuracy of 72.64%. The increase in accuracy can be seen as support for our claim. However, larger datasets are needed to perform a more in-depth analysis.

5 Conclusion

In this paper, we address the problem of micro-expression recognition using a ConvLSTM-based network. The proposed shallow network can capture spatiotemporal information more effectively than other neural networks, such as CNNs and RNNs. We demonstrate the claim by comparing the proposal with the state-of-the-art comprising of hand-crafted, shallow CNN and fine-tuned deep CNN+LSTM networks. The proposed ConvLSTM-based network beats the state-of-the-art, thus emphasising its significance in capturing spatiotemporal information for micro-expression recognition. The proposed ConvLSTM based network also reports reasonably uniform accuracies across its classes in the face of the unbalanced SAMM dataset. The state-of-the-art performs poorly in such scenarios, as their performance is skewed by correct recognition of the class with the highest number of sequences.

The biggest challenge we still face is the absence of a large balanced dataset. Capturing a large dataset that allows us to explore deep ConvLSTM based networks can be a possible future direction. Since the downsampling of the input video sequences is a part of our pre-processing, the proposal performs micro-expression recognition with fewer frames than the state-of-the-art. Thus, its evaluation using low fps videos will also be a part of our future work.

References

1. Svetieva, E., Frank, M.G.: Empathy, emotion dysregulation, and enhanced micro-expression recognition ability. Motiv. Emot. **40**, 309–320 (2016)
2. Pan, H., Xie, L., Wang, Z., Liu, B., Yang, M., Tao, J.: Review of micro-expression spotting and recognition in video sequences. Virtual Reality Intell. Hardw. **3**(1), 1–17 (2021)
3. Yan, W.J., Wu, Q., Liang, J., Chen, Y.H., Fu, X.: How fast are the leaked facial expressions: the duration of micro-expressions. J. Nonverbal Behav. **37**, 217–230 (2013)

4. Polikovsky, S., Kameda, Y., Ohta, Y.: Facial micro-expressions recognition using high speed camera and 3D-gradient descriptor. In: International Conference on Imaging for Crime Detection and Prevention. IET (2009)
5. Li, X., Pfister, T., Huang, X., Zhao, G., Pietikäinen, M.: A spontaneous micro-expression database: inducement, collection and baseline. In: 10th IEEE International Conference and Workshops on Automatic Face and Gesture Recognition (2013)
6. Yan, W.J., Wu, Q., Liu, Y.J., Wang, S.J., Fu, X.: CASME database: a dataset of spontaneous micro-expressions collected from neutralized faces. In: 10th IEEE International Conference and Workshops on Automatic Face and Gesture Recognition (FG) (2013)
7. Yap, C.H., Kendrick, C., Yap, M.H.: SAMM long videos: a spontaneous facial micro- and macro-expressions database. In: 15th IEEE International Conference on Automatic Face and Gesture Recognition (2020)
8. Moilanen, A., Zhao, G., Pietikäinen, M.: Spotting rapid facial movements from videos using appearance-based feature difference analysis. In: 22nd International Conference on Pattern Recognition (2014)
9. Davison, A.K., Yap, M.H., Lansley, C.: Micro-facial movement detection using individualised baselines and histogram-based descriptors. In: IEEE International Conference on Systems, Man, and Cybernetics (2015)
10. Li, J., Soladie, C., Seguier, R.: Local temporal pattern and data augmentation for micro-expression spotting. IEEE Trans. Affect. Comput. (2020)
11. Zhao, G., Pietikainen, M.: Dynamic texture recognition using local binary patterns with an application to facial expressions. IEEE Trans. Pattern Anal. Mach. Intell. **29**, 915–928 (2007)
12. Wang, Y., See, J., Phan, R.C.W., Oh, Y.H.: LBP with six intersection points: reducing redundant information in LBP-top for micro-expression recognition. In: Asian Conference on Computer Vision (2014)
13. Huang, X., Zhao, G., Hong, X., Zheng, W., Pietikäinen, M.: Spontaneous facial micro-expression analysis using spatiotemporal completed local quantized patterns. Neurocomputing **175**, 564–578 (2016)
14. Xu, F., Zhang, J., Wang, J.Z.: Microexpression identification and categorization using a facial dynamics map. IEEE Trans. Affect. Comput. **8**, 254–267 (2017)
15. Liong, S.T., Phan, R.C.W., See, J., Oh, Y.H., Wong, K.: Optical strain based recognition of subtle emotions. In: International Symposium on Intelligent Signal Processing and Communication Systems (ISPACS) (2014)
16. Liong, S.T., See, J., Wong, K., Phan, R.C.W.: Less is more: micro-expression recognition from video using apex frame. Signal Process. Image Commun. **62**, 82–92 (2018)
17. Takalkar, M.A., Xu, M.: Image based facial micro-expression recognition using deep learning on small datasets. In: International Conference on Digital Image Computing: Techniques and Applications (DICTA) (2017)
18. Li, Q., Yu, J., Kurihara, T., Zhan, S.: Micro-expression analysis by fusing deep convolutional neural network and optical flow. In: 5th International Conference on Control, Decision and Information Technologies (CoDIT) (2018)
19. Wang, C., Peng, M., Bi, T., Chen, T.: Micro-attention for micro-expression recognition. Neurocomputing **410**, 354–362 (2020)
20. Khor, H.Q., See, J., Liong, S.T., Phan, R.C., Lin, W.: Dual-stream shallow networks for facial micro-expression recognition. In: IEEE International Conference on Image Processing (ICIP) (2019)

21. Gan, Y.S., Liong, S.T., Yau, W.C., Huang, Y.C., Tan, L.K.: OFF-ApexNet on micro-expression recognition system. Signal Process. Image Commun. **74**, 129–139 (2019)
22. Reddy, S.P.T., Karri, S.T., Dubey, S.R., Mukherjee, S.: Spontaneous facial micro-expression recognition using 3D spatiotemporal convolutional neural networks. In: International Joint Conference on Neural Networks (IJCNN) (2019)
23. Khor, H.Q., See, J., Phan, R.C.W., Lin, W.: Enriched long-term recurrent convolutional network for facial micro-expression recognition. In: 13th IEEE International Conference on Automatic Face Gesture Recognition (2018)
24. Shi, X., Chen, Z., Wang, H., Yeung, D.Y., Wong, W.K., Woo, W.C.: Convolutional LSTM network: a machine learning approach for precipitation nowcasting. In: Advances in Neural Information Processing Systems, vol. 28 (2015)

Adversarial Machine Learning Towards Advanced Vision Systems

ADVFilter: Adversarial Example Generated by Perturbing Optical Path

Lili Zhang(✉) and Xiaodong Wang

College of Computer, National University of Defense Technology, Changsha, China
{zhanglili18,xdwang}@nudt.edu.cn

Abstract. Deep Neural Networks (DNNs) have achieved great success in many applications, and they are taking over more and more systems in the real world. As a result, the security of DNN system has attracted great attention from the community. In typical scenes, the input images of DNN are collected through the camera. In this paper, we propose a new type of security threat, which attacks a DNN classifier by perturbing the optical path of the camera input through a specially designed filter. It involves many challenges to generate such a filter. First, the filter should be input-free. Second, the filter should be simple enough for manufacturing. We propose a framework to generate such filters, called ADVFilter. ADVFilter models the optical path perturbation by thin plate spline, and optimizes for the minimal distortion of the input images. ADVFilter can generate adversarial pattern for a specific class. This adversarial pattern is universal for the class, which means that it can mislead the DNN model on all input images of the class with high probability. We demonstrate our idea on MNIST dataset, and the results show that ADVFilter can achieve up to 90% success rate with only 16 corresponding points. To the best of our knowledge, this is the first work to propose such security threat for DNN models.

Keywords: Adversarial example · Deep neural networks · Security threat · Physical attack

1 Introduction

Deep Neural Networks (DNNs) [1] have achieved tremendous success in many real world applications. DNN models are taking over the control of more and more systems, which are traditionally considered to be operated only by humans, such as automatic pilot system. Therefore, the security [2] of DNN is directly related to the safety of the physical world, including the safety of human life and property. The community have conducted extensive research on DNN security [2].

An important direction of DNN security is adversarial example attack [3,4]. A deliberate perturbation on the input data can mislead the DNN model to incorrect predictions. For computer vision applications, the DNN model usually uses the camera to obtain the input image. Surprisingly, most cameras are

Y. Zheng et al. (Eds.): ACCV 2022, LNCS 13848, pp. 33–44, 2023.
https://doi.org/10.1007/978-3-031-27066-6_3

physically accessible to malicious users, so that it is difficult to detect abnormality even when installing or refitting a filter on the camera. Traditional security research only focuses on the electronic part of the DNN system, while this paper points out that the optical part can also trigger adversarial attacks.

Compared with traditional adversarial example attack, adversarial filter should conquer more difficulties. First, traditional adversarial example usually handles a specific input image, while adversarial filter is input independent. Clearly, we cannot preset the input image of the camera, so that adversarial filter should mislead the DNN model on the whole data distribution of a certain class. Second, traditional adversarial example can apply pixel-level modification, while adversarial filter should be simple enough for manufacturing. Under the view of adversarial example, it means that adversarial filter has much fewer dimensions to operate than the pixel-level adversarial example. Consequently, the difficulty of generating adversarial filter is much higher.

To conquer such difficulties, we propose a novel framework to generate adversarial filters, called ADVFilter. ADVFilter models the optical path perturbation by thin plate spline (TPS). We use as few corresponding points as possible in TPS method to control the complexity of the result, and optimizes for the minimal distortion of the input images. For a given DNN model, ADVFilter can generate adversarial pattern for a specific class. This pattern acts as a virtual filter to perturb all input images of the DNN model. The adversarial pattern is universal for the selected class, which means that it can mislead the DNN model on all input images of the class with high probability. We demonstrate our idea on MNIST dataset, and the results show that ADVFilter can achieve up to 90% success rate with only 16 corresponding points. To the best of our knowledge, this is the first work to propose such security threat for DNN models.

To summarize, we list our contributions as follows:

1. We propose the security threat for DNN model by perturbing the optical path, and demonstrate that this threat can lead to serious consequences.
2. We propose novel framework to generate adversarial filters, named ADVFilter. ADVFilter can generate universal adversarial filters, which attacks a DNN model in an input-independent way. Moreover, ADVFilter adopts as few corresponding points as possible to simplify the result.
3. We conduct experiments to verify the idea by MNIST dataset. The results show that ADVFilter can achieve up to 90% success rate with only 16 corresponding points, which constitutes a space with only 64 dimensions.

The rest of this paper is organized as follows. We introduce the related works in Sect. 2. We overview the design of ADVFilter in Sect. 3. Next, we present the details of ADVFilter in Sect. 4 and show the experiment results in Sect. 5. Finally, we conclude this work in Sect. 6.

2 Related Work

The research on the security of neural networks has attracted significant efforts [3–7], Early work focused on generating adversarial examples in different ways to explore the border of DNN security. Another research direction is to attack black-box model in an efficient way. This section mainly discusses the research closely related to our work.

Some papers [8–12] propose to generate universal adversarial examples that are image-agnostic. The adversarial patterns generated by these methods can be applied to different images. These patterns can convert a given image into an adversarial example. These methods only consider the color space of the pixels, so that they cannot solve the problem of adversarial filters. Compared with these works, ADVFilter only adopts several corresponding points to generate universal adversarial pattern. The dimension of optimization space is much smaller than that of traditional methods, so our problem is correspondingly more difficult.

Some researchers explore image transformation to generate adversarial examples [13–15]. These methods generate adversarial examples by rotations and translations. They only obtain the adversarial ability on single target image, which cannot be generalized to our scenario. In contrast, ADVFilter generates a perturb pattern that can be applied to any image of a certain class. The problem of ADVFilter is more general and therefore more difficult to solve.

Some studies break the boundary between physics and cyber space, so as to generate adversarial examples in the physical world. Early attempt simply prints adversarial examples and fool DNN classifier through camera input [16]. Study [17] also show that deliberately constructed glasses can mislead face recognition systems. Besides, printable adversarial patches can control the model prediction to a predefined target class [18]. Moreover, some studies [19–22] explore robust adversarial examples in a predefined distribution. They show the ability to fool DNN model under a continuous viewpoints in three dimensional physical world. Adv t-shirt [23] generates a t-shirt with special designed pattern that can evade person detectors. This work adopts TPS to model the non-rigid surface of the t-shirt. Adv t-shirt shows the potential of TPS-based technique to generate physical adversarial examples. Compared with this study, ADVFilter deals more general cases that should fit all potential inputs of the target model.

In a word, perturbing optical path is a new type of DNN attack, which cannot be realized by simply expanding the existing technology.

3 ADVFilter Overview

This section introduces the design of the ADVFilter, and shows the architecture of the framework.

We show the architecture of ADVFilter in Fig. 1. In the figure, the part surrounded by dotted lines is a standard DNN prediction pipeline. The model collects images through an input camera, and outputs the prediction results. In order to facilitate the algorithm description, we only discuss the case of white

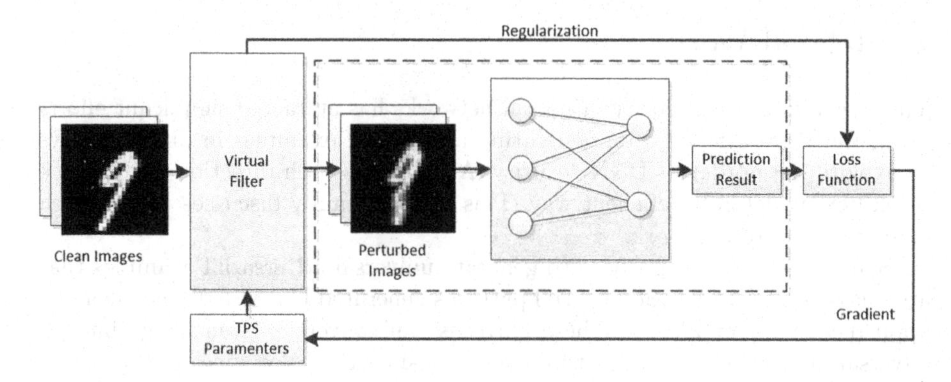

Fig. 1. The architecture of ADVFilter

box attack in this section. We will discuss how to extend ADVFilter to black box attack in the following section. In order to evaluate the effect of a malicious filter on the camera, we design the virtual filter module, which is the core of the whole system. After adding the virtual filter, all input images of the DNN model will be universally perturbed by the virtual filter. In the figure, the clean images are the images of the physical world, and the perturbed images are the images perturbed by the virtual filter. Under this configuration, the DNN model can only get the perturbed images as the input.

We model the effect of the adversarial filter by TPS. TPS is a physical analogy involving the bending of a thin sheet of metal, which is widely used as the non-rigid transformation model in image alignment and shape matching. It achieves several advantages to adopt TPS for adversarial filter generation. First, TPS produces smooth surface, which is consistent with the characteristics of optical path change. Second, TPS adopts several discrete corresponding points to control the entire transformation over the whole image, which facilities the manufacturing of the adversarial filter. Third, TPS has closed form solutions for both warping and parameter estimation, which can be integrated to current deep learning frameworks. The mathematical details of the virtual filter module are discussed in the next section.

In the view of the learning architecture, the goal of generating adversarial filter is to get an appropriate set of parameters for the virtual filter model, i.e. the parameters of the TPS transformation. The goal of the optimization is two folded. On one hand, the result should mislead the target DNN model, so that the loss function contains the prediction result of the DNN model. On the other hand, the overall distortion of the image should be as low as possible. We add several regularization terms to the loss function to limit the deformation of the images. Finally, the parameters of TPS are optimized by gradient information. The final loss function of ADVFilter consists of three parts:

$$Loss = \lambda_1 L_{\text{CrossEntropy}} + \lambda_2 L_{\text{radius}} + \lambda_3 L_{\text{distortion}} \tag{1}$$

In the loss function, $L_{\text{CrossEntropy}}$ is standard cross entropy loss to control the output label, expressed as:

$$L_{\text{CrossEntropy}} = -\sum y_i \log(p_i) \tag{2}$$

where y_i denotes each element of the one-hot vector of target class, and p_i denotes each element of the predicted probability vector of the target model. The cross entropy of a batch is the mean of each cross entropy loss for each data in the batch.

Besides, L_{radius} and $L_{distortion}$ are two regularization terms to control the visibility of the adversarial filter, λ_1, λ_2 and λ_3 are hyper parameters to adjust the attack ability and distortion. These regularization terms are closely related to TPS implementation. We leave the detail design of this part to the following section.

4 ADVFilter Design

In this section, we formally define the problem of generating adversarial filters and show the mathematical details of ADVFilter design. We also adopt white-box attack model for clarity and discuss the black-box case in the end of this section.

We first introduce the formal definition of adversarial examples. Let set X and set Y denote the possible input and output of the target model, respectively. Given a particular input $x \in X$ and a target class $y \in Y$, we can obtain the model prediction $P(y \mid x)$, as well as the gradient $\nabla_x P(y \mid x)$. We use a function M to denote the predict result with maximum probability. For example, $M(x) = y$ means that label y is the most likely result under input x. Traditional adversarial example is to find an input x' within the ε-ball of the initial input x, i.e. $\|x' - x\| < \varepsilon$ that is classified as another class y_t, i.e. $M(x') = y_t$ and $y_t \neq y$.

The definition of adversarial filter is on the whole input set rather than a single image. An adversarial filter is defined as a transformation F with parameter θ on the whole input set X. For any input $x \in X$, the result of the transformation $F(x;\theta)$ is misclassified by the target model, i.e. $M(F(x;\theta)) = y_t$ and $y_t \neq y$. Hence, generating adversarial filter is to find an appropriate transformation $F(x;\theta)$ and optimize parameter θ.

ADVFilter adopts TPS as the transformation function. In two dimensional case, the TPS fits a mapping function between two corresponding point-sets $\{p_i\}$ and $\{p_i'\}$ one by one. The result is to minimize the following energy function:

$$E(F) = \sum_{i=1}^{K} \|p_i' - F(p_i)\|^2 \mid \tag{3}$$

where K is the number of corresponding points. We do not consider the smoothness variant of standard TPS, so that the mapping results accurately coincide with the target corresponding points. Given the corresponding point-sets $\{p_i\}$ and $\{p_i'\}$, the TPS has closed-form solution for the optimal mapping.

Clearly, the parameter space for standard TPS is $\theta = \{p_i; p'_i\}$. To simplify the expression in deep learning framework, we convert the TPS parameters into an equivalent form. We preserve the source corresponding point-sets $\{p_i\}$ and rewrite the target corresponding point-sets $\{p'_i\}$ in a relative form. That is,

$$p'_i = p_i + r_i R(\alpha_i) \tag{4}$$

where r_i denotes the distance between the two corresponding points, $R(\alpha) = \begin{bmatrix} \cos(\alpha) \\ \sin(\alpha) \end{bmatrix}$ denotes the relative angle between the two corresponding points. The parameter space of the problem is $\theta = \{p_i, r_i, \alpha_i\}$. After the transformation, the radius regularization term of the loss function is as follow:

$$L_{\text{radius}} = \left(\sum r_i^2\right)^{1/2} \tag{5}$$

Moreover, we also set an upper bound value r_{max} for all radius values and clip the maximum absolute value of each r_i to r_{max}.

The distortion regularization term is defined as the distance between the original image and the transformed image as follow:

$$L_{\text{distortion}} = \left[\sum (x_i - F(x_i; \theta))^2\right]^{1/2} \tag{6}$$

Note that all TPS transformations share the same parameter θ, which is the most significant difference between this work and existing researches.

The parameter space of the optimization is $\theta = \{p_i, r_i, \alpha_i\}$, where p_i denotes the locations of the source corresponding points, r_i and α_i denote the distance and relative angles between the corresponding points. We adopt Adam optimizer with learning rate 0.1 to search the optimal solution.

Fig. 2. Initial distribution of the corresponding points

The initialization of corresponding point $\{p_i\}$ is in grid form. Let $n = k^2$, $k \in N$ denote the total number of corresponding points. We set n to be a square number for simplicity. Suppose the size of the input image is $h \times w$, then the distribution of all k^2 points is a two-dimensional grid with an interval of $\frac{h}{k+1} \times \frac{w}{k+1}$, as shown in Fig. 2. This design ensures that the TPS transformation can capture the vulnerability of any position of the input image.

For black-box attack case, ADVFilter can leverage off-the-shelf gradient estimation scheme, such as zero order optimization [7] or natural evolution strategies [24]. The cost of these algorithms depends heavily on the dimension number of the entire search space. Note that the search space of TPS is very low. For example, if we adopt 16 corresponding points, the dimension of the whole search space is only 16*4 = 64, which is far less than the dimensions of pixel color space. As a result, the overhead of gradient estimation is correspondingly low.

5 Experiments

We conduct several experiments to verify the correctness and the efficiency of ADVFilter.

5.1 Experiment Setup

We conduct all the experiments on a laptop computer with Intel Core i7-8550U and 16G RAM. The software platform for deep learning is Tensorflow 1.15.4.

The target dataset is the MNIST database, which is a data collection of handwritten digits. In the process of generating the adversarial pattern, we record all the local optimal solutions. The criteria include two aspects: the success rate of the attacking and the average l2 distance between the clean images and the adversarial images in the tested batch. Based on this vector criterion, we record all results with lower average l2 distance or higher success rate. To facilitate comparison, we fix the average l2 distance to a predefined value and compare the success rate in different settings.

For all the following tests, the batch size is 100, which means we randomly sample 100 images from the dataset. In each iteration, we clip the distances between all corresponding points to 0.05. That is, we set $r_{max} = 0.05$ to control the distortion of the adversarial images.

5.2 Correctness Test

This section demonstrates how adversarial filter attacks the DNN model. We conduct an untargeted attack with all images labeled "9" in the dataset. The number of corresponding points is 9, which are arranged in 3*3 grid initially.

Figure 3 shows the experiment results. The first row of the images are the clean images, which are the input of the adversarial filter. The second row of the images are the adversarial images, which are output of the adversarail filter. The colored dots are the corresponding points of the TPS conversion. Clearly, the same color indicates the corresponding relationship of points. After several iterations, all corresponding points are far away from their initial grid position, and moves to the sensitive position to change the features of the target images.

Note that all the adversarail images follow the same conversion pattern, which is the major difference between our work with existing ones. Comparing the clean images with the adversarail images, we found that most areas of the images had

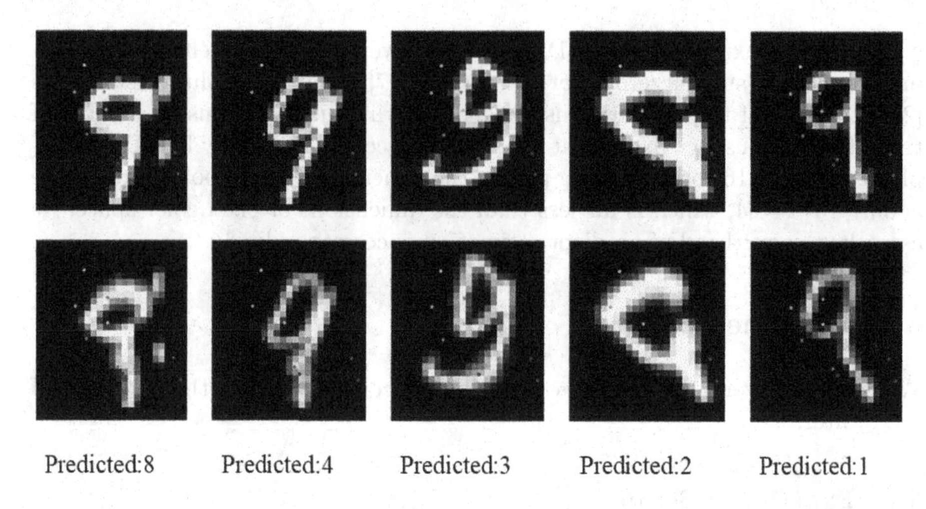

Fig. 3. The clean images and the adversarial images in MNIST dataset

only a small amount of modification, and only the lower right part is deflected sharply to the right. Through the visual judgment of human eyes, it is difficult to distinguish between the clean images and adversarail images. Therefore, we will still label these adversarial images as "9". In contrast, these adversarail images are all misclassified by the DNN classifier.

5.3 Comparison with SOTA Universal Adversarial Examples

This section compare our ADVFilter with existing universal adversarial examples [8–11]. We perform the experiment with the same configuration as in Sect. 5.2. That is, we use the same DNN model, the same dataset, and the same source image set. Under this configuration, we generate pixel-level universal adversarial example (PUAE), as shown in Fig. 4. The perturbation of all adversarial examples follows the same pattern. After adding the perturbation, we clip the color value of each pixel to the legal range.

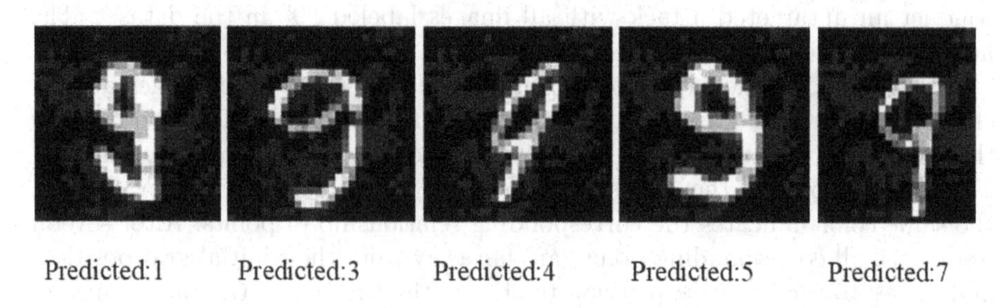

Fig. 4. The adversarial examples generated by perturbing color space

It is difficult to make a quantitative comparison between PUAE and ADV-Filter, because there are inherent difference in the techniqual roadmap. We first discuss the similarities between PUAE and ADVFilter. First, both PUAE and ADVFilter generate a universal adversarial pattern, which is independent of specific input image. Second, they all achieved a relatively high attack success rate, i.e. 90% in our experiment. Finally, the adversarial examples generated by both methods are not easily perceptible to humans, while the invisibility are defined in completely different ways. Next, we discuss the differences between PUAE and ADVFilter. First of all, the dimension of freedom for manipulating color space is much larger than that of TPS. As a result, the problem of generating an adversarial filter is much more difficult. Second, The result of ADVFilter only change the position distribution of pixels, hence the distribution of the color space is similar to the clean image. In contrast, PUAE significantly changes the color distribution of the image. Last, in terms of physical manufacturing, adv-filter does not need pixel level alignment, thus to accept higher manufacturing errors. In contrary, PUAE requires pixel-level alignment accuracy, which is more challenging.

5.4 The Number of Corresponding Points

This section discusses the impact of the number of corresponding points on the success rate. We conduct an untargeted attack with all images labeled "9" in the dataset. We repeat the experiments with different number of corresponding points and record the success rate under different settings, as shown in Fig. 5.

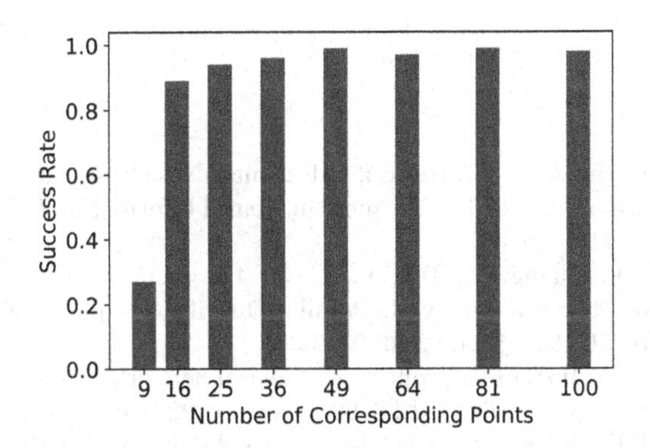

Fig. 5. The success rate under various numbers of corresponding points

In order to evenly distribute all corresponding points in the image, we set the number of corresponding points to be square number. The test range covers from three square to ten square, i.e. from 9 to 100. We conclude from the figure that the success rate will increase approximately monotonically with the increase of the number of the corresponding points in the first half. However, there is an upper limit of this increase trend. After reaching a specific value, i.e. 36 in this test, the success rate will remains stable.

This experiment shows that universal adversarial examples can be generated in a very low dimensional subspace. For example, in this experiment, we use only 16 corresponding points to achieve the success rate of about 90%. Even if we ignore the restriction between the dimensions of the corresponding points, this means that our estimated dimension is higher than the actual one. The total dimension of this subspace is only $16 * 4 = 64$.

6 Conclusion and Future Work

In this paper, we propose a new type of threat for DNN system, which is to generate adversarial example by perturbing optical path. One possible way to apply this idea is to add a specially designed filter to the input camera of the target DNN model. We define the problem of generating such an adversarial filter, and propose a framework, named ADVFilter. ADVFilter models the optical path perturbation by thin plate spline, and optimizes for the minimal distortion of the input images. The generated adversarail filter can mislead the DNN model on all input images of the class with high probability, which shows that the threat can lead to serious consequences. The future work is to physically produce a filter and test the attack effect in the physical world.

Acknowledgements. We are particularly grateful to Inwan Yoo who implements TPS in Tensorflow and shares the code on https://github.com/iwyoo/tf_ThinPlateSpline.

References

1. Krizhevsky, A., Sutskever, I., Hinton, G.E.: ImageNet classification with deep convolutional neural networks. In: Advances in Neural Information Processing Systems 2012, pp. 1097–1105 (2012)
2. Wang, X., Li, J., Kuang, X., Tan, Y., Li, J.: The security of machine learning in an adversarial setting: a survey. J. Parallel Distrib. Comput. **130**, 12–23 (2019). https://doi.org/10.1016/j.jpdc.2019.03.003
3. Szegedy, C., et al.: Intriguing properties of neural networks. Presented at the ICLR (2014). http://arxiv.org/abs/1312.6199. Accessed 22 Aug 2019
4. Goodfellow, I.J., Shlens, J., Szegedy, C.: Explaining and harnessing adversarial examples. Presented at the ICLR (2015). http://arxiv.org/abs/1412.6572. Accessed 22 Aug 2019
5. Chen, P.-Y., Zhang, H., Sharma, Y., Yi, J., Hsieh, C.-J.: ZOO: zeroth order optimization based black-box attacks to deep neural networks without training substitute models. In: Proceedings of the 10th ACM Workshop on Artificial Intelligence and Security (AISec), pp. 15–26 (2017). https://doi.org/10.1145/3128572.3140448
6. Ilyas, A., Engstrom, L., Athalye, A., Lin, J.: Black-box Adversarial Attacks with Limited Queries and Information (2018). http://arxiv.org/abs/1804.08598. Accessed 18 Aug 2019

7. Tu, C.-C., et al.: AutoZOOM: Autoencoder-Based Zeroth Order Optimization Method for Attacking Black-box Neural Networks (2019). http://arxiv.org/abs/1805.11770. Accessed 18 Aug 2019
8. Moosavi-Dezfooli, S.-M., Fawzi, A., Fawzi, O., Frossard, P.: Universal Adversarial Perturbations, p. 9 (2017)
9. Li, Y., Bai, S., Xie, C., Liao, Z., Shen, X., Yuille, A.: Regional homogeneity: towards learning transferable universal adversarial perturbations against defenses. In: Vedaldi, A., Bischof, H., Brox, T., Frahm, J.-M. (eds.) ECCV 2020. LNCS, vol. 12356, pp. 795–813. Springer, Cham (2020). https://doi.org/10.1007/978-3-030-58621-8_46
10. Zhang, C., Benz, P., Imtiaz, T., Kweon, I.S.: Understanding Adversarial Examples From the Mutual Influence of Images and Perturbations, p. 10 (2020)
11. Baluja, S., Fischer, I.: Learning to attack: adversarial transformation networks. In: The Thirty-Second AAAI Conference on Artificial Intelligence (AAAI) 2018, p. 9 (2018)
12. Li, M., Yang, Y., Wei, K., Yang, X., Huang, H.: Learning universal adversarial perturbation by adversarial example. In: Proceedings of the AAAI Conference on Artificial Intelligence, vol. 36, pp. 1350–1358 (2022). https://doi.org/10.1609/aaai.v36i2.20023
13. Engstrom, L., Tran, B., Tsipras, D., Schmidt, L., Madry, A.: Exploring the Landscape of Spatial Robustness (2019). http://arxiv.org/abs/1712.02779. Accessed 28 Apr 2022
14. Alaifari, R., Alberti, G.S., Gauksson, T.: ADef: an Iterative Algorithm to Construct Adversarial Deformations (2019). http://arxiv.org/abs/1804.07729. Accessed 28 Apr 2022
15. Xiao, C., Zhu, J.-Y., Li, B., He, W., Liu, M., Song, D.: Spatially Transformed Adversarial Examples (2018). http://arxiv.org/abs/1801.02612. Accessed 28 Apr 2022
16. Kurakin, A., Goodfellow, I., Bengio, S.: Adversarial examples in the physical world, arXiv:1607.02533 Cs Stat (2016). http://arxiv.org/abs/1607.02533. Accessed 22 Aug 2019
17. Sharif, M., Bhagavatula, S., Bauer, L., Reiter, M.K.: Accessorize to a crime: real and stealthy attacks on state-of-the-art face recognition. In: Proceedings of the 2016 ACM SIGSAC Conference on Computer and Communications Security - CCS 2016, Vienna, Austria, pp. 1528–1540 (2016)
18. Brown, T.B., Mané, D., Roy, A., Abadi, M., Gilmer, J.: Adversarial Patch, arXiv:1712.09665 Cs (2017). http://arxiv.org/abs/1712.09665. Accessed 22 Aug 2019
19. Athalye, A., Engstrom, L., Ilyas, A., Kwok, K.: Synthesizing Robust Adversarial Examples (2018). http://arxiv.org/abs/1707.07397. Accessed 28 July 2019
20. Athalye, A., Carlini, N., Wagner, D.: Obfuscated Gradients Give a False Sense of Security: Circumventing Defenses to Adversarial Examples (2018). http://arxiv.org/abs/1802.00420. Accessed 18 Aug 2019
21. Eykholt, K., et al.: Robust physical-world attacks on deep learning visual classification. In: IEEE/CVF Conference on Computer Vision and Pattern Recognition (CVPR), Salt Lake City, UT, USA, pp. 1625–1634 (2018). https://doi.org/10.1109/CVPR.2018.00175
22. Wang, D., et al.: FCA: learning a 3D full-coverage vehicle camouflage for multi-view physical adversarial attack. In: Proceedings of the AAAI Conference on Artificial Intelligence, vol. 36, pp. 2414–2422 (2022). https://doi.org/10.1609/aaai.v36i2.20141

23. Xu, K., et al.: Adversarial T-shirt! evading person detectors in a physical world. In: Vedaldi, A., Bischof, H., Brox, T., Frahm, J.-M. (eds.) ECCV 2020. LNCS, vol. 12350, pp. 665–681. Springer, Cham (2020). https://doi.org/10.1007/978-3-030-58558-7_39
24. Wierstra, D., Schaul, T., Peters, J., Schmidhuber, J.: Natural evolution strategies. J. Mach. Learn. Res. **15**, 949–980 (2014). https://doi.org/10.1109/CEC.2008.4631255

Enhancing Federated Learning Robustness Through Clustering Non-IID Features

Yanli Li[1(✉)], Abubakar Sadiq Sani[2], Dong Yuan[1], and Wei Bao[3]

[1] School of Electrical and Information Engineering, The University of Sydney, Sydney, NSW 2006, Australia
{yanli.li,dong.yuan}@sydney.edu.au

[2] School of Computing and Mathematical Sciences, Faculty of Engineering and Science, University of Greenwich, London SE10 9LS, UK
S.Sani@greenwich.ac.uk

[3] School of Computer Science, The University of Sydney, Sydney, NSW 2006, Australia
wei.bao@sydney.edu.au

Abstract. Federated learning (FL) enables many clients to train a joint model without sharing the raw data. While many byzantine-robust FL methods have been proposed, FL remains vulnerable to security attacks (such as poisoning attacks and evasion attacks) because of its distributed nature. Additionally, real-world training data used in FL are usually Non-Independent and Identically Distributed (Non-IID), which further weakens the robustness of the existing FL methods (such as Krum, Median, Trimmed-Mean, etc.), thereby making it possible for a global model in FL to be broken in extreme Non-IID scenarios.

In this work, we mitigate the vulnerability of existing FL methods in Non-IID scenarios by proposing a new FL framework called Mini-Federated Learning (Mini-FL). Mini-FL follows the general FL approach but considers the Non-IID sources of FL and aggregates the gradients by groups. Specifically, Mini-FL first performs unsupervised learning for the gradients received to define the grouping policy. Then, the server divides the gradients received into different groups according to the grouping policy defined and performs byzantine-robust aggregation. Finally, the server calculates the weighted mean of gradients from each group to update the global model. Owning the strong generality, Mini-FL can utilize the most existing byzantine-robust method. We demonstrate that Mini-FL effectively enhances FL robustness and achieves greater global accuracy than existing FL methods when against the security attacks and in Non-IID settings.

Keywords: Federated Learning (FL) · Byzantine-robust aggregation · Untargeted model attack

1 Introduction

Federated Learning (FL) is an emerging distributed learning paradigm that enables many clients to train a machine learning model collaboratively while

Y. Zheng et al. (Eds.): ACCV 2022, LNCS 13848, pp. 45–59, 2023.
https://doi.org/10.1007/978-3-031-27066-6_4

keeping the training data decentralized and users' privacy protected [13]. Generally speaking, FL contains three steps: 1) a server broadcasts the current global model to selected clients; 2) each client locally trains the model (called local model) and sends back the local model updates[1]; and 3) the server updates the global model by aggregating the local model updates received through a particular aggregation algorithm (AGR).

However, the distributed nature of training data makes FL vulnerable to various attacks (such as poisoning attacks) by malicious attackers and untrusted clients. Poisoning attack, which seeks to damage the model and generate misbehaviour, draws the most important threats to FL security. Through poisoning in different training stages, poisoning attacks can lead the global model to show an indiscriminate accuracy reduction (called untargeted attack) or attacker-chosen behaviour on a minority of examples (called targeted attack) [13]. One popular defence solution against the untargeted attack is introducing the byzantine-robust aggregation rule [3,4,11,20] on the server to update the global model. By comparing the client's model updates, these aggregation rules can find and discard the statistical outliers and prevent the suspected model uploaded from poisoning the global model. Although most of the studies [3,20] are designed and evaluated in an Independent and Identically Distributed (IID) setting. Assuming each client's data follows the same probability distribution, the training data in real-world FL applications are usually Non-IID due to location, time, and user clusters reasons, which make the existing byzantine-robust FL methods show little effectiveness and even fully break when facing the state-of-the-art attack [9].

The most common sources of Non-IID are a client corresponding to a particular location, a particular time window, and/or a particular user cluster [13,15]. In terms of location, various kinds of locations factors drive the most impact on the Non-IID of a dataset. For instance, the mammal's distributions are different due to the geographic location [12], customer profiles are different due to various city locations [18], and emoji usage patterns are different due to the demographic locations [13]. In terms of a time window, people's behaviour and objects' features can be very different at different times. For instance, the images of the parked cars sometimes are snow-covered due to the seasonal effects, and people's shopping patterns are different due to the fashion and design trends. In terms of a particular user, different personal preferences can result in a dataset Non-IID. For instance, [5] shows students from different disciplines have very different library usage patterns.

In this paper, we design a new FL framework, namely Mini-FL framework, to mitigate the research gap. Mini-FL considers the main source of Non-IID and identifies Geo-feature, Time-feature, and User feature as the alternative grouping features. Based on the grouping feature selected, the server defines the grouping principle through performing unsupervised learning. In each iteration, the server first assigns the received gradients to different groups and then performs byzantine-robust aggregation, respectively. Finally, the server aggregates the aggregation outcomes (called group gradient) from each group to update the

[1] In this work, we combined use "model update" and "gradient" with same meaning.

global model in each iteration. We use Krum [3], Median [20], and Trimmed-mean [20] as the byzantine-robust aggregation rule to evaluate our Mini-FL on the various dataset from different Non-IID levels. Our results show that Mini-FL effectively enhances the security of existing byzantine-robust aggregation rules and also reaches a high level of accuracy (without attack) in the extreme Non-IID setting. We also provide a case study to further demonstrate the effectiveness of Mini-FL in the real world.

Our contributions are summarized as follows:

- We propose the group-based aggregation method and identify three features (i.e., Geo-feature, Time-feature, and User-feature) as the grouping principles.
- We propose the Mini-FL framework to enhance the robustness of existing FL methods. Our results show these methods can achieve byzantine robustness through the Mini-FL framework even in an extreme Non-IID setting.

2 Related Work

2.1 Poisoning Attacks on Federated Learning

Poisoning attacks generally indicate the attack type that crafts and injects the model during training time. These attacks include data poisoning attacks [2] and model poisoning attacks [6,8,9,12,18] which are performed by poisoning the training data owned and gradients, respectively. The model poisoning attack directly manipulates gradients, which can bring higher attack impacts to FL.

Based on the adversary's goals, the attacks can be further classified into untargeted attacks [6,8,9,12,18] (model downgrade attacks) and targeted attacks [10,16] (backdoor attacks). In untargeted attacks, the adversary aims to reduce the global model's accuracy and entirely 'break' the model by participating in the learning task. In contrast, target attacks maintain the global model's overall accuracy but insert 'back door' in minority examples. These back-doors can result in a wrong reaction when the attacker-chosen action event occurs. For instance, [10] can force GoogLeNet to classify a panda as a gibbon by adding an imperceptibly small vector on the panda image; the Faster RCNN can not detect the 'stop' sign that added small perturbations [16]. As the untargeted draws lead to security threats for FL, we consider the setting of **untargeted model poisoning attacks** in this study which shows as follows:

"Reverse attack" [6] and "Random attack" [8] poison the global model by uploading a reverse gradient and a random gradient. "Partial drop attack" [8] replaces the gradient parameter as a 0 with a given probability and subsequently uploads the crafted gradient to poison the global model. "Little is enough attack" [1] and "Fall of empires attack" [19] leverage the dimension curse of machine learning and upload the crafted gradient by adding perturbation on the mean of the gradient owned (based on the capability). "Local model poisoning attack" [9] is a state of art attack. It infers the convergence direction of the gradients and uploads the scaled, reverse gradient to poison the global model.

2.2 Byzantine-Robust Aggregation Rules for FL

The FL server can effectively average and aggregate the local models received in non-adversarial settings [17]. However, linear combination rules, including averaging, are not byzantine resilient. In particular, a single malicious worker can corrupt the global model and even prevent global model convergence [3]. Therefore, the existing byzantine-robust aggregation rules have been designed to replace the averaging aggregation and address byzantine failures. Next, we discuss the popular byzantine-robust aggregation rules.

Krum [3]**:** Krum discards the gradients that are too far away from benign gradients. In particular, for each gradient received, Krum calculates the sum Euclidean distance of a number of the closest neighbours as the score. The gradient with the lowest score is the aggregation outcome and becomes the new global model in this iteration. As the number of the closest neighbors selected influences the score, Krum requires the number of attackers.

Trimmed-Mean and Median [20]**:** Trimmed-mean is a coordinate-wise aggregation rule which aggregates each model parameter, respectively. Specifically, for a given parameter, the server firstly sorts the parameter from all gradients received. Then, the server discards a part of the largest and smallest values and finally averages the remaining gradients as the corresponding parameter of the new global model in this iteration. The Median method is another coordinate-wise aggregation rule. In the Median method, the server firstly sorts the parameter from all gradients received and selects the median as the corresponding parameter of the new global model in this iteration.

Bulyan [11]**:** Bulyan can be regarded as a combination of Krum and Trimmed-mean. Specifically, Bulyan first selects a number of gradients by performing Krum (the gradient is then removed from the candidate pool once selected). Then Bulyan performs Trimmed-mean in the gradients selected to update the global model.

FLTrust [4]**:** FLTrust considers both the directions and magnitudes of the gradients. Particularly, the server collects a clean dataset and owns a corresponding model; in each iteration, FLTrust first calculates the cosine similarity between the gradient received and owned. The higher cosine similarity gradient gains a higher trust score and consequently participates in the weighted average with a higher proportion. Instead of directly participating in the aggregation, each gradient is normalized by the gradient server owned before the weighted average.

Table 1 illustrates the robustness of the existing FL methods/proposed Mini-FL methods against different attacks under the IID/Non-IID settings. Since these attacks (i.e., untargeted attack) aim to reduce the model's global accuracy indiscriminately, we use the global testing accuracy to evaluate the robustness of FL methods.

Table 1. The robustness of the existing FL methods against poisoning attacks

	"Reverse", "Random"		"Partial"		"Little", "Fall"		"Local"	
	IID	Non-IID	IID	Non-IID	IID	Non-IID	IID	Non-IID
Average	×	×	×	×	×	×	×	×
Krum	✓	×	✓	×	O	×	×	×
Trimmed-Mean	O	×	×	×	O	×	O	O
Median	✓	O	✓	O	✓	O	O	×
Bulyan	✓	O	✓	O	✓	O	O	O
FL-Trust	✓	O	✓	O	✓	O	✓	O
Mini Krum	✓	✓	✓	✓	✓	✓	✓	✓
Mini T-Mean	✓	O	✓	×	✓	✓	✓	×
Mini Median	✓	✓	✓	✓	✓	✓	✓	✓

Non: Non-IID, ✓: effective, O: partially effective, ×: ineffective.

3 Problem Setup

3.1 Adversary's Objective and Capability

Adversary aims to reduce the model's global accuracy or 'fully break' the global by uploading the malicious gradients; this is also known as untargeted model poisoning attacks or model downgrade attacks [10,13,16]. We consider the adversary's capability and knowledge from three dimensions: the adversary amount, the malicious client's distribution, and the knowledge of aggregation rule. We assume the adversary controls some clients, called malicious clients, and we keep the setting of the adversary number of each existing FL method that Krum: $2f + 2 < n$, Trimmed-Mean: $2f < n$ and Trimmed-Median: $2f < n$, where f is the number of attackers, n denotes the number of all clients. The adversary knows the local training data on malicious clients and can arbitrarily send crafted local model updates to the server in each iteration. To guarantee the generality, we assume the distribution of malicious clients and benign clients are similar. Furthermore, we assume the adversary knows the aggregation rules but does not know the grouping principle.

3.2 Defense Objective and Capability

We aim to develop the FL framework to achieve byzantine robustness against untargeted attacks and embody the data minimization principle. Specifically, the new framework does not need clients to upload further information beyond local model updates. The server plays the defender's role and has access to the information naturally brought with the gradients uploaded (e.g., IP, Timestamp, etc.). We notice some byzantine-robust aggregation rules need to know the upper bound of the malicious clients [3,20]; we follow these settings but don't leak further information of malicious clients; specifically, the defender does not know the distribution of malicious clients.

4 Mini-FL Design and Analysis

4.1 Overview of Mini-FL

In our Mini-FL, the server assigns the model updates received into different groups and executes byzantine-robust aggregation accordingly. Specifically, Mini-FL follows the general FL framework but adds a new step (i.e., Grouped model aggregation) before the Global model update. Furthermore, a prepossessing step: Grouping principle definition is introduced before the training task starts. Figure 1 illustrates the Mini-FL framework.

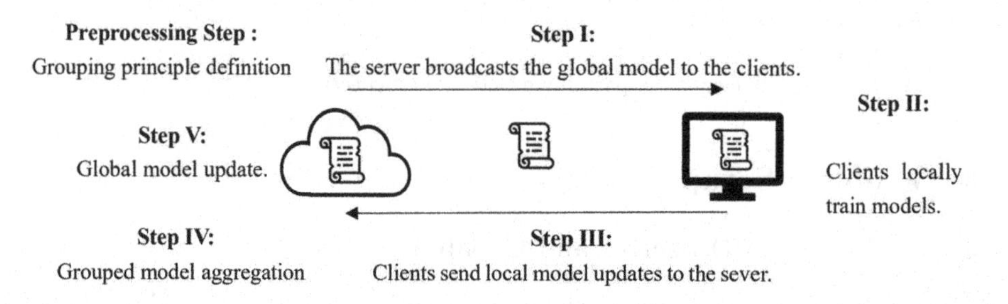

Fig. 1. Illustration of the Mini-FL framework.

To craft the malicious gradient and avoid being excluded by byzantine-robust aggregation rules, the adversary commonly statistically analyzes the gradient owned and calculates (or infers) the range of the benign gradients. By restricting the crafted gradient under this range, the attackers can effectively hide their gradients in benign gradients and subsequently attack the global model. However, because most federated learning models are trained through Non-IID data, the gradients uploaded naturally tend to be clustered due to location, time and user clusters reason. Thus, Mini-FL firstly defined the groups and then execute byzantine-robust aggregation accordingly. The similar behaviour of each group brings a smaller gradient range and therefore results in a smaller attack space. Finally, the server aggregates the outcome from each group and updates the global model to finish the current iteration.

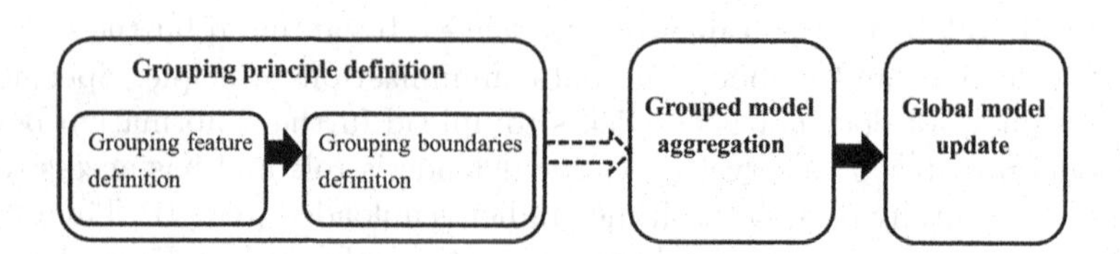

Fig. 2. Illustration of the Mini-FL aggregation rule.

4.2 Mini-FL Framework

Our Mini-FL considers leveraging the Non-IID nature of federated learning to define groups and execute byzantine-robust aggregation accordingly. Figure 2 illustrates the Mini-FL aggregation rule.

Grouping Principle Definition. Before the learning task starts, the server defines the grouping principle (i.e., prepossessing step), which includes "grouping feature definition" and "grouping boundaries definition"; the grouping principle could only be defined before the learning task starts or is required to be updated.

- **Grouping feature definition**: The existing research [13] believes the major sources of Non-IID are due to each client corresponding to a particular geographic location, a particular time window, and/or a particular user. For instance, [12] demonstrates the real-world example of skewed label partitions: geographical distribution of mammal pictures on Flickr, [13] illustrates the same label can also look very different at different times(e.g., seasonal effects, fashion trends, etc.).
  Considering the major source of Non-IID and the features naturally carried in server-client communication, we identify textbfGeo-feature (e.g., IP address), Time-feature (e.g., Timestamp), and User-feature (e.g., User ID) of the local model update as the based grouping feature to maintain the principle of focused collection and guarantee the effectiveness of clustering.
  When defining the grouping feature, the server firstly regroups the gradient collection C by Geo-feature; the collection C should accumulate the gradients received in a few iterations to maintain the generality. Then, we execute the 'elbow method' [14] to detect the number for clustering and subsequently get the SSE (i.e., Sum of the Squared Errors, which reflected the grouping effectiveness). By repeating the first two steps through replacing the Geo-feature with Time-feature and User-feature, we can find the feature F with the lowest SSE. Finally, we select that feature F acts as the grouping feature and the corresponding elbow point as the number of groups.
- **Grouping boundaries definition**: Once the grouping feature has been defined, we cluster the collection regrouped through unsupervised learning. In this research, we use the K-means algorithm to execute the unsupervised learning; the "elbow" point is assigned to the algorithm as the number of groups. By analyzing the gradient's feature value in different groups, the grouping boundaries could be defined.

Grouped Model Aggregation. According to the grouping principle, the server divides the gradients received into different groups and executes byzantine-robust aggregation respectively. The mini-FL framework has strong generality and can utilize most existing byzantine-robust aggregation rules. In this research, we used 'Krum,' 'Trimmed-mean,' and 'Median' for aggregation in this research, and the detail of the experiments are studied in Sect. 5.

Global Model Update. The server calculates the weighted mean of grouped gradients (i.e., outcome from each group) and updates the global model to finish this iteration.

4.3 Security Enhancement Analysis

In this section, we analyze the security enhancement of Mini-FL from 'information asymmetry' perspective.

As discussed in Sect. 2, most existing byzantine-robust aggregation rules can effectively detect and discard the malicious gradient if it is far (based on Euclidean distance) from benign gradients. To guarantee the attack effectiveness and avoid being excluded by the byzantine-robust aggregation rules, a common perturbation strategy is determining the attack direction and then scaling the crafted gradient to stay close with benign gradients. Depending on different knowledge, the adversary can precisely or generally infer the statistics (e.g., max, min, mean, and Std (Standard Deviation)) of the benign gradients and subsequently scale the crafted gradient; Table 2 illustrates the scalier of gradient crafted in different attacks.

Table 2. Illustration of the scalier of gradient crafted in different attacks.

Attack	Crafted gradients range
"Little" [1]	$(\mu - z\sigma, \mu + z\sigma)$ μ: mean, z: scalar (set 0$\sim$1.5 in research), σ: Std.
"Fall" [19]	$(-z\mu, -z\mu)$ μ: mean, z: scalar (0$\sim$10 in research), σ: Std.
"Local" [9]	$(\mu + 3\sigma, \mu + 4\sigma)$ when the adversary has partial knowledge. or (μ-4σ, μ-3σ) depends on the gradient direction $(Wmax, z * Wmax)$ when the adversary has full knowledge. or $(z * Wmin, Wmin)$ depends on the gradient direction μ: mean, z: scalar(set 2 in research), σ: Std, $Wmax/Wmin$: the max/min gradient value at that iteration

However, Mini-FL defines the grouping principles and clusters the gradients received **only** on the server-side. The information asymmetry makes the adversary hardly infer the members of different groups, much less calculate the relevant statistical parameters to scale the crafted gradients and bypass the defense of Mini-FL.

5 Evaluation

5.1 Experimental Setup

Dataset. We evaluate our Mini-FL framework on the MNIST [7]. To simulate the dataset pattern in the real world, we set different Non-IID degrees when distributing training data. Suppose we have m groups of clients and l different data labels; we set training data size as s and assign $p * s$ training examples with label l to the client group m with probability p, then we randomly select and assign other $s-(p * s)$ training data to m groups. As the parameter p controls the distribution of training data on clients, we call p the Non-IID degree. To further embody the source of each Non-IID distribution, we assign a feature (i.e., Geo, Time, or User feature) for each item of local model updates.

MNIST-1.0: The MNIST [7] (Modified National Institute of Standards and Technology) database is an extensive database of handwritten digits that includes 60,000 training images and 10,000 testing images. To simulate people's different handwriting habits in different countries [13], we divide clients into five groups; each group owns one unique IP range (reflect different countries) and training examples with two different labels (reflect different handwriting habits). We use MNIST-1.0 ($p = 1.0$) to simulate the extreme Non-IID situation (Non-IID degree $= 1.0$). In other words, each group only has two different unique labels of training examples in MNIST-1.0.

MNIST-0.75 and MNIST-0.5: We use MNIST-0.75 and MNIST-0.5 to evaluate the effectiveness of Mini-FL in different Non-IID degrees. MNIST-0.75 and MNIST-0.5 have similar settings as MNIST-1.0, but the Non-IID degree p is 0.75 and 0.5, respectively.

Evaluated Poisoning Attacks. Mini-FL provides a new framework to enhance the security of FL and the excellent generalization enables Mini-FL can introduce most existing byzantine-robust aggregation rules. We introduce Krum [3], Trimmed-mean [20], Median [20] in experiments, respectively, and select the following poisoning attacks to evaluate the effectiveness of Mini-FL; we have not introduced FL-Trust in Mini-FL as FL-Trust does not fit extreme Non-IID scenarios - Krum adapted attacks can achieve 90% attack success rate when the root dataset's bias probability is over 0.6 [4].

"Reverse Attack" [6]**:** "Reverse attack" poisons the global model through uploading the reverse gradient. We follow the setting in [6] and set the attack multiple as 100.

"Random Attack" [8]**:** "Random attack" poisons the global model through uploading a random gradient.

"Partial Drop Attack" [8]**:** "Partial drop attack" masks the gradient parameter as 0 with probability p. As the parameter naturally carries a few 0 in our training tasks, we enhance the attack strength by replacing the mask 0 as -1 and setting p as 0.8 in experiments.

"Little is Enough Attack" [1]**:** "Little is enough attack" leverages the dimension curse of ML and upload the crafted gradient where gradient $= \mu + \mathrm{z} * \sigma$; here, μ and σ are the mean and standard deviation of the gradients respectively. z is the attack multiple, and we set z as 1.035, 1.535, and 2.035.

"Fall of Empires Attack" [19]: "Fall of empires attack" uploads the crafted gradient where gradient = $-z * \mu$. Here, μ is the mean of gradients and z is the attack multiple; we set z as 1 and 10.

"Local Model Poisoning Attack" [9]: "Local model poisoning attack" is a state of art attack. It infers the convergence direction of the gradients and uploads the scaled, reverse gradient to poison the global model. We follow the default setting in [9] for the local model poisoning attack.

Evaluation Metrics. Since these attacks (i.e., untargeted attack) aim to reduce the model's global accuracy indiscriminately, we use the testing accuracy to evaluate the effectiveness of our Mini-FL. In particular, we use a part of data owned as testing examples and test the model's global accuracy each iteration. The testing accuracy reflects the model's robustness against byzantine attacks; in other words, it is more robust if the model has a higher testing accuracy. We further use the existing FL methods with the original framework as the baseline to compare against.

FL System Setting. Without other specific notifications, we use the setting as follows.

Global model setting: As this study does not aim to improve the model accuracy through crafting the model, we use a general model for training MNIST. This model consists of a dense layer (28 * 28) and a softmax layer (10).

Learning parameters: We set the learning rate as 0.01, the batch size as 128, and the epoch as 50. We set the global iterations as 300. As some byzantine-robust methods (Krum in this study) require the parameter M for the upper bound of the number of malicious clients, we follow the setting in [3] that the server knows the exact number of all malicious clients. However, since Mini-FL defines groups and performs aggregation accordingly, Mini-FL further requires the malicious clients m of each group when introducing Krum. To maintain the generality, We set m belong with the group size:

$$m = \frac{n_{group}}{N_{global}} M$$

Here, n_{group} is the client number of the group (i.e., group size) and N_{global} is the total number of clients. In other words, we do not give any privilege to Mini-FL, and Mini-FL can only use the proportion to infer the number of malicious clients in each group.

Clients & data setting: We assume 20 clients participate in the learning task in each iteration, and 25% of clients are malicious. In Mini-FL, gradients are assigned in different groups as they carry different features. To simulate the Non-IID setting in the real world, we assign different numbers of clients to different groups; subsequently, the larger group has more malicious clients. Table 3 illustrates the setting detail for the MNIST (Non-IID degree = 1.0).

Table 3. Illustration of the setting (Client & Data) for the MNIST.

	Group1	Group2	Group3	Group4	Group5
Training labels	1, 2	3, 4	5, 6	7, 8	9, 0
Client ID	C1, C6 C11, C16, C19	C2, C7 C12, C17, C20	C3, C8 C13, C18	C4, C9 C14	C5, C10 C15
Attackers	C1, C11	C2, C12	C3	None	None

5.2 Experimental Results

The results show Mini-FL achieves better robustness than the existing FL methods. Figures 3 and 4 illustrate the global accuracy of the existing FL methods/Mini-FL methods under different Non-IID degrees. When increasing the Non-IID degree, the results show that most Mini-FL methods can maintain a similar global accuracy under the same attack, while the existing FL methods witness decreasing global accuracy. For instance, Mini-median stably maintains around 90% global accuracy against various attacks and Non-IID settings. In contrast, Median achieves around 85% global accuracy against various attacks in MNIST-0.5 but drops global accuracy to 64.62%, 26.51%, 55.79%, and 53.68% in MNIST-1.0 under "little attack", "fall attack", "random attack", and "drop attack", respectively.

Mini-FL Achieves the Defense Objectives: Recall that the defense objectives include two parts (see Sect. 3): **achieving byzantine robustness against untargeted attacks** and **maintaining the data minimization principle of FL.** The experimental results show our Mini-FL framework achieves these goals.

Table 4. The global accuracy of different FL/Mini-FL methods under different Non-IID degrees and non-attack setting

	MNIST-1.0	MNIST-0.75	MNIST-0.5
Avg	88.89%	90.04%	90.38%
Krum	88.25%	77.27%	87.47%
Mini Krum	89.51%	90.03%	90.09%
Median	53.60%	80.45%	86.30%
Mini median	90.34%	90.26%	90.47%
Trimmed-mean	86.88%	88.29%	89.24%
Mini trimmed-mean	90.38%	90.47%	90.51%

First, Mini-FL achieves similar global accuracy as FedAvg (average aggregation rule) in the non-attack setting, but most existing byzantine robust FL methods have a decreased accuracy. For instance, FedAvg and all Mini-FL methods (i.e., Mini-Krum, Mini-Median, Mini-Trimmed mean) achieve over 90% global accuracy on MNIST-0.75 while Krum, Median, Trimmed mean get 77.27%, 80.45%, 88.29%, respectively. Table 4 illustrates the global accuracy of different FL/Mini-FL methods under different Non-IID degrees and non-attack settings. The result shows the Mini-FL framework increases the accuracy for existing

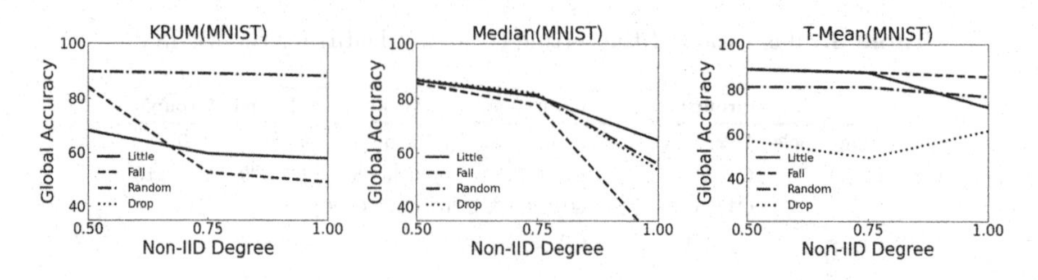

Fig. 3. The robustness of existing FL-methods under different Non-IID levels.

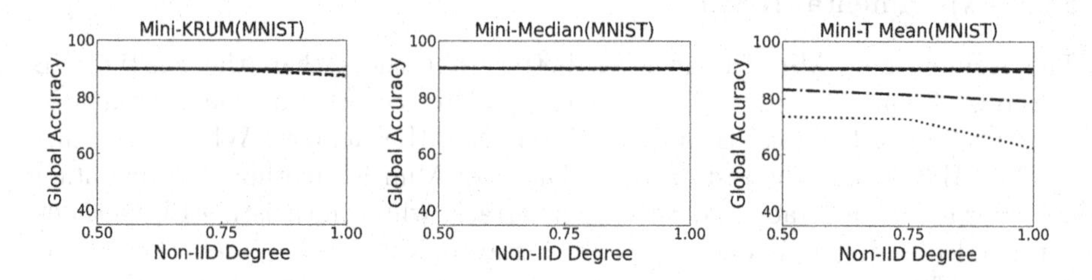

Fig. 4. The robustness of Mini FL-methods under different Non-IID levels.

Table 5. The global accuracy of FL/Mini-FL methods under different Non-IID degrees and non-attack setting

	Average	Krum	Mini Krum	T-Median	Mini T-Median	T-Mean	Mini T-Mean
(a) MNIST-1.0							
Little (2.035)	74.71%	74.92%	89.62%	53.76%	90.37%	52.83%	89.76%
Little (1.035)	84.42%	57.78%	89.70%	64.62%	90.34%	71.95%	90.40%
Fall (10)	23.73%	77.34%	88.38%	54.98%	90.10%	61.54%	90.18%
Fall (1)	78.23%	48.97%	87.61%	26.51%	89.95%	85.41%	89.41%
Random	80.19%	88.03%	89.68%	55.79%	90.37%	76.80%	79.04%
Partial Drop	61.65%	88.05%	87.33%	53.68%	90.42%	61.47%	62.33%
Local	78.62%	n/d	n/d	2.85%	89.77%	64.51%	87.32%
(b) MNIST-0.75							
Little (2.035)	66.89%	83.84%	89.74%	81.25%	90.22%	61.05%	89.94%
Little (1.035)	89.49%	59.55%	89.81%	80.65%	90.34%	87.33%	90.41%
Fall (10)	85.33%	77.27%	89.77%	79.15%	89.98%	64.09%	90.23%
Fall (1)	89.63%	52.43%	89.74%	77.59%	90.10%	87.51%	89.95%
Random	74.85%	88.93%	89.74%	81.28%	90.24%	80.85%	81.28%
Partial Drop	69.99%	88.83%	89.77%	81.87%	90.31%	49.36%	72.53%
Local	85.16%	n/d	n/d	62.31%	90.00%	75.90%	88.78%
(c) MNIST-0.5							
Little (2.035)	79.15%	88.71%	90.02%	86.56%	90.36%	80.83%	90.34%
Little (1.035)	89.88%	68.06%	90.01%	86.51%	90.35%	88.80%	90.41%
Fall (10)	88.38%	89.66%	90.03%	85.88%	90.38%	71.64%	90.35%
Fall (1)	90.11%	84.03%	89.99%	85.48%	90.41%	88.77%	90.30%
Random	78.43%	89.67%	90.02%	86.60%	90.36%	80.92%	83.08%
Partial Drop	72.06%	89.66%	90.00%	86.86%	90.43%	57.11%	73.48%
Local	86.37%	n/d	n/d	80.68%	90.28%	76.48%	89.81%

FL methods in the non-attack scenario. This is because benign gradients could be very different in the Non-IID setting, which may be regarded as malicious gradients and discarded by the existing FL method. As Mini-FL performs the aggregation by groups, it could comprehensively collect features from different groups and guarantee global accuracy.

Second, our Mini-FL shows better robustness and stability than most existing FL methods against different attacks and under different Non-IID settings. Specifically, most Mini-FLs can maintain the unattacked global accuracy even facing a state of art attack and under an extreme Non-IID setting; on the contrary, existing FL methods immensely decrease global accuracy and even be fully broken. For instance, Mini-median achieves 89.77% global accuracy in MNIST-1.0 under 'local attack,' while Median drops global accuracy from 53.60% to 2.85%. Table 5 illustrates the global accuracy of FL/Mini-FL methods under different Non-IID degrees and different attacks.

Moreover, the result shows that although the Mini-trimmed mean improves the robustness for the trimmed mean method, it achieves lower global accuracy than other Mini-Fl methods. For instance, Mini-trimmed mean achieves 62.33% global accuracy under drop attack in MNIST1.0 while other Mini-FL methods get around 90%. This is because the original FL method (Trimmed mean ($\beta = 20\%$)) draws a larger attack surface than Krum and Median as Trimmed mean ($\beta = 20\%$) accept and aggregates 80% gradients received while Krum and Median accept only one gradient.

Third, Mini-FL maintains the principles of focused collection and data minimization of FL. All of the information used for grouping (i.e., IP address, response time, and client ID) are naturally carried by the gradients when uploading. Mini-FL neither asks clients to upload their information further nor digs their features through reverse engineering, which provides the same privacy protection as the existing FL methods.

6 Discussion and Future Work

Mini-Krum and Bulyan: Mini-Krum and Bulyan [11] are different, although both of them rely on performing Krum and mean/trimmed methods. Specifically, Mini-Krum performs Krum by group and generates the weighted average as the global model. In contrast, Bulyan globally performs Krum n times to select n gradients and performs Trimmed-mean to generate the global model. As Bulyan does not consider the Non-IID setting of FL, it faces a similar degraded performance as other FL methods in Non-IID scenarios.

Non-IID sources: As Geo-feature, Time-feature and User-feature are the most common source of Non-IID in the real world, we select these three features as the grouping feature in this research, but we note that the Non-IID source could be more complicated and even be a combination in some cases [13]. We leave investigating further to explore more possibilities of Non-IID sources and improve the Mini-FL method.

7 Conclusion

We evaluated the robustness of existing FL methods in different Non-IID settings and proposed a new framework called Mini-FL to enhance Federated Learning robustness. The main difference between Mini-FL and existing FL methods is that Mini-FL considers FL's Non-IID nature and performs the byzantine tolerant aggregation in different groups. Our evaluation shows that Mini-FL effectively enhances existing FL methods' robustness and maintains a stable performance against untargeted model attacks and different Non-IID settings.

References

1. Baruch, G., Baruch, M., Goldberg, Y.: A little is enough: circumventing defenses for distributed learning. In: Advances in Neural Information Processing Systems, vol. 32 (2019)
2. Biggio, B., Nelson, B., Laskov, P.: Poisoning attacks against support vector machines. arXiv preprint arXiv:1206.6389 (2012)
3. Blanchard, P., El Mhamdi, E.M., Guerraoui, R., Stainer, J.: Machine learning with adversaries: byzantine tolerant gradient descent. In: Advances in Neural Information Processing Systems, vol. 30 (2017)
4. Cao, X., Fang, M., Liu, J., Gong, N.Z.: FLtrust: byzantine-robust federated learning via trust bootstrapping. arXiv preprint arXiv:2012.13995 (2020)
5. Collins, E., Stone, G.: Understanding patterns of library use among undergraduate students from different disciplines. Evid. Based Libr. Inf. Pract. **9**(3), 51–67 (2014)
6. Damaskinos, G., El-Mhamdi, E.M., Guerraoui, R., Guirguis, A., Rouault, S.: Aggregathor: byzantine machine learning via robust gradient aggregation. In: Proceedings of Machine Learning and Systems, vol. 1, pp. 81–106 (2019)
7. Deng, L.: The MNIST database of handwritten digit images for machine learning research [best of the web]. IEEE Signal Process. Mag. **29**(6), 141–142 (2012)
8. El-Mhamdi, E.M., Guerraoui, R., Guirguis, A., Hoang, L.N., Rouault, S.: Genuinely distributed byzantine machine learning. In: Distributed Computing, pp. 1–27 (2022)
9. Fang, M., Cao, X., Jia, J., Gong, N.: Local model poisoning attacks to {Byzantine-Robust} federated learning. In: 29th USENIX Security Symposium (USENIX Security 2020), pp. 1605–1622 (2020)
10. Goodfellow, I.J., Shlens, J., Szegedy, C.: Explaining and harnessing adversarial examples. arXiv preprint arXiv:1412.6572 (2014)
11. Guerraoui, R., Rouault, S., et al.: The hidden vulnerability of distributed learning in byzantium. In: International Conference on Machine Learning, pp. 3521–3530. PMLR (2018)
12. Hsieh, K., Phanishayee, A., Mutlu, O., Gibbons, P.: The non-IID data quagmire of decentralized machine learning. In: International Conference on Machine Learning, pp. 4387–4398. PMLR (2020)
13. Kairouz, P., et al.: Advances and open problems in federated learning. Found. Trends® Mach. Learn. **14**(1–2), 1–210 (2021)
14. Kodinariya, T.M., Makwana, P.R.: Review on determining number of cluster in k-means clustering. Int. J. **1**(6), 90–95 (2013)

15. Li, Q., Diao, Y., Chen, Q., He, B.: Federated learning on non-IID data silos: an experimental study. In: 2022 IEEE 38th International Conference on Data Engineering (ICDE), pp. 965–978. IEEE (2022)
16. Lu, J., Sibai, H., Fabry, E.: Adversarial examples that fool detectors. arXiv preprint arXiv:1712.02494 (2017)
17. McMahan, B., Moore, E., Ramage, D., Hampson, S., Arcas, B.A.: Communication-efficient learning of deep networks from decentralized data. In: Artificial Intelligence and Statistics, pp. 1273–1282. PMLR (2017)
18. Moreno-Torres, J.G., Raeder, T., Alaiz-Rodríguez, R., Chawla, N.V., Herrera, F.: A unifying view on dataset shift in classification. Pattern Recogn. **45**(1), 521–530 (2012)
19. Xie, C., Koyejo, O., Gupta, I.: Fall of empires: breaking byzantine-tolerant SGD by inner product manipulation. In: Uncertainty in Artificial Intelligence, pp. 261–270. PMLR (2020)
20. Yin, D., Chen, Y., Kannan, R., Bartlett, P.: Byzantine-robust distributed learning: towards optimal statistical rates. In: International Conference on Machine Learning, pp. 5650–5659. PMLR (2018)

Towards Improving the Anti-attack Capability of the RangeNet++

Qingguo Zhou[1], Ming Lei[1], Peng Zhi[1], Rui Zhao[1], Jun Shen[2], and Binbin Yong[1](✉)

[1] School of Information Science and Engineering, Lanzhou University, Tianshui South Road 222, Lanzhou 730000, Gansu, China
yongbb@lzu.edu.cn

[2] School of Computing and Information Technology, University of Wollongong, Wollongong, NSW, Australia

Abstract. With the possibility of deceiving deep learning models by appropriately modifying images verified, lots of researches on adversarial attacks and adversarial defenses have been carried out in academia. However, there is few research on adversarial attacks and adversarial defenses of point cloud semantic segmentation models, especially in the field of autonomous driving. The stability and robustness of point cloud semantic segmentation models are our primary concerns in this paper. Aiming at the point cloud segmentation model RangeNet++ in the field of autonomous driving, we propose novel approaches to improve the security and anti-attack capability of the RangeNet++ model. One is to calculate the local geometry that can reflect the surface shape of the point cloud based on the range image. The other is to obtain a general adversarial sample related only to the image itself and closer to the real world based on the range image, then add it into the training set for training. The experimental results show that the proposed approaches can effectively improve the RangeNet + +'s defense ability against adversarial attacks, and meanwhile enhance the RangeNet++ model's robustness.

Keywords: Adversarial attacks · Adversarial defenses · Semantic segmentation · RangeNet++ · Local geometry · Adversarial samples

1 Introduction

As part of the 3D point cloud scene understanding, semantic segmentation is a very important task in the field of autonomous driving. Semantic segmentation assigns a class label to each data point in the input modality, where a data point can be a pixel from a camera or a 3D point acquired from a radar. A stable semantic segmentation model can greatly improve the safety of autonomous driving systems. However, most semantic segmentation models perform poorly when facing the adversarial attacks and they have fairly weak adversarial defenses.

In recent years, there are many researches on applying adversarial attacks and adversarial defenses to various practical scenarios. Weng et al. [3] used adversarial attacks to find flaws in image QR codes. Wu et al. [4] summarized methods for image generation and

Y. Zheng et al. (Eds.): ACCV 2022, LNCS 13848, pp. 60–70, 2023.
https://doi.org/10.1007/978-3-031-27066-6_5

Fig. 1. Velodyne HDL-64E laser scan from KITTI dataset [1] with semantic information from RangeNet++. Best viewed in color, each color represents a different semantic class [2].

editing based on adversarial networks. Deng et al. [5] proposed a hippocampal segmentation method that combines residual attention mechanism with generative adversarial networks. However, there have been few studies on adversarial attacks and adversarial defenses of point cloud semantic segmentation models.

The essence of adversarial attack is to generate adversarial disturbances that can mislead model judgments and add them to samples. Therefore, how to generate aggressive and versatile adversarial samples is the focus of this research, while adversarial defense mainly focuses on improving the robustness of the model so that the model can still work normally in the face of disturbance. Among them, adding adversarial samples to the training set is a commonly used method for adversarial defense.

As one of the best point cloud semantic segmentation methods in the field of autonomous driving, the basic concept of RangeNet++ is to convert the point cloud into a range image, and then put it into a 2D convolutional network for learning, so as to obtain the label of the point. An example of RangeNet++ is shown in Fig. 1. However, the semantic information about the point cloud contained in the range image is too simple, which is not enough to support the model's good performance under adversarial attacks. Therefore, we can enrich the semantic information of the point cloud by obtaining the local geometric features of the points, so that the model has better robustness.

In this paper, we improve the security and anti-attack capability of the RangeNet++ model in two ways.

One is to add adversarial samples to the training set. We propose a method to obtain perturbation based on the surrounding local information of each point in the range image. Considering the fact that the semantic segmentation model classifies each point based on the feature differences between different objects, our method of generating adversarial samples calculates the average feature difference of each point and all points around it based on the range image. By adding the feature difference between each point and its surrounding points, the feature difference will become blurred, which makes it difficult for the semantic segmentation model to obtain the unique feature values of different objects and ensures the effectiveness of adversarial samples.

The other method is designed in such a way that instead of directly extracting local features for each point in the point cloud, we will calculate the difference between the normal vectors of different triangular patches formed by each point in the image and the surrounding points according to the range image to obtain the local geometry features. Such features reflect the topographic features of the 3D point cloud surface, thereby they will be increasing the semantic information of the point cloud.

In summary, contribution of our work are as follows:

i) A method for generating local features of point cloud based on range image and reflecting the geometric shape of point cloud.
ii) A method for generating general adversarial samples based on range image that are closer to the real world and adding them to the training set.
iii) Adversarial attack against RangeNet++ added to our method and original RangeNet++.

The rest of this paper is organized as follows. Section 2 presents a brief introduction to related works of local features of point clouds and adversarial examples. In Sect. 3, we describe the methodology of local feature extraction, and the generation of adversarial samples are also detailed. In Sect. 4, experimental results and the analysis are presented. Section 5 concludes this paper.

2 Related Work

2.1 Local Features of Point Clouds

Point cloud features are mainly divided into single point features, local features and global features. Single point features only focus on their own details, ignoring global information, and global features cannot describe the details sufficiently. In contrast, local features strike a balance between these two aspects.

Rusu et al. [6, 7] proposed to use point feature histograms (PFH) to obtain the shape information of the local surface. Using this method to describe point cloud features can extract rich information, but the computational complexity is high. It is also computationally expensive, especially for feature extraction tasks on large point cloud data.

In order to solve this problem, the Fast Point Feature Histograms (FPFH) method [8] reduces the computational complexity of the algorithm at the expense of reducing feature information by simplifying the interconnection characteristics of the PFH center point neighborhood, but it is simply affected by the point cloud sampling density. The feature distribution will be quite different when facing different sampling scales.

The Viewpoint Feature Histogram (VFH) [9] operator uses the angle between the viewpoint direction and the normal to the sampling point as a feature. However, VFH loses the property of rigidly maintaining the relative positional relationship between point clouds, and has poor performance in point cloud segmentation and registration.

Splash descriptor [10] describes local features by calculating the angle component between the normal vector of the key point and the normal vector of all points in the neighborhood.

Inspired by these works, we take the points in the range image as the key points, calculate the normal vectors of several triangular patches formed between itself and other key points in the neighborhood, and take the cosine value of any included angle between two normal vectors as the local feature of the point cloud. This feature can satisfactorily describe the surface shape of the point cloud in the neighborhood of the key points, better utilizing the morphological features of the point cloud.

2.2 Generation of Adversarial Examples

The key to adversarial sample generation is to find suitable and effective adversarial maneuvers. The fast gradient sign method (FGSM) proposed by Ian J et al. [11] maximizes the loss function to obtain the corresponding perturbation by backpropagating the gradient. Based on this, the DeepFool method proposed by Moosavi et al. [12] aimed at the selection of artificially specified perturbation coefficients in FGSM. This method firstly obtains good-performance adversarial samples by calculating the minimum necessary perturbation. Algorithms such as FGSM and DeepFool are all focusing on adversarial samples with different perturbation degrees for each adversarial sample generated by a single sample. In a work based on DeepFool, Moosavi et al. [13] extended it to propose a noise independent of the input image, known as Universal Adversarial Perturbations (UAP), which could destroy most natural images and would become a universal adversarial noise that could be trained offline and generate interference to the corresponding output of a given model online. These methods all rely heavily on the model to be attacked, and the adversarial disturbances generated for a certain model often perform poorly for other models and have poor scalability.

Xiang et al. [14] firstly proposed an adversarial sample generation method for 3D point clouds, and several new algorithms to generate adversarial point clouds for PointNet. However, on the one hand, the direct manipulation based on 3D point cloud will result in a large amount of computation. On the other hand, it is mainly used for adversarial sample generation of a single object and does not work well in point cloud segmentation where multiple objects exist.

On the basis of the above work, we herein propose a model-independent adversarial perturbation generation method for the semantic segmentation model. By calculating the difference between each point and its neighboring points in the image, the perturbation is generated to make the points in the image similar, so that the boundaries of each object become blurred in the process of semantic segmentation then the effective adversarial sample can be obtained.

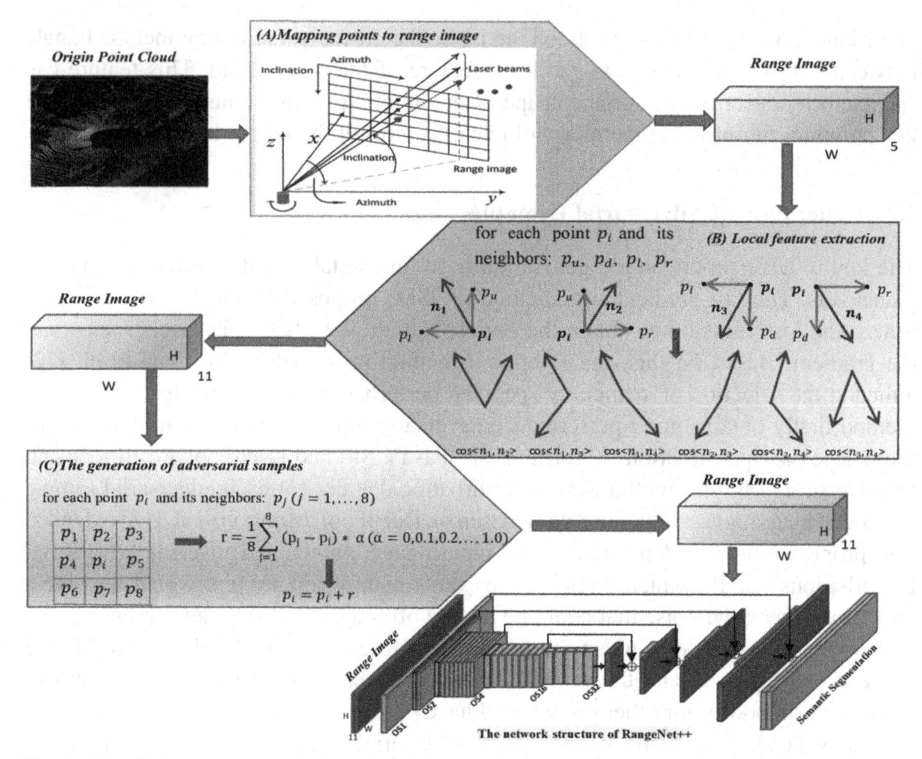

Fig. 2. The flow of our model. (A) convert the point cloud into a range image according to the method in RangeNet++, (B) extract local features based on the range image, (C) generate adversarial samples based on the range image, and put the new range image into the network of RangeNet++.

3 Methods

Our work is an improvement of the RangeNet++ model, which is carried out on the basis of range image, so we will first review the range image processing component in RangeNet++ in this section, and then introduce the local feature extraction method and adversarial samples generation method.

Our model is as follows: (A) convert the point cloud into a range image according to the method in RangeNet++; (B) extract local features based on the range image; (C) generate adversarial samples based on the range image, and then put the new range image into the network of RangeNet++. See Fig. 2 for details.

3.1 Mapping of Points to Range Image

For the points in the point cloud, the contained information of each of them includes coordinates *(x, y, z)* and remission. To obtain accurate semantic segmentation of point clouds, RangeNet++ converts each point cloud into a depth representation. To achieve this, it converts each point $p_i = (x, y, z)$ to spherical coordinates by mapping *Π: R^3*

$\rightarrow$ *R^2,* and finally to image coordinates, we denote the height and width of the range image by *H, W,* while using *u, v* to represent the mapping of each point to the range image, then this mapping can be defined as:

$$\begin{pmatrix} u \\ v \end{pmatrix} = \begin{pmatrix} \frac{1}{2}\left[1 - \arctan(y, x)\pi^{-1}\right]W \\ \left[1 - \left(\arcsin\left(zr^{-1}\right) + f_{down}\right)f^{-1}\right]H \end{pmatrix} \tag{1}$$

where, $r = \sqrt{x^2 + y^2 + z^2}$ represents the depth of each point, and $f = f_{up} + f_{down}$ is the vertical field-of-view of the sensor. With this approach, we get a list of *(u, v)* tuples containing a pair of image coordinates for each p_i. For each p_i, we extract its distance *r*, its *x*, *y*, and *z* coordinates, and its remission, and store them in the image, ultimately creating a $[5 \times h \times w]$ tensor.

3.2 Local Feature Extraction

It should be noted that, after we obtain the mapping of each point on the range image, for the multiple points contained in each pixel, RangeNet++ selects the minimum depth point to represent this pixel, since there are multiple points mapped to the same pixel.

In order to ensure the proposed time complexity advantage of the RangeNet++ model, we did not choose to directly extract local features for each point in the point cloud, but relied on the corresponding points that their index could be used to easily find in the range image so as to quickly obtain its neighbor points, thus completing the extraction of local features.

Specifically, we first extract the neighborhood points of each point p_i in the range image in four directions: up, down, left, and right, namely p_u, p_d, p_l, p_r. With p_i and the four points around it, we can get four vectors:

$$\begin{cases} \boldsymbol{n}_{ui} = p_u - p_i \\ \boldsymbol{n}_{di} = p_d - p_i \\ \boldsymbol{n}_{li} = p_l - p_i \\ \boldsymbol{n}_{ri} = p_r - p_i \end{cases} \tag{2}$$

Then, we use "×" to denote the Vector product. Meanwhile, we can get four planes, and each of which includes p_i and two non-collinear neighbor points: (p_i, p_l, p_u), (p_i, p_r, p_u), (p_i, p_l, p_d), (p_i, p_r, p_d). We use their normal vectors to represent them:

$$\begin{cases} \boldsymbol{n}_1 = \boldsymbol{n}_{li} \times \boldsymbol{n}_{ui} \\ \boldsymbol{n}_2 = \boldsymbol{n}_{ri} \times \boldsymbol{n}_{ui} \\ \boldsymbol{n}_3 = \boldsymbol{n}_{li} \times \boldsymbol{n}_{di} \\ \boldsymbol{n}_4 = \boldsymbol{n}_{ri} \times \boldsymbol{n}_{di} \end{cases} \tag{3}$$

$\boldsymbol{n}_1, \boldsymbol{n}_2, \boldsymbol{n}_3, \boldsymbol{n}_4$ represent the directions of the four planes. Then, we calculate the angle between these normal vectors, here we use the cosine of the angle to represent the angle:

$$\begin{cases} \theta_1 = cos < \boldsymbol{n}_1, \boldsymbol{n}_2 > \\ \theta_2 = cos < \boldsymbol{n}_1, \boldsymbol{n}_3 > \\ \theta_3 = cos < \boldsymbol{n}_1, \boldsymbol{n}_4 > \\ \theta_4 = cos < \boldsymbol{n}_2, \boldsymbol{n}_3 > \\ \theta_5 = cos < \boldsymbol{n}_2, \boldsymbol{n}_3 > \\ \theta_6 = cos < \boldsymbol{n}_3, \boldsymbol{n}_4 > \end{cases} \tag{4}$$

Finally, we add these 6 values to the *[5, H, W]* tensor created in the previous section to get a *[11, H, W]* tensor and put it into the RangeNet++ semantic segmentation convolutional network for learning.

3.3 The Generation of Adversarial Samples

Our work on adversarial sample generation is based on range image. The basic concept is as follows: firstly, obtain the difference between each point in the range image and each point in the surrounding neighborhood, and then multiply it by a perturbation parameter, which determines the strength of the adversarial attack, so as to obtain an adversarial perturbation, and generate adversarial samples.

For the semantic segmentation model, its essence is to find the feature differences between different objects. The smaller the differences are, the more blurred the different objects become, and the semantic segmentation model is more difficult to play a role.

Therefore, in the first step, we need to calculate the difference between each point p_i in the range image and its surrounding points $p_j (j = 1, 2, \ldots, 8)$:

$$d_j = p_j - p_i \tag{5}$$

When getting the difference value from each surrounding point, we average these difference values to generate the average difference value $\overline{d_i}$:

$$\overline{d_i} = \frac{1}{8} \sum\nolimits_{j=1}^{8} d_j \tag{6}$$

$\overline{d_i}$ can be used as an adversarial perturbation for the semantic segmentation model.

In order to ensure that the adversarial samples are effective, we must have some control over the adversarial perturbation, so that it cannot be observed. Therefore, we set the perturbation parameter α, which is in the range [0.0, 1.0]. For the difference value obtained in Eq. (6), we multiply it by α as the final perturbation value. Then, we set the perturbation parameter α in the range [0.0, 1.0]. For the difference value obtained in Eq. (6), we multiply it by α as the final perturbation value:

$$r_i = \alpha * \overline{d_i} \tag{7}$$

Finally, the perturbation generated by each point is added to the sample, and the corresponding adversarial sample is obtained.

3.4 Adversarial Attacks Against Models

In the previous subsection, we obtained adversarial examples based on range image. Using these adversarial examples, we design adversarial attacks against RangeNet++ and our model. The adversarial attack mainly consists of two parts. The difference between the two parts lies in the selection of the perturbation parameters α. In the first part, for each range image, we randomly select the perturbation parameters in the range of [0, 1] to achieve a more comprehensive attack. In the second part, we set the same perturbation parameters to conduct a more targeted attack on the model.

Table 1. IoU [%] before and after adversarial attack.

Approach	RangeNet++			RangeNet++ + Local Feature			RangeNet++ + Local Feature + Adversarial Samples		
State	Before	After	Diff	Before	After	Diff	Before	After	Diff
Car	80.2	60.1	**20.1**	79.1	67.1	**12.0**	80.2	79.8	**0.4**
Bicycle	16.4	7.5	**8.9**	15.9	7.2	**8.7**	16.6	12.4	**4.2**
Motorcycle	34.5	10.6	**23.9**	34.2	20.3	**13.9**	30.6	29.8	**0.8**
Truck	3.5	1.7	**1.8**	2.0	0.9	**1.1**	4.8	3.5	**1.3**
Other-vehicle	21.3	3.7	**17.6**	22.0	11.4	**10.6**	21.0	20.6	**0.4**
Person	15.1	10.8	**4.3**	12.3	8.8	**3.5**	13.9	14.1	**−0.2**
Bicyclist	35.4	16.8	**18.6**	33.8	25.5	**8.3**	33.9	33.6	**0.3**
Motorcyclist	0.0	0.0	**0.0**	0.0	0.0	**0.0**	0.0	0.0	**0.0**
Road	89.8	73.0	**16.8**	91.0	77.1	**13.9**	90.1	90.3	**−0.2**
Parking	49.7	27.7	**22.0**	52.0	28.1	**23.9**	48.9	48.7	**0.2**
Sidewalk	76.2	49.6	**26.6**	77.9	54.7	**23.2**	76.6	76.9	**−0.3**
Other-ground	0.0	0.1	**−0.1**	0.1	0.5	**−0.4**	0.0	0.0	**0.0**
Building	76.5	60.9	**15.6**	77.2	73.0	**4.2**	77.1	77.0	**0.1**
Fence	46.1	22.4	**23.7**	44.6	25.8	**18.8**	48.7	48.5	**0.2**
Vegetation	75.0	50.6	**24.4**	76.0	63.4	**12.6**	75.9	76.0	**−0.1**
Trunk	37.8	34.3	**3.5**	39.5	37.3	**2.2**	32.0	32.6	**−0.6**
Terrain	61.5	38.5	**23.0**	64.9	45.5	**19.4**	64.1	64.1	**0.0**
Pole	27.6	24.3	**3.3**	31.6	28.5	**3.1**	29.6	29.2	**0.4**
Traffic-sign	27.2	20.7	**6.5**	29.4	26.3	**3.1**	25.3	25.2	**0.1**
MeanIoU	40.7	27.0	**13.7**	41.2	31.7	**9.5**	40.5	40.1	**0.4**
MeanAcc	83.4	67.7	**16.6**	85.0	74.5	**10.5**	84.9	84.9	**0.0**

4 Experiments and Results

Dataset: We use SemanticKITTI as experiment dataset, which is a sub-dataset of KITTI in the direction of semantic segmentation. It annotates all sequences in the KITTI Vision Odometry Benchmark and provides dense point-by-point annotations for the full 360° field of view of the automotive LiDAR used in [2]. It includes a total of 22 sequences, of which sequences 00-10 are the training set and sequences 11-21 are the test set.

Parameters: To get objective and fair results, we train RangeNet++ and our model with the same parameters. We choose the sequences 00-05 as the training set, 07-10 as the test set, and 06 as the validation set. Meanwhile, we train each model for 100 epochs. We employ random perturbation parameters in the first experiment, i.e. $\alpha \in [0.0, 1.0]$, and specific perturbation parameters in the second experiment, i.e. $\alpha \in \{0.2, 0.4, 0.6, 0.8, 1.0\}$.

To ensure consistency, we chose the same evaluation metric as in RangeNet++. That is, we use the mean intersection over union (mIoU) over all classes [6], with the following formula:

$$mIoU = \frac{1}{D}\sum_{d=1}^{D} \frac{TP_d}{TP_d + FP_d + FN_d} \tag{8}$$

where TP_d, FP_d and FN_d correspond to the number of true positive, false positive, and false negative predictions for class d and D is the number of classes.

4.1 Adversarial Attacks with Randomly Perturbed Parameters

The first experiment aims to demonstrate the effectiveness of the two methods proposed in this paper, that is, to improve the security and anti-attack capability of RangeNet++ in the face of adversarial attacks. Therefore, we compare the defense ability of the original RangeNet++ with the RangeNet++ with only local features added and the model containing local features and adversarial samples in the face of adversarial attacks.

Table 1 shows the changes of the IoU of these three models before and after being attacked. As can be seen from Table 1, compared to the original RangeNet++, the RangeNet++ added to our method performs better in the face of adversarial attacks. More specifically, the RangeNet++ that only contains local features is improved compared to the original RangeNet++ and the RangeNet++ that includes both methods have a higher improvement, which fully shows that our two methods are both effective in improving the defense capability of the RangeNet++.

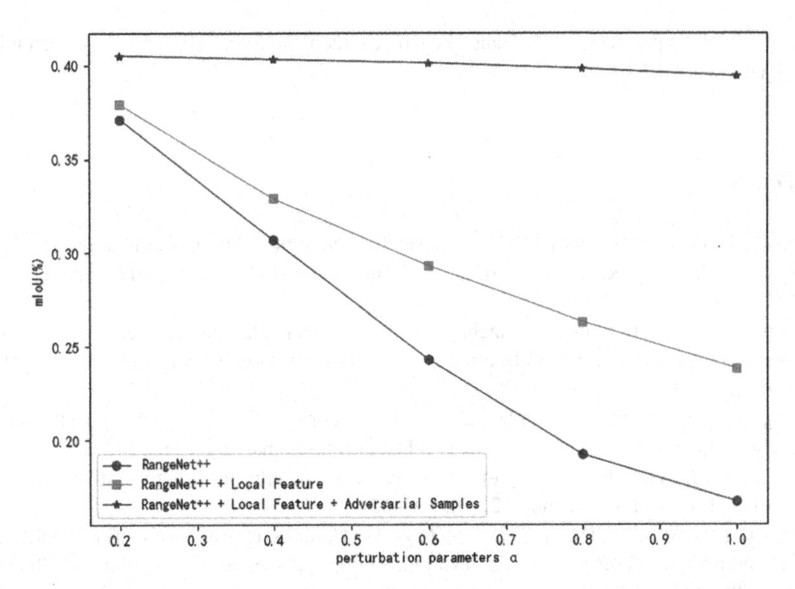

Fig. 3. Experimental results: mIoU with different perturbation parameters.

4.2 Adversarial Attack with Specified Perturbation Parameters

The second experiment demonstrates the effectiveness of adversarial attacks based on the adversarial samples proposed in this paper. Figure 3 shows the variation of mIoU for the three models under different parameters, which represent adversarial attacks of different strengths.

As can be seen from Fig. 3, as the perturbation parameter increases, the mIoU of the three models gradually decreases, which proves the effectiveness of adversarial samples. At the same time, we can find that our method has better performance in the face of different levels of adversarial attacks, which further illustrates that our method can improve the defense ability of the RangeNet++.

5 Conclusions

In this work, we proposed two methods to improve the security and anti-attack capability of the semantic segmentation model RangeNet++. One was to extract the local geometric features of each point reflecting the topographic features of the point cloud based on the range image. The other was to generate adversarial examples against the semantic segmentation model based on the range image. And then we put them into the training set. Through experiments, we demonstrated the effectiveness of these two methods. Moreover, the adversarial attack based on our designed adversarial samples also had good performance for the RangeNet++ model.

Acknowledgments. This work was partially supported by the Gansu Provincial Science and Technology Major Special Innovation Consortium Project (No. 21ZD3GA002), the name of the innovation consortium is Gansu Province Green and Smart Highway Transportation Innovation

Consortium, and the project name is Gansu Province Green and Smart Highway Key Technology Research and Demonstration.

References

1. Behley, J., et al.: SemanticKITTI: a dataset for semantic scene understanding of LiDAR sequences. In: Proceedings of the IEEE/CVF International Conference on Computer Vision (ICCV) (2019)
2. Milioto, A., Vizzo, I., Behley, J., Stachniss, C.: RangeNet++: fast and accurate LiDAR semantic segmentation. In: IEEE/RSJ International Conference on Intelligent Robots and Systems (IROS) (2019)
3. Weng, H., et al.: Towards understanding the security of modern image captchas and underground captcha-solving services. Big Data Mining Anal. **2**(2), 118–144 (2019)
4. Wu, X., Xu, K., Hall, P.: A survey of image synthesis and editing with generative adversarial networks. Tsinghua Sci. Technol. **22**(6), 660–674 (2017)
5. Deng, H., Zhang, Y., Li, R., Hu, C., Feng, Z., Li, H.: Combining residual attention mechanisms and generative adversarial networks for hippocampus segmentation. Tsinghua Sci. Technol. **27**(1), 68–78 (2022)
6. Rusu, R.B., Marton, Z.C., Blodow, N., et al.: Learning informative point classes for the acquisition of object model maps. In: 2008 10th International Conference on Control, Automation, Robotics and Vision, pp. 643–650. IEEE (2008)
7. Rusu, R.B., Blodow, N., Marton, Z.C., et al.: Aligning point cloud views using persistent feature histograms. In: 2008 IEEE/RSJ International Conference on Intelligent Robots and Systems, pp. 3384–3391. IEEE (2008)
8. Rusu, R.B., Blodow, N., Beetz, M.: Fast point feature histograms (FPFH) for 3D registration. In: 2009 IEEE International Conference on Robotics and Automation, pp. 3212–3217. IEEE (2009)
9. Rusu, R.B., Bradski, G., Thibaux, R., et al.: Fast 3D recognition and pose using the viewpoint feature histogram. In: 2010 IEEE/RSJ International Conference on Intelligent Robots and Systems, pp. 2155–2162. IEEE (2010)
10. Stein, F., Medioni, G.: Structural indexing: efficient 3-D object recognition. IEEE Trans. Pattern Anal. Mach. Intell. **14**(2), 125–145 (1992)
11. Goodfellow, I.J., Shlens, J., Szegedy, C.: Explaining and harnessing adversarial examples. In: International Conference on Learning Representations, California (2015)
12. Moosavi-Dezfooli, S.-M., Fawzi, A., Frossard, P.: DeepFool: a simple and accurate method to fool deep neural networks. In: IEEE Conference on Computer Vision and Pattern Recognition, Nevada, pp. 2574–2582. IEEE (2016)
13. Moosavi-Dezfooli, S.-M., Fawzi, A., Fawzi, O., Frossard, P.: Universal adversarial perturbations. In: IEEE Conference on Computer Vision and Pattern Recognition, Hawaii, pp. 86–94. IEEE (2017)
14. Xiang, C., Qi, C.R., Li, B.: Generating 3D adversarial point clouds. In: 2019 IEEE/CVF Conference on Computer Vision and Pattern Recognition (CVPR), pp. 9128–9136 (2019). https://doi.org/10.1109/CVPR.2019.00935

Computer Vision for Medical Computing

Ensemble Model of Visual Transformer and CNN Helps BA Diagnosis for Doctors in Underdeveloped Areas

Zhenghao Wei(✉)

School of Computer Science and Engineering, Sun Yat-Sen University, Guangzhou 510006, China
weizhh8@mail2.sysu.edu.cn

Abstract. The diagnosis of Biliary Atresia (BA) is still complicated and high resource consumed. Though sonographic gallbladder images can be used as an initial detection tool, lack of experienced experts limits BA infants to be treated timely, resulting in liver transplantation or even death. We developed a diagnosis tool by ViT-CNN ensemble model to help doctors in underdeveloped area to diagnose BA. It performs better than human expert (with 88.1% accuracy versus 87.4%, 0.921 AUC versus 0.837), and still has an acceptable performance on severely noised images photographed by smartphone, providing doctors in clinical facilities with outdated Ultrasound instruments a simple and feasible solution to diagnose BA with our online tool

Keywords: Biliary atresia · Visual transformer · Medical image processing · Ensemble model

1 Introduction

Biliary atresia (BA) is a pediatric disease affects both intrahepatic and extrahepatic bile ducts, which leads to pathological jaundice and liver failure in early infancy [1,2]. Though BA only has a prevalence rate of about 1 in 5000-19,000 infants all over the world [3,4], it is the most common cause for liver transplantation in infants below 1 year old [5]. If not treated timely, the disease will progress to end-stage liver cirrhosis rapidly, and liver transplantation will be necessary. Receiving Kasai portoenterostomy (KPE) surgery before age 2 months can largely extend the infant's native liver survival time [6]. Therefore diagnosis in an early stage is essential. However, it is still hard to distinguish BA from common causes of cholestasis [7], and diagnostic methods of BA including assay of serum matrix metalloproteinase-7 and screening the direct bilirubin concentration [7,8], are not feasible for medical facilities with underdeveloped conditions especially in rural regions.

In the rural area of developing Asian country like China and India, ultrasound examination is the main approach to diagnose BA in jaundiced infants. Though the method has a specificity over 90% [9], the lack of experienced doctor who are

Y. Zheng et al. (Eds.): ACCV 2022, LNCS 13848, pp. 73–89, 2023.
https://doi.org/10.1007/978-3-031-27066-6_6

capable to make diagnosis and perform operation in undeveloped region, delays the appropriate treatment and the patient may miss the optimum therapeutic time. Thus, an AI-assisted diagnosis mechanism is vital to enable clinical staffs in undeveloped region to make preliminary diagnosis timely before the condition worsen.

Deep learning methods, specifically, the convolutional neural networks (CNNs), have been widely used in Biomedical Image Analysis tasks, and have been proved to be superior or comparable to human experts in tasks like diagnosis of lung cancer or Covid-19 [10,11]. In 2021, teams from Sun Yat-Sen University (SYSU) and The First Affiliated Hospital of SYSU developed a deep learning Model based on Deep Residual Networks, to help diagnose BA, as well as a smartphone application which can help doctors especially in rural area, to diagnose rare disease like BA [12].

Though the model outperforms human experts, both in accuracy and sensitivity, the worst case is still unconsidered. In the case when sonographic system is not connected to Internet and exporting images is impermissible, taking a photo by smartphone may be the simplest solution. Moiré patterns and other noise can catastrophically reduce the accuracy of the classification model, and CNN is proved to be effective in such denoising tasks as Moiré photo restoration [13,14]. Visual Transformer (ViT) is another state of the art (SOTA) deep learning model for image processing, which holds better performance in multi-task than CNN [15], and is also utilized in Biomedical Image Processing (BIP) tasks, like MRI Segmentation [16] and Covid-19 diagnosis [17]. We combined the two techniques and utilize both the high accuracy of ViT models and the denoising ability of CNNs.

In summary, it is of great necessity to develop a diagnosis system of BA to assist unexperienced clinical staff in rural area to make preliminary diagnosis, and the system should have the ability to eliminate noises like Moiré patterns. In this study, we developed an ensemble deep learning model (EDLM) which contains four ViT models and two CNN models. It outperforms human experts in normal cases and have a better performance in noised dataset than ensemble CNN models or ViT models. Our main contributions can be concluded as follow:

1. We proposed an EDLM of both ViT and CNN to diagnose BA for infants aged below 60 days. The model is an image classification model which classifies sonographic gallbladder images, and it contains 4 ViT models (ViT-base, Swin Transformer-base, CvT-24 and CSwin Transformer-base) and 2 CNN models (ResNet-152 and SENet). Our model has a better accuracy and AUC score than human expert (with 88.1% accuracy versus 87.4%, 0.921 AUC versus 0.837).
2. Due to the imbalanced data (as a rare disease, BA-positive sample is lesser), we compared several strategies to enhance the specificity and sensitivity of the model.
3. We discussed several designs of the EDLM, compared the ViT-CNN ensemble model with ensemble CNN model, ensemble ViT model and single models, and two kinds of voting strategy. And we proved our mechanism performs better.

4. In the noised case, our ensemble model evidently outperforms others. And we adopted several few-shot tuning methods to enhance the model's ability to classify noised sonographic images photographed by cellphone.

2 Related Works

2.1 Conventional BA Diagnosis Methods

BA is such a rare and complicated disease, as well as the infantile patients can't afford surgical exploration, an accurate diagnosis of BA is still a challenge. There are several clinical symptoms of BA including persistent jaundice, pale stools, and dark urine [3], for the conjugated hyperbilirubinemia due to cholestasis. However, neonatal cholestasis can be caused by many other diseases. Though screen the direct bilirubin concentration [9,18] or stool color [4] can yield sensitivities over 97%, and serum gamma-glutamyl transferase (GGT) can be considered to distinguish BA from PFIC, bile acid synthesis or metabolism disorders, but mechanical bile duct obstruction, paucity of interlobular bile ducts or cystic fibrosis are still possible [19]. The specificity of method beyond is not satisfying (that means other disease still can't be excluded).

Those BA-suspected infants need further examinations including ultrasonography (US), hepatobiliary scintigraphy and magnetic resonance cholangiography (MRCP). Hepatobiliary scintigraphy has a low specificity as 70.4% and is time-consuming and radiant [20]. MRCP is also limited by the small body size of infants and is also deprecated for the need of sedation [21]. Therefore, US has been recommended as the preferred imaging tool for the initial detection of BA [3]. Several direct features reflecting the abnormalities of biliary system can be the criterion of BA, including Gallbladder abnormalities [20,22], Triangular Cord sign [23,24], and porta hepatic cyst [25,26], but any single feature can't guarantee extremely high specificity and sensitivity at the same time. Even some infants may still have equivocal US results, that US-guided percutaneous cholecystocholangiography (PCC) may be needed [27]. In conclusion, the diagnosis of BA is a process full of complexity and uncertainty, that's why human experts mentioned in Zhou, W *et al.*[12] have an AUC lower than 0.85 (Fig. 1).

2.2 CNN Models

The convolutional neural networks (CNN) has been the mainstream of Computer Vision (CV) domain since the invention of AlexNet [29], though come up decades ago [30]. In the last decade, deeper and more effective CNN models have been proposed and achieved enormous success, like VGG [31], GoogleNet [32], ResNet [33], DenseNet [34], PNASNet [35], and EfficientNet [36], etc.

ResNet and its variants are still the solid backbone architectures of CV, which introduces the Residual mechanism. Se-ResNet or SENet is an evolved model with a new mechanism called Squeeze-and-Excitation (SE) [37], allows the network to perform feature recalibration. That means it can use global information

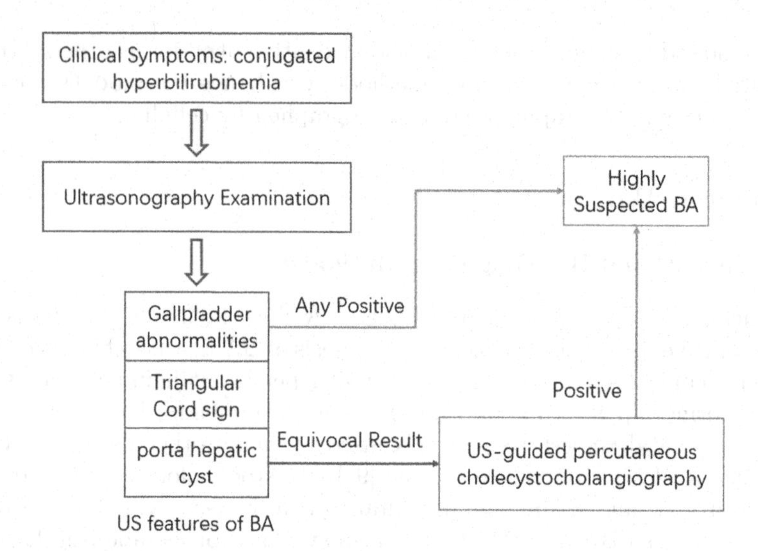

Fig. 1. A sketch map of the conventional BA diagnosis process. To learn a more detailed diagnostic decision flow chart can refer to [28]

to reemphasize valuable local features while restrains the less important ones. It can be likened to attention mechanism, which has a global receptive field on its input. Therefore, it performs better on classification tasks, and it is adopted by the BA-diagnosis application of Zhou, W *et al.*[12] as the main network.

2.3 Visual Transformer Models

Visual Transformer (**ViT**) is firstly proposed by Google in 2021 [15], is the SOTA technique in image classification tasks. Since 2017, Transformer has been widely used in natural language processing (NLP) tasks [38], and pretrained models based on Transformer architecture like BERT [39], GPT [40] and UniLM [41] are still the SOTA techniques. Transformer consists of several Encoder blocks and decoder blocks, based on self-attention mechanism, which provide the capability to capture non-local features or encode dependencies between distant pixels [42]. It can be concluded as that CNN performs better on the extraction of local features and Transformer concentrates on non-local features. ViT imported the Transformer method into CV and proposed an approach to embedded patches of image into sequential input which is similar to inputs of NLP models. It outperforms all SOTA CNN models on all popular image classification benchmarks [15], but it doesn't have an equivalent performance on other downstream tasks like object detection and instance segmentation, due to its lack of inductive bias compared to CNNs.

DeiT is a follow-up model of ViT, which didn't change the network architecture but imported Knowledge Distillation method to transfer inductive bias from CNN teacher model to the student model [43].

Swin Transformer is one of the best performed CV backbone in multitask [44]. It proposed an improved sliding window method (shifted window) to eliminate ViT's shortage caused by fixed scale. Swin Transformer utilized both the hierarchical design and the shifted window approach to transcend the former state of art models in several tasks, as well as reduce the computational complexity. Comparing to ViT, it holds less parameters and have a better speed-accuracy trade-off.

CvT is a combination of ViT and CNN, like Swin Transformer, it contains a hierarchy of Transformers to capture features of different receptive fields but utilizes convolutional layers to produce token embedding [45]. These changes introduce desirable properties of CNN, such as inductive bias.

CSwin Transformer is another evolved model of ViT proposed by Microsoft, the same as Swin Transformer [46]. Instead of the Shifted Window Attention (SWA) mechanism of Swin Transformer, it introduced a Cross-Shaped Window (CSwin) mechanism. SWA allows information exchange through nearby windows, but the receptive field is still enlarged quite slowly, therefore Swin Transformer needs a dozen of Transformer blocks in the pyramid structure. CSwin has a lower computational cost while can achieve the global receptive filed in less steps. CSwin Transformer is still the CV backbone models with highest performance.

3 Methods and Data Materials

3.1 EDLM of CNNs

Due to the Condorcet's Jury Theorem [47], ensemble classification model made up of individual models which have probabilities of being correct greater than 1/2, has a probability of being correct higher than the individual probabilities [48]. Hybrid or ensemble machine learning models have been popular since decades and can trace their history back to random forest [49] and bootstrap aggregating [50].

Our first attempt is to train k different models on k-fold split trainsets. That means we randomly separate the internal dataset into k (*e.g.*, five) complementary subsets, then in the k cases, different single subset is selected as the internal validation set, and other $k-1$ subsets will be combined as the training set. On each training set, a CNN model is trained, and for different models, their performances are evaluated on different validation sets (the remained one subset). It is similar to k-fold cross validation, but every model is kept (in k-fold cross validation, only the best model will be selected, with others abandoned). The ensemble model predicts the label of test data by calculating the average of outputs (predicted score of each category) of k models, and then do SoftMax to identify the final prediction label (Fig. 2).

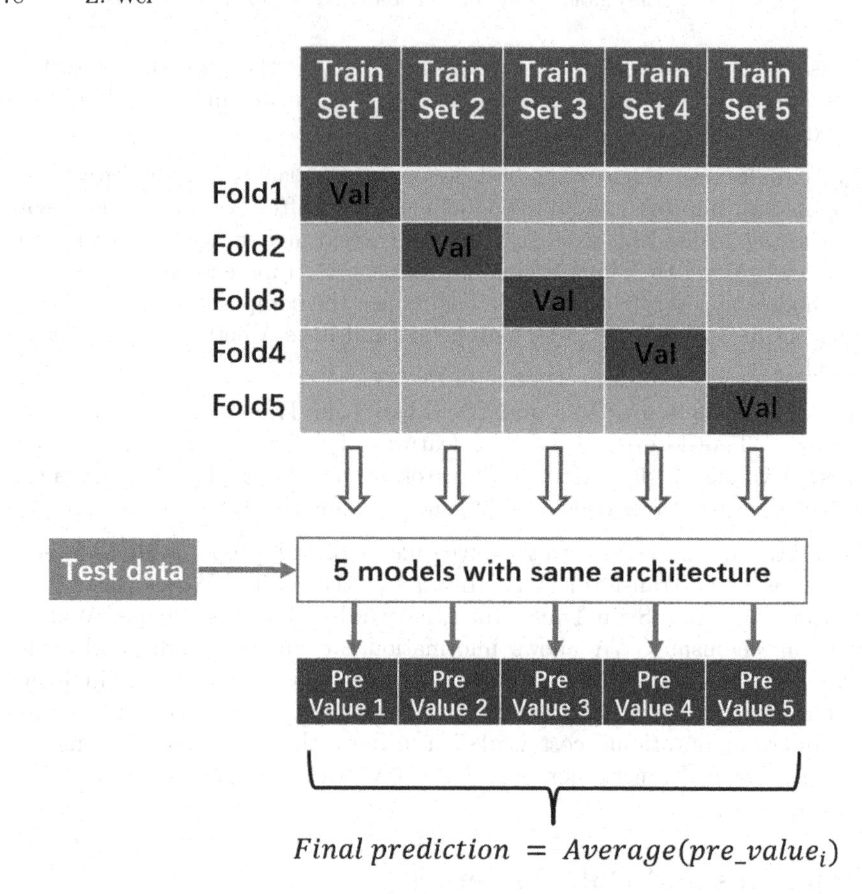

Fig. 2. The framework of an ensemble model. The form upside shows that the training set was divided into complementary subsets (means they are disjoint and the union of them are the universal set), only one of the subsets is selected as validation set in the training of each model. Differ to the classic case of Condorcet's Jury, neither simple majority rule nor unanimity rule is adopted, but we calculate the average of SoftMax outputs of the models (a real value instead of 0–1 value).

To construct the ensemble model, several CNN architectures were considered. Including ResNet [33], DenseNet [34], SeNet [37] and EfficientNet [36]. Zhou, W *et al.*[12] utilized SeNet-154 as the base classification of the EDLM, but we found it not a best choice, due to that the sonographic images are greyscale images de facto, and the motivation of SeNet is to enhance ResNet from the aspect of channel relationship. So, the SE mechanism may not function, and we can see, as a result, SeNet-154 didn't perform better than ResNet-152 (83.57% versus 85.60% on accuracy). Consequently, we adopted EfficientNetB6 as the base CNN model

3.2 ViT-CNN EDLM

Though ensemble model based on single model can promote the performance indeed, those models with same architecture have extremely high correlation, which conflicts with the assumption of independence of the Condorcet's Jury Theorem [47]. That means positive correlation between base models will decrease the accuracy of the ensemble model, and diversity of models will conversely improve the accuracy [51]. Therefore, choosing models with entirely different architectures may perform better than single architecture ensemble model.

As is concluded in Sect. 2.3, Visual Transformer has been becoming the new backbone architecture of computer vision, and it has a very different mechanism from CNNs. Convolutional mechanism endows CNNs the characteristics of translation invariance and inductive bias, while Transformer gives ViT models stronger ability of extracting global features in a patch, and they gain inductive bias by other approaches. So that if we combine CNNs with ViT models, they may concentrate on very different patterns and operates in different ways, and the diversity of the ensemble model will be guaranteed.

Other than EDLM in Sect. 3.1, since base models vary in architecture, it has no necessity to ensure models with different architectures trained on different dataset to produce diversity. Thus, models of same architecture will share a group of k-fold generated sets, while training sets of models with different architectures have no relevancy. The prediction rule is still the average strategy like in Sect. 3.1.

Several Visual Transformer methods were considered in our attempts, including ViT (the origin google version instead of the abbreviation of Vision Transformer), DeiT, Swin Transformer, CvT, and CSwin Transformer. As is shown below, single CSwin Transformer performed best in all these models (Fig. 3).

3.3 Datasets

It should be mentioned that, besides the lack of experienced doctors, the sonographic machines are usually not connected to the internet, and export images is not feasible. Therefore, taking a photo by cellphone and upload to remote diagnosis system might be a temporary expedient.

Thus, we adopted two datasets, besides the sonographic gallbladder images provided on https://zenodo.org/record/4445734, we generated more noised pictures, which are photographed by several doctors with different models of smartphones. Differ from experiments in Zhou, W *et al.*[12], we want to explore the generalization ability of the model more, if we only take one type into account, there might be some bias unconsidered.

The first dataset consists of a 3705 images internal training dataset (internal validation set is also split from it), and an 842 images external validation dataset, all the images are segmented. The second dataset has 24150 pictures for the internal dataset, and 840 for the external validation one. 3659 pics of the internal set are original sonographic gallbladder images, and the others are reproduced picture took by smartphone, based on foresaid original images. Images in

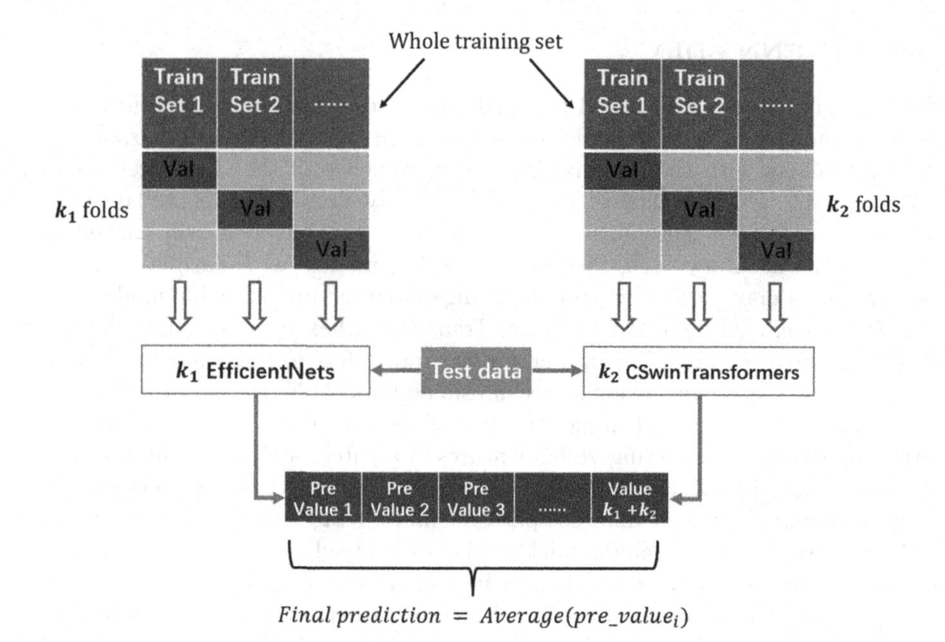

Fig. 3. The framework of a ViT-CNN EDLM (*i.e.*, two architectures, namely CSwin Transformer and EfficientNet). The difference is that, for each architecture, several models are trained on k-fold cross validation rule independently. For models of an architecture, they monopolize a whole training set, because there are enough diversity from different architectures, and models with same architecture are required to gain patterns.

the internal set were took by five doctors, and there were two new coming doctors producing the external data. that is because we expect the model to gain adequate generalization ability, and the immunity against random noise (Fig. 4).

3.4 Data Processing

What should be considered at first is, that the positive samples are much less than the negative ones, due to BA is such a rare disease, as is mentioned, has a low incidence about 1 in 5000-19,000 infants [3]. There are only about 23% of the samples are BA positive in both datasets, so some measures should be applied to solve the imbalance problem. We tried several approaches, like resampling, down sampling, modifying class weights in the loss function, or k fold cross validation. As a result, we notice that resampling is the most practical strategy. Because we have adopted an EDLM method, k fold cross validation is redundant; down sampling made a precision loss, deservedly; changing class weights might be an efficient strategy, but we found it performs worse than simply resampling. It may blame the ensemble model, for in the trainset of a single model, some key samples of a pattern may miss, while in the resampling case, it is ensured that the trainset are more likely to contain all key samples.

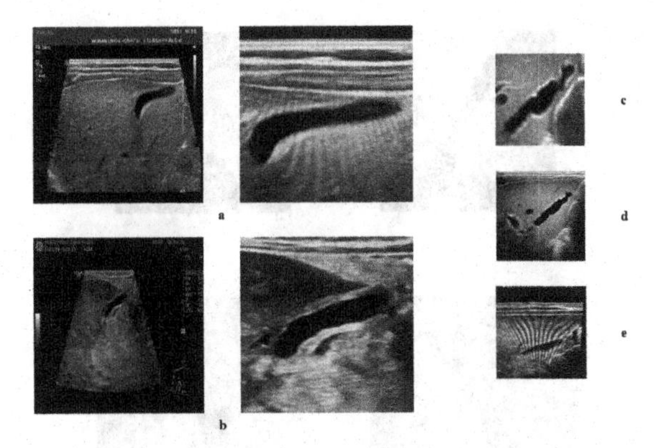

Fig. 4. **a**: an example of the second dataset, the left one is an original sonographic gallbladder image of a non-BA infant, and the right one is the reproduced picture of the left. **b**: BA patient, the same as **a**, original image on the left and reproduced on the left. **c**: a sample of the first dataset, differ from the second one, pictures are pre-segmented. **d**: smartphone reproduced picture in Zhou, W *et al.*[12]. **e**: smartphone reproduced picture of the second dataset. It's obvious that noise in **d** isn't severe, and the performance on **e** can better illustrate the generalization ability of the model.

Secondly, if we aim at improving the generalization ability of the model, even the 24k training dataset won't be sufficient, adopting some data augmentation techniques is necessary. Random rotation, random horizontal flip are adopted, resize and random crop is also applied to the original set, and all the images are turned into greyscale, for color has no significance in sonographic images (Fig. 5).

And whether training data need masks, is also worth discussing. At first, the hospital provided manually annotated images, with a mask showed the focus, but finally we found the mask an obstacle to improve the model performance. From our perspective, the mask quality is unsatisfactory, but doctors who made the annotation are not to blame. As is mentioned in Sect. 2.3, multiple features are considered in the US diagnosis of BA, when the doctor label the images, he may only focus on one of those features, and omit some. Thus, ignoring the mask information and just train models on the original data could perform better.

4 Experiments and Results

4.1 Experimental Settings

There are two tasks in the experiment, one is the diagnosing BA on original sonographic gallbladder images, the other one is on noised pictures photographed by smartphone. We solved the two problems step by step.

Firstly, we trained the classification model for original images. Models with different architectures were trained on different 5-fold split dataset, that means

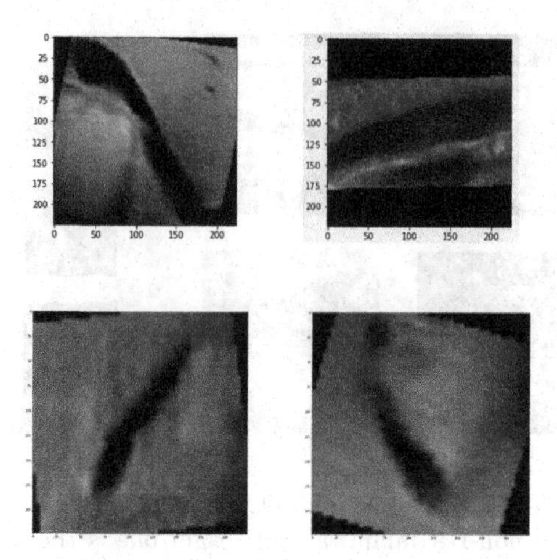

Fig. 5. Samples after data augmentation.

for one architecture, five models were trained at once, the training set was divided into five subsets as discussed in Sect. 3.1, and for each model, different subsets were used as validation set, remained four sets were used as the training cohort.

For every architecture, models are trained by transfer learning, the network loaded pretrained weight on ImageNet-1k classification dataset as the initial weight. And we fine-tuned the model on our training cohort. The loss function was Cross Entropy loss, and we tried weighted Cross Entropy as mentioned in Sect. 3.4, but we found it not better than resampling. The model was trained for 200 epochs and evaluated on internal validation set for every 5 epochs. If the evaluation loss didn't decrease for next 40 epochs, then the training progress would be forced to an early termination, and the parameter model with the lowest loss (*i.e.*, 40 epochs before) will be saved.

At the step of comparing different architectures, like k-fold cross validation, we chose the model with best performance on the test sets in every k-fold training as the representative. But after we chose the base architectures of the ensemble model, the five models will be adopted together to construct the ensemble model.

To control the number of models in different ensemble models consistent, all the ensemble models in the comparison consists of six base models. The ViT-CNN ensemble model has three SWin Transformer Base and three EfficientNet B6, trained on 3-fold cross training cohort.

For the second problem, we also utilized transfer learning, but it was the weight of ensemble model trained previously to be loaded. For each base model, we fine-tuned the model on the noised data, with augmentation mentioned in Sect. 3.4, for 30 epochs, evaluated every 5 epochs, and the best model in evaluation would be selected.

4.2 Results

As is shown below, for the single models, ViT family outperforms CNNs, and CSwin Transformer is the best single model. All the ensemble models have a better performance than human expert, but the ViT-CNN ensemble model do the best. CSwin Transformer has an 87.86% accuracy on the original images (while precision is 88.07% and recall is 87.74%, areas under the receiver operating characteristic curve of 90.78%), and 80.5% accuracy on the smartphone-took images (while precision is 80.49% and recall is 81.32%, areas under the receiver operating characteristic curve of 80.83%).

For ensemble models, ViT-CNN ensemble model has 88.11% accuracy on the original images (while precision is 88.35% and recall is 87.98%, areas under the receiver operating characteristic curve of 92.90%), and 81.11% accuracy on the smartphone-took images (while precision is 82.33% and recall is 81.71%, areas under the receiver operating characteristic curve of 81.04%) (Figs. 6 and 7).

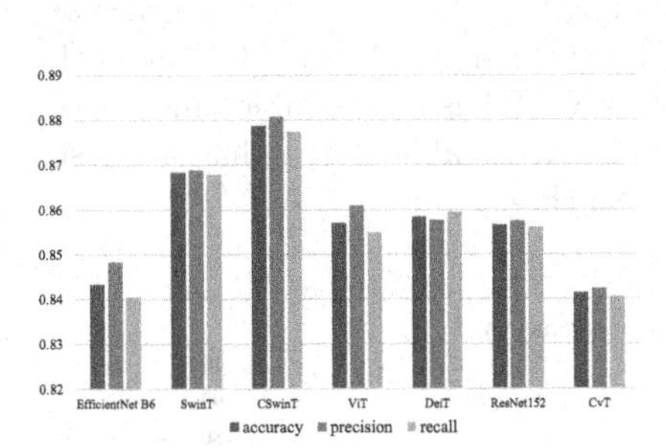

Fig. 6. Comparison between single models.

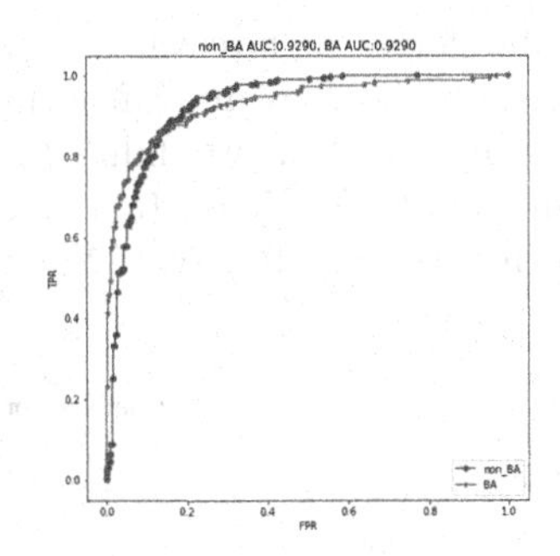

Fig. 7. The areas under the receiver operating characteristic curve (AUC) of ViT-CNN ensemble model.

Other than human experts, the deep learning models have precision and recall very similar, while the human experts have a recall obviously lower than the precision. The phenomenon indicates that our rebalance strategy works, that the deep learning models didn't performs worse on positive samples than negative ones. But human experts are still limited by the rarity of the disease, they can't recognize the atypical samples so that they have a lower sensitivity than machine.

5 Discussion

5.1 Comparison Between Methods

As has been discussed in Sect. 2.3, ViT models have been proved to be better backbone architectures than CNN models like ResNet or EfficientNet. In the

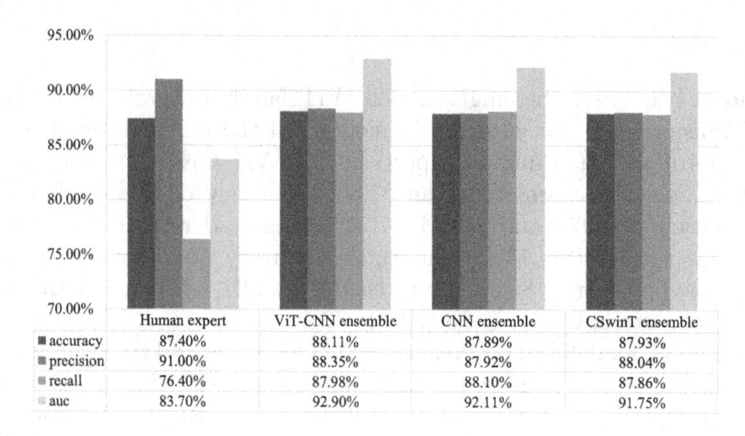

	Human expert	ViT-CNN ensemble	CNN ensemble	CSwinT ensemble
accuracy	87.40%	88.11%	87.89%	87.93%
precision	91.00%	88.35%	87.92%	88.04%
recall	76.40%	87.98%	88.10%	87.86%
auc	83.70%	92.90%	92.11%	91.75%

Fig. 8. Comparison between different ensemble models and human expert.

test of different architectures, generally, ViT models did better, and the order of performance is: CSwin Transformer, Swin Transformer, DeiT, ViT, CNNs. CvT is an exception that we didn't find a version of CvT has comparable amount of parameters to ViT-base (about 80M). So it is predictable that it has a worse performance than ResNet152 and other models (Fig. 9).

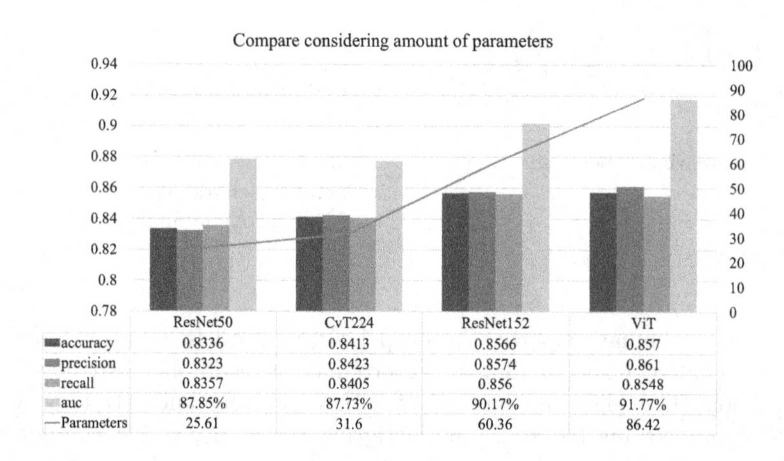

	ResNet50	CvT224	ResNet152	ViT
accuracy	0.8336	0.8413	0.8566	0.857
precision	0.8323	0.8423	0.8574	0.861
recall	0.8357	0.8405	0.856	0.8548
auc	87.85%	87.73%	90.17%	91.77%
—Parameters	25.61	31.6	60.36	86.42

Fig. 9. Comparison considered difference in amount of parameters between models

However, it does not absolutely mean that ensemble ViT model will outperform ensemble CNN model. Though single CNN models have a worse performance than ViT models, as is shown in Fig. 8, ensemble CNN model has a comparable or better performance than ensemble ViT model. It can be interpreted as that in classification tasks, CNNs are more sensitive to training data. In each fold of training cohort, the performance of CNNs are limited by the lack of some images, but in the ensemble case, the diversity produced by different data, will improve the performance of ensemble model.

The ViT-CNN ensemble model has the best performance as we expected, with the same number of base models as other ensemble models. The diversity of model architectures may account for the better performance, it matches the assumption of independence of the Condorcet's Jury Theorem.

5.2 Ensemble Strategy: Simple Majority or Average

Though ensemble model has become a popular technique in improving the upper bound of model performance, few investigators take the difference between classic Condorcet's Jury and EDLM into account.

Assume an ensemble model is made up of several individual models whose prediction score of an input are independent and identically distributed (*i.e.*, suppose it is a random variable and we know the priori probability distribution), *e.g.*, a Gaussian distribution $N(\mu, \sigma^2)$, the difference of two kinds of voting strategy will be obvious. In the simple majority case, the probability of a single model misclassifies a positive sample (*i.e.*, $\mu > 0.5$, and prediction score is less than 0.5) is

$$\Pr\left(prediction\ score < 0.5\right) = \int_0^{0.5} \frac{1}{\sqrt{2\pi}\sigma} e^{-\frac{(x-\mu)^2}{2\sigma^2}}, \mu > 0.5. \tag{1}$$

For convenience, we record $p_1 = \Pr\left(prediction\ score < 0.5\right)$, and the prediction score of a single model is S_i. Suppose the ensemble model has k base models, the voting game turns into a Bernoulli experiment $b(k, p_1)$. The probability of majority voting predicts incorrectly is the probability of the number of cases $S_i < 0.5$ is less than $k/2$ for $0 < i \leq k$ (k is odd), like this:

$$\begin{gathered}\Pr\left(simple\ majority\ voting\ is\ wrong\right) = \\ p_1^k + C_k^1 p_1^{k-1}\left(1-p_1\right) + \ldots C_k^{\frac{k-1}{2}} p_1^{\frac{k+1}{2}}\left(1-p_1\right)^{\frac{k-1}{2}}\end{gathered} \tag{2}$$

And the average strategy is much simpler. The average output of the ensemble model can be represented as $\bar{X} = (X_1 + X_2 + \ldots + X_k)/k, X_i \sim N(\mu, \sigma^2)$, and all X_i are independent and identically distributed. Thus, $\bar{X} \sim N(\mu, \frac{\sigma^2}{k})$, that means more base models there are, more accurate the prediction is (Fig. 10).

Let us quantitatively compare these two strategies. Assume the output of a base model correspond to $N\left(0.6, 0.1\right)$ and there are $k = 5$ base models, and $p_1 = 15.87\%$ can be easily calculated (having a 1σ deviation). Then

$\Pr\left(simple\ majority\ voting\ is\ wrong\right) = 0.0311$, and the probability of the average case is wrong is the probability of having a $\sqrt{5}\sigma$ deviation,

$$\Pr\left(average\ score\ is\ wrong\right) = \int_0^{0.5} \frac{\sqrt[4]{k}}{\sqrt{2\pi}\sigma} e^{-\frac{k(x-\mu)^2}{2\sigma^2}} = 0.0127 \tag{3}$$

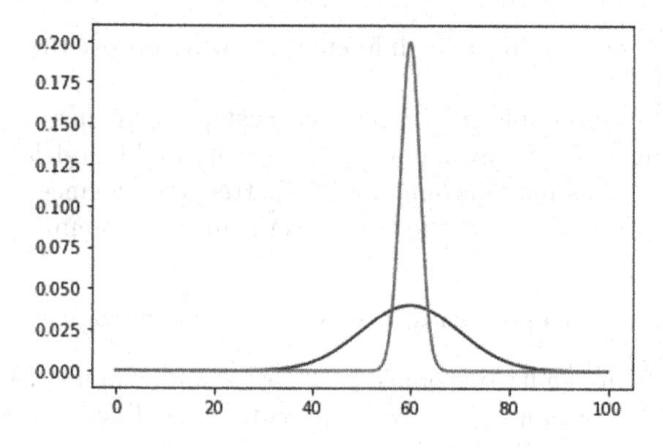

Fig. 10. PDF of X_i and $\bar{X}$.

It is obvious that the average strategy is less possible to mistake than the simple majority strategy.

References

1. Wang, J., et al.: Liver immune profiling reveals pathogenesis and therapeutics for biliary atresia. Cell. **183**(7), pp. 1867–1883.e26 (2020). https://www.sciencedirect.com/science/article/pii/S0092867420314550
2. Asai, A., Miethke, A., Bezerra, J.: Pathogenesis of biliary atresia: defining biology to understand clinical phenotypes. Nat. Rev. Gastroenterol. Hepatol. **12**, 05 (2015)
3. Hartley, J.L., Davenport, M., Kelly, D.A.: Biliary atresia. The Lancet **374**(9702), 1704–1713 (2009)
4. Hsiao, C.-H., et al.: Universal screening for biliary atresia using an infant stool color card in Taiwan. Hepatology **47**(4), 1233–1240 (2008)
5. Ohi, R.: Surgical treatment of biliary atresia in the liver transplantation era, pp. 1229–1232 (1998)
6. Serinet, M.-O., et al.: Impact of age at Kasai operation on its results in late childhood and adolescence: a rational basis for biliary atresia screening. Pediatrics **123**(5), 1280–1286 (2009)
7. Lertudomphonwanit, C., et al.: Large-scale proteomics identifies MMP-7 as a sentinel of epithelial injury and of biliary atresia. Sci. Transl. Med. 9(417), eaan8462 (2017)
8. Harpavat, S., et al.: Diagnostic yield of newborn screening for biliary atresia using direct or conjugated bilirubin measurements. Jama **323**(12), 1141–1150 (2020)
9. Humphrey, T.M., Stringer, M.D.: Biliary atresia: US diagnosis. Radiology **244**(3), 845–851 (2007)
10. Qi, X., et al.: Machine learning-based CT radiomics model for predicting hospital stay in patients with pneumonia associated with SARS-COV-2 infection: a multicenter study. MedRxiv (2020)
11. Ardila, D., et al.: End-to-end lung cancer screening with three-dimensional deep learning on low-dose chest computed tomography. Nat. Med. **25**(6), 954–961 (2019)

12. Zhou, W., et al.: Ensembled deep learning model outperforms human experts in diagnosing biliary atresia from sonographic gallbladder images. Nat. Commun. **12**(1), 1–14 (2021)
13. Zhang, K., Zuo, W., Chen, Y., Meng, D., Zhang, L.: Beyond a gaussian denoiser: Residual learning of deep CNN for image denoising. IEEE Trans. Image Process. **26**(7), 3142–3155 (2017)
14. Sun, Y., Yu, Y., Wang, W.: Moiré photo restoration using multiresolution convolutional neural networks. IEEE Trans. Image Process. **27**(8), 4160–4172 (2018)
15. Dosovitskiy, A., et al.: An image is worth 16x16 words: Transformers for image recognition at scale. arXiv:2010.11929 (2021)
16. Gu, Y., Piao, Z., Yoo, S.J.: STHardNet: Swin transformer with HardNet for MRI segmentation. Appl. Sci. **12**(1), 468 (2022)
17. Zhang, L., Wen, Y.: A transformer-based framework for automatic covid19 diagnosis in chest CTS. In: Proceedings of the IEEE/CVF International Conference on Computer Vision, pp. 513–518 (2021)
18. Harpavat, S., Garcia-Prats, J.A., Shneider, B.L.: Newborn bilirubin screening for biliary atresia. N. Engl. J. Med. **375**(6), 605–606 (2016)
19. Feldman, A.G., Sokol, R.J.: Recent developments in diagnostics and treatment of neonatal cholestasis. In: Seminars in Pediatric Surgery, vol. 29, no. 4, p. 150945. Elsevier (2020)
20. Zhou, L., Shan, Q., Tian, W., Wang, Z., Liang, J., Xie, X.: Ultrasound for the diagnosis of biliary atresia: a meta-analysis. Am. J. Roentgenol. **206**(5), W73–W82 (2016)
21. Kim, M.-J., et al.: Biliary atresia in neonates and infants: triangular area of high signal intensity in the porta hepatis at t2-weighted MR cholangiography with us and histopathologic correlation. Radiology **215**(2), 395–401 (2000)
22. Farrant, P., Meire, H., Mieli-Vergani, G.: Ultrasound features of the gall bladder in infants presenting with conjugated hyperbilirubinaemia. Br. J. Radiol. **73**(875), 1154–1158 (2000)
23. Lee, H.-J., Lee, S.-M., Park, W.-H., Choi, S.-O.: Objective criteria of triangular cord sign in biliary atresia on us scans. Radiology **229**(2), 395–400 (2003)
24. Park, W.-H., Choi, S.-O., Lee, H.-J., Kim, S.-P., Zeon, S.-K., Lee, S.-L.: A new diagnostic approach to biliary atresia with emphasis on the ultrasonographic triangular cord sign: comparison of ultrasonography, hepatobiliary scintigraphy, and liver needle biopsy in the evaluation of infantile cholestasis. J. Pediat. Surg. **32**(11), 1555–1559 (1997)
25. Koob, M., Pariente, D., Habes, D., Ducot, B., Adamsbaum, C., Franchi-Abella, S.: The porta hepatis microcyst: an additional sonographic sign for the diagnosis of biliary atresia. Eur. Radiol. **27**(5), 1812–1821 (2017)
26. Caponcelli, E., Knisely, A.S., Davenport, M.: Cystic biliary atresia: an etiologic and prognostic subgroup. J. Pediat. Surg. **43**(9), 1619–1624 (2008)
27. Zhou, L.-Y., et al.: Percutaneous us-guided cholecystocholangiography with microbubbles for assessment of infants with us findings equivocal for biliary atresia and gallbladder longer than 1.5 cm: a pilot study. Radiology **286**(3), 1033–1039 (2018)
28. Zhou, W., Zhou, L.: Ultrasound for the diagnosis of biliary atresia: from conventional ultrasound to artificial intelligence. Diagnostics **12**(1), 51 (2021)
29. Krizhevsky, A., Sutskever, I., Hinton, G.E.: Imagenet classification with deep convolutional neural networks. In: Advances in Neural Information Processing Systems, vol. 25 (2012)

30. LeCun, Y., Bottou, L., Bengio, Y., Haffner, P.: Gradient-based learning applied to document recognition. Proc. IEEE **86**(11), 2278–2324 (1998)
31. Simonyan, K., Zisserman, A.: Very deep convolutional networks for large-scale image recognition. CoRR, abs/1409.1556 (2015)
32. Szegedy, C., et al.: Going deeper with convolutions. In: Proceedings of the IEEE conference on computer vision and pattern recognition, pp. 1–9 (2015)
33. He, K., Zhang, X., Ren, S., Sun, J.: Deep residual learning for image recognition. In: Proceedings of the IEEE Conference on Computer Vision and Pattern Recognition, pp. 770–778 (2016)
34. Huang, G., Liu, Z., Van Der Maaten, L., Weinberger, K.Q.: Densely connected convolutional networks. In: Proceedings of the IEEE Conference on Computer Vision and Pattern Recognition, pp. 4700–4708 (2017)
35. Liu, C., Zoph, B., Neumann, M., Shlens, J., Hua, W., Li, L.-J., Fei-Fei, L., Yuille, A., Huang, J., Murphy, K.: Progressive neural architecture search. In: Ferrari, V., Hebert, M., Sminchisescu, C., Weiss, Y. (eds.) ECCV 2018. LNCS, vol. 11205, pp. 19–35. Springer, Cham (2018). https://doi.org/10.1007/978-3-030-01246-5_2
36. Tan, M., Le, Q.: Efficientnet: rethinking model scaling for convolutional neural networks. In: International Conference on Machine Learning, pp. 6105–6114. PMLR (2019)
37. Hu, J., Shen, L., Sun, G.: Squeeze-and-excitation networks. In: Proceedings of the IEEE Conference on Computer Vision and Pattern Recognition, pp. 7132–7141 (2018)
38. Vaswani, A., et al.: Attention is all you need. In: Advances in Neural Information Processing Systems, vol. 30 (2017)
39. Kenton, J. D. M.-W. C., Toutanova, L.K.: BERT: pre-training of deep bidirectional transformers for language understanding. In: Proceedings of NAACL-HLT, pp. 4171–4186 (2019)
40. Radford, A., et al.: Language models are unsupervised multitask learners. OpenAI Blog **1**(8), 9 (2019)
41. Dong, L., et al.: Unified language model pre-training for natural language understanding and generation. In: Advances in Neural Information Processing Systems, vol. 32 (2019)
42. Wang, X., Girshick, R., Gupta, A., He, K.: Non-local neural networks. In: Proceedings of the IEEE Conference on Computer Vision and Pattern Recognition, pp. 7794–7803 (2018)
43. Touvron, H., Cord, M., Douze, M., Massa, F., Sablayrolles, A., Jégou, H.: Training data-efficient image transformers & distillation through attention. In: International Conference on Machine Learning, pp. 10:347–10:357. PMLR (2021)
44. Liu, Z., et al.: Swin transformer: hierarchical vision transformer using shifted windows. In: Proceedings of the IEEE/CVF International Conference on Computer Vision, pp. 10:012–10:022 (2021)
45. Wu, H., et al.: CVT: introducing convolutions to vision transformers. In: Proceedings of the IEEE/CVF International Conference on Computer Vision, pp. 22–31 (2021)
46. Dong, X., et al.: CSwin transformer: a general vision transformer backbone with cross-shaped windows. In: Proceedings of the IEEE/CVF Conference on Computer Vision and Pattern Recognition, pp. 12:124–12:134 (2022)
47. Boland, P.J.: Majority systems and the condorcet jury theorem. J. R. Statist. Soc. Ser. D (The Statist.) **38**(3), 181–189 (1989)

48. Ardabili, S., Mosavi, A., Várkonyi-Kóczy, A.R.: Advances in machine learning modeling reviewing hybrid and ensemble methods. In: Várkonyi-Kóczy, A.R. (ed.) INTER-ACADEMIA 2019. LNNS, vol. 101, pp. 215–227. Springer, Cham (2020). https://doi.org/10.1007/978-3-030-36841-8_21
49. Ho, T.K.: Random decision forests. In: Proceedings of 3rd International Conference on Document Analysis and Recognition, vol. 1, pp. 278–282. IEEE (1995)
50. Breiman, L.: Bagging predictors. Mach. Learn. **24**(2), 123–140 (1996)
51. Kaniovski, S., Zaigraev, A.: Optimal jury design for homogeneous juries with correlated votes. Theory Decis. **71**(4), 439–459 (2011)

Understanding Tumor Micro Environment Using Graph Theory

Kinza Rohail[1], Saba Bashir[1], Hazrat Ali[2], Tanvir Alam[2], Sheheryar Khan[3], Jia Wu[4], Pingjun Chen[4], and Rizwan Qureshi[1,4](✉)

[1] National University of Computer and Emerging Sciences, Karachi, Pakistan
engr.rizwanqureshi786@gmail.com
[2] College of Science and Engineering, Hamad Bin Khalifa University, Doha, Qatar
[3] School of Professional Education and Executive Department, The Hong Kong Polytechnic University, Hong Kong, China
[4] Department of Imaging Physics, MD Anderson Cancer Center, The University of Texas, Houston, TX, USA

Abstract. Based over the historical data statistics of about past 50 years from National Cancer Institute's Surveillance, the survival rate of patients affected with Chronic Lymphocytic Leukemia (CLL) is about 65%. Neoplastic lymphomas accelerated Chronic Lymphocytic Leukemia (aCLL) and Richter Transformation - Diffuse Large B-cell Lymphoma (RT-DLBL) are the aggressive and rare variant of this cancer that are subjected to less survival rate in patients and becomes worse with age of the patients. In this study, we developed a framework based over Graph Theory, Gaussian Mixture Modeling and Fuzzy C-mean Clustering, for learning the cell characteristics in neoplastic lymphomas along with quantitative analysis of pathological facts observed with integration of Image and Nuclei level analysis. On H&E slides of 60 hematolymphoid neoplasms, we evaluated the proposed algorithm and compared it to four cell level graph-based algorithms, including the global cell graph, cluster cell graph, hierarchical graph modeling and FLocK. The proposed method achieves better performance than the existing algorithms with mean diagnosis accuracy of 0.70833.

Keywords: Hematolymphoid cancer · Graph theory · Digital pathology · Fuzzy clustering

1 Introduction

Hematolymphoid Cancer is a type of primary cancer associated with blood, bone marrow and lymphoid organs. This type of cancer is subjected to high mortality rate. It is very challenging task to distinguish the healthy cells with tumor affected ones since no standard way is yet proposed to distinguish these cells into biological subtypes for diagnosing hematolymphoid tumors on cells. The diagnosis of cancer using histopathology depends upon the topological structure and

Y. Zheng et al. (Eds.): ACCV 2022, LNCS 13848, pp. 90–101, 2023.
https://doi.org/10.1007/978-3-031-27066-6_7

phenotyping of histology based entities like nuclei, tissue areas and cells. Characterization and understanding of these cellular morphology based over image level analysis, nuclei level analysis and entity based analysis is becoming popular and still a challenge when dealing with complex structures [1]. Graphical representations using nodes and edges help to analyze network of cancer cells more deeply and detect the tumor areas more efficiently. Nucleus is represented by a node within original image and edges represent cellular interactions for defining nodes similarity.

1.1 Lymphoid Neoplastic Cells

The proposed work focused on three lymphoid neoplastic cells associated with hematolymphoid cancer, named as Richter transformation-diffuse large B-cell lymphoma (RT-DLBL), chronic lymphocytic leukemia/small lymphocytic lymphoma (CLL/SLL), and Accelerated aCLL/SLL, discussed below [1–3]:

1. **Chronic Lymphocytic Leukemia (CLL)** is a low grade B lymphoid neoplasm that originates in lymphocytes cells (white blood cells) present in bone marrow and then mix with blood. This type of lymphoma grows slowly and patients affected with this type have mild symptoms in initial stages.
2. **Accelerated Chronic Lymphocytic Leukemia (aCLL)** is the aggressive variant of chronic lymphocytic leukemia (CLL). The diagnosis of this type is more challenging when we deal with biopsy specimens.
3. **Richter Transformation - Diffuse Large B-cell Lymphoma (RT-DLBL)** is the rare variant and grows fast in lymphocytes cells. It becomes more aggressive with age of patients.

In this paper, we present a framework based mainly on Graph theory for understanding Tumor Micro Environment (TME) related to Hematolymphoid Cancer based over three neoplastic lymphomas CLL, aCLL and RT. The rest of the paper is organized as follows; Sect. 2 discusses the existing Graph based methods, Sect. 3 discusses proposed research methodology used in this study. Section 4 presents the implementation details, results, comparison and analysis and Sect. 5 concludes this work with future prospects.

2 Literature Review

Most Graph based approaches are applied for the diagnosis of Breast Cancer using Digital Pathology. For understanding the Tumor Micro Environment, based over Graph representations and entity based analysis on cells, nuclei and tissues. This area has not been much explored.

HACT-NET [10] handled the hierarchical structure and shape of tissue in Tumor Micro Environment on the basis histological entities like nuclei, tissues, cells etc. The major limitation of this study is that it is restricted to Breast Cancer yet. It has to be explored to other cancer types and other imaging modalities. Graph-Based Spatial Model [11] constructed topological tumor graphs to diagnose stromal phenotypes in Melanoma. The main limitation was the limited access of clinical cohorts for treating the patients after getting immunotherapy.

Spatial Analysis [12] presents a survey to learn various methods to learn spatial heterogeneity of cellular patterns in Tumor Micro Environment. From the survey, it is concluded that an automated and quantitative analysis of spatial analysis is required for dealing with large clinical cancer data that are subjected to complex spatial patterns. Graph based analysis [16] learns the complex relationship of cancer cells and components of Tumor Micro Environment for diagnosing breast tumors using Graph approach and mathematical morphology.

Cell Spatial Graph [9] decoded cellular and clonal phenotypes in cellular morphology and characterize spatial architectures. They used the concept of local and global graphs for understanding the profile orchestration and interaction of cellular components. It achieves better understanding of intratumoral heterogeneity in Digital Pathology. However, the proposed scheme was unable to handle complex architectures when dealing with critical neoplastic cases and fails to integrate with image level and nuclei level analysis. In this paper, we further improve [9] and integrate nuclei level information.

3 Research Methodology

In this paper, we have presented a framework based over Graph Theory for the diagnosis of Hematolymphoid Cancer using three neoplastic lymphomas. This Section presents the details of the dataset and the brief description of methods used (Fig. 1).

3.1 Proposed Framework

The proposed framework used in this research for the diagnosis of lymphocytic pathology images can be explained as follows:

1. Cell feature extraction for local graphs on the basis of Intensity, Morphology and Region. In this step, 24 features were extracted for each neoplastic cell. These features exhibit the morphological, regional and intensity based patterns in nuclei cells. Intensity based features were related to mean, deviation, range and boundary. The morphological based features were related

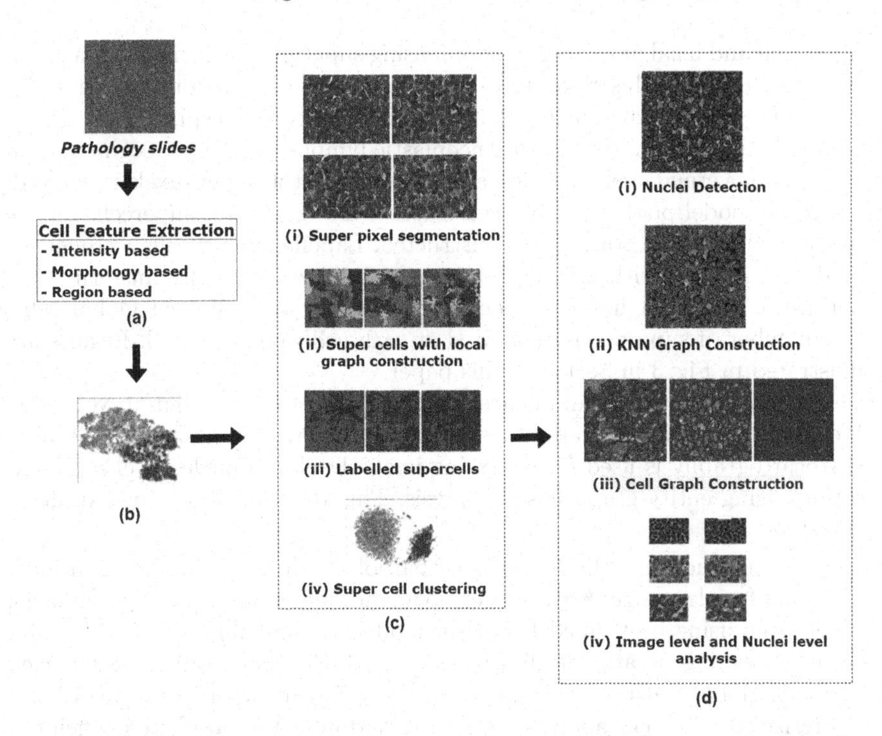

Fig. 1. The illustration of proposed framework used in this research for the diagnosis of lymphocytic pathology images. (a) Cell feature extraction for local graphs on the basis of Intensity, Morphology and Region. (b) Identification of healthy and tumor affected local cells using Fuzzy C Means Clustering. (c) Super cell and local graph construction using super pixel algorithm - Gaussian Mixture Model. (d) Cell graph construction and cancer diagnosis using Graph Neural Networks.

to shape and structure of nuclei in pathology slides such as circularity, elliptical deviation, orientation, axis length, perimeter and equiv diameter. The regional based features were related to boundary saliency, mass displacement, solidity and weighted centroid. Laplacian score method is used for eliminating the redundant features. Using multiple pass adaptive voting, nuclei was segmented on each image and then overlaid over original images. 10 features were selected for local graph construction.

2. Identification of healthy and tumor affected local cells using Fuzzy C Means Clustering. In this step, cells were clustered into tumor affected and healthy cells for each neoplastic cell and build a cell classifier model based on cell types. Results are illustrated in Fig. 2 in Sect. 4 of this paper.

3. Super cell and local graph construction using super pixel algorithm - Gaussian Mixture Model. In this step, super pixel segmentation was done at four scales 8×8, 14×14, 20×20 and 26×26. Highest number of superpixels were generated in CLL as compared to other neoplastic lymphomas. These segmentation results are overlaid over the original ones based on the super pixels generated. Then the model pools the supercell features and generates supercells on the basis of which local graphs are constructed. Labelled super cells were generated and overlaid with original images. Again, fuzzy c means clustering was performed to distinguish between healthy and tumor affected global cells. The number of superpixels generated at each cell were obtained. Results are illustrated in Fig. 3 in Sect. 4 of this paper.
4. Cell graph construction and cancer diagnosis using Graph Neural Networks. We focused on global graph construction, Python based library known as Histocartography is used for analyzing image level and nuclei level representations using entity-graph based analysis. This step is further divided into 3 sub steps: .
   - (i) Constructing Cell Graph using pathology slides - in which identification of nodes, edge, features per node and detection of patch level nuclei is done using pretrained HoverNet model. Highest number of nodes and edges were generated at CLL by the model. For characterizing the nuclei, node global features were extracted using ResNet with patch size 72 and resize size 224. For analyzing the intra and inter tumoral heterogeneity in nuclei with the help of edges, K-Nearest Neighbor graphs were constructed with k=5 and graph properties were obtained with outputs of number of nodes (CLL-5543, aCLL-3579 and RT-2376), edges (CLL-27715, aCLL-17645 and RT-11880) and features extracted per node (514). Results are illustrated in Fig. 4 in Sect. 4 of this paper.
   - (ii) Classification and Analysis of Cell Graphs - For classifying the tumor areas in neoplastic cells RT, aCLL and CLL using Cell Graphs, Graph Neural Network (GNN) was trained with node dimensions 514 and number of classes 3. The model was able to give highest relative node importance to CLL as compared to aCLL and RT. For the analysis of graph representation, GraphGradCAM modified version for feature attribution was integrated with Graph Neural Network, and was used for feature attribution. Node importance was extracted for neoplastic cells on the basis of constructed cell graphs. Results are illustrated in Fig. 4 in Sect. 4 of this paper.
   - (iii) Analysis of Results - For understanding the analysis of shape and size of nuclei and tumor cells in the images, quantitative analysis was

conducted for nuclei analysis along with patch level analysis for each cell defining the importance of values generated by model for area, contrast, crowdness based over nuclei level analysis.

Quantitative analysis was conducted in which pathological facts were observed and important scores were evaluated by the model for neoplastic cells. Following pathological facts were observed in this analysis:
(i) "aCLL is bigger in size than CLL" - aCLL, an aggressive form of neoplastic lymphoma, is bigger in size than CLL, as analyzed in importance value (0.8033) of Area feature.
(ii) "RT is more solid in shape as compared to CLL that has gas bubbles" - as analyzed in importance value (1.3161) of GLCM contrast feature.
(iii) "RT is faster and grows rapidly as compared to aCLL" - as analyzed in importance value (1.2123) of crowdness.

Image level and nuclei level analysis of neoplastic cells was conducted for understanding patch level nuclei in neoplastic lymphomas in which 20 most important nuclei of cells were visualized and then random nuclei were visualized. Results are illustrated in Fig. 4 in Sect. 4 of this paper.

3.2 Dataset Description

A dataset of digital pathology slides of The University of Texas MD Anderson Cancer Center (UTMDACC) were used in this research. We have 20 Digital Pathology slides for each neoplastic cells CLL, aCLL and RT-DLBL and are affected with Hematolymphoid Cancer. Each slide is associated with one patient.

3.3 Description of Methods Used in This Framework

Fuzzy C means clustering algorithm (FCM) - a type of soft clustering in which we assign each data point with a value of likelihood or probability belonging to a particular cluster. This algorithm is used for identifying the healthy cells versus the tumor affected cancer cells in CLL, aCLL and RT.

Gaussian Mixture Model (GMM) - A super pixel based algorithm used for segmenting images for identifying cellular regions and structural patterns. The main properties of this model are that it is able to handle pixels not identically distributed and Eigen decomposition is used for defining the covariance matrix [5]. This algorithm is used for segmenting the images based over super pixels that are further used for making supercells on the basis which local graphs are constructed.

Graph Neural Network - A type of Artificial Neural Network that deals with Graph based structure data. This architecture is well suited for analyzing the interactions between cells. Since, the cellular data does not lie in a grid format. Here Node can be considered as nuclei and edges as interactions between cells.

4 Results and Discussion

In this Section, we provide the results of proposed framework for diagnosing cancer using Digital Pathology and compared the results of this study with Hierarchical Graph Modelling study related to Hematolymphoid Cancer. Figure 2 shows Fuzzy C-mean Clustering results. Figure 3 shows generation of super pixels and super-cell constructed with local graphs. Figure 4 shows cell graph construction with node importance, and image and nuclei level visualization of 20 most important nuclei.

Fig. 2. FCM Clustering results of healthy (represented by blue color) versus tumor affected local cells (represented by green color) in CLL, aCLL and RT. (Color figure online)

4.1 Discussion and Comparison of This Study with Related Existing Hierarchical Graph Modeling Study

The results of the proposed framework are compared with the results of Hierarchical Graph Modeling study [22] and other three graph methods. The comparative analysis is provided in Table 1. In Hierarchical Graph Modeling study, a multi scale framework was proposed for examining the tissue in lymphoid neoplasms accurately. Same lymphoid neoplastic cells were used in this study.

Hierarchical phenotyping was integrated with graph modeling for characterizing the spatial architecture in TME using digital pathology [87]. The proposed approach was able to decode the cellular and clonal hierarchy in the Tumor Micro environment (TME). Also, it extracted both the local and global cellular interactions and their intratumoral heterogeneity. The major limitation was

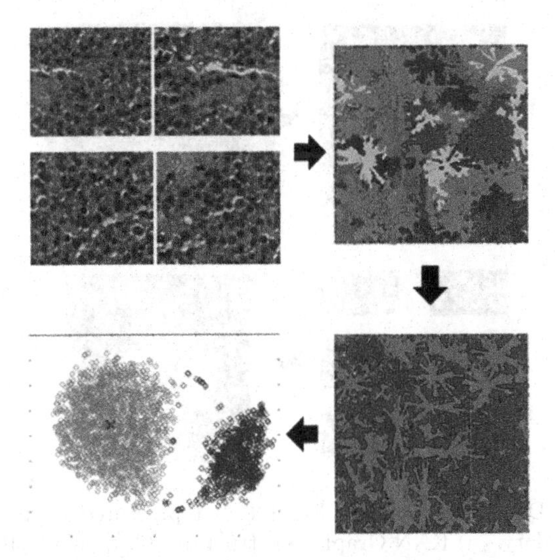

Fig. 3. (i) Gaussian Mixture Model - Super pixel segmented results at four scales 8×8, 14×14, 20×20 and 26×26 scales for CLL, aCLL and RT cells, super pixels labelling is represented by red color - complex labelling CLL with 20255 (8×8), 8254 (14×14), 3340 (20×20) and 1788 (26×26), moderate in aCLL with 17818 (8×8), 6532 (14×14), 3361 (20×20) and 1907 (26×26) and low in RT with 17035 (8×8), 6033 (14×14), 3187 (20×20) and 1796 (26×26) number of superpixels generated at each scale - Highest number of superpixels were generated in CLL as compared to other neoplastic lymphomas. (ii) On basis of segmented results, Super cells were constructed with Local Graph Construction (CLL, aCLL, RT - moderate variation in CLL as compared to Complex in aCLL and RT. (iii) Labelled Super cells constructed with Local Graph Construction (CLL, aCLL, RT - better graphs constructed in aCLL and RT as compared to CLL. (iv) FCM Clustering results of healthy (represented by blue color) verses tumor affected super cells (represented by green color) in CLL, aCLL and RT.

that it was not able to capture neoplastic cell characteristics when dealing with critical cases.

Comparing our results with Hierarchical Graph Modeling ones, Fuzzy algorithm affected the clustering results and allowed to control the level of fuzziness by the Fuzziness parameter helping better classification of cell types at local and global level in TME. Whereas in Hierarchical Graph study, spectral clustering was used which lacked to cover the cell characteristics at cluster boundaries. The use of Gaussian Mixture allowed better segmentation of tumor affected areas and then super-pixels are generated, local graphs were constructed. For global

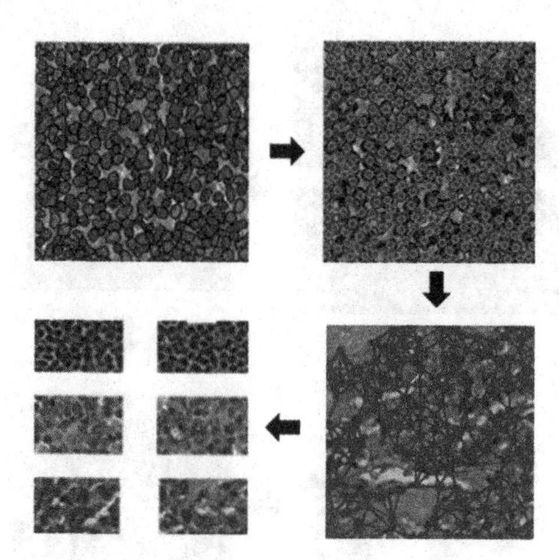

Fig. 4. (i) Nuclei Detection (CLL, aCLL, RT) - represented by purple circles with black outline boundary. (ii) KNN Graph Construction (CLL, aCLL, RT) - graphs constructed with blue color with yellow labelling. (iii) Cell Graphs Construction with node importance, represented by blue color (CLL, aCLL, RT) - high variation in CLL, moderate variation in RT and least variation was seen in aCLL.(iv) Quantitative Analysis of Importance values - pathological facts were observed and important scores were evaluated by a model for neoplastic cells. Image level and Nuclei level analysis (CLL, aCLL, RT) - 20 most important nuclei of cells were visualized and then random nuclei were visualized. (Color figure online)

graph analysis, we initially detected patch level nuclei at global level to measure the intra and inter tumoral heterogeneity and obtained the graph properties. Then with the help of Cell Graphs constructed by Graph Neural Network, we measured the relative node importance for each cell.

The hierarchical graph modeling lacks in integration of image and nuclei level analysis. Also the model was not able to cover complex spatial patterns when dealing with critical neoplastic cases. In Hierarchical Graph Modelling study, Delaunay triangulation was used for global graphs construction that only covered the edge and node information but lacked in covering nuclei level details at global level. The three sub steps workflow used in the four step of our proposed framework has solved the problem with detection of patch level nuclei at global level.

Quantitative analysis of pathological facts helped to observe the shape and structure of tumor with the help of integrated image and nuclei level analysis. The results show the effectiveness of the proposed approach, for understanding and characterizing the cellular morphology and spatial architecture both at global as well as local level. The integration of image and nuclei level analysis have allowed better graph representations of pathological slides for the diagnosis of Hematolymphoid Cancer.

Table 1. Comparison with other graph based methods

Method	Accuracy	AUC(CLL)	AUC(aCLL)	AUC(RT-DLBL)
GCG [26]	0.436 ± 0.037	0.421 ± 0.054	0.730 ± 0.027	0.770 ± 0.023
LCG [27]	0.471 ± 0.042	0.555 ± 0.049	0.669 ± 0.050	0.763 ± 0.032
Flock [28]	0.601 ± 0.045	0.545 ± 0.054	**0.816 ± 0.025**	0.847 ± 0.022
Graph Modeling [9]	0.703 ± 0.030	0.915 ± 0.009	0.724 ± 0.033	0.866 ± 0.028
Proposed	**0.7088 ± 0.035**	**0.922 ± 0.0187**	0.7457 ± 0.0396	**0.913 ± 0.0251**

5 Conclusion and Future Work

Using the concept of local and global graph theory for cancer diagnosis using neoplastic lymphoid cells proved to be efficient method for characterizing spatial architecture and better understanding of cellular morphology. Integrating this concept with Graph Neural Networks has allowed getting more detailed analysis of pathological images at image and nuclei level. Image level analysis allows identifying complex microscopic structural patterns around tumor cells and nuclei level analysis helps to get better learning of cellular architectures more deeply. Entity based analysis of these graph representations allow to extract biological insights more deeply in tissue images. The hierarchical graph approach provides better results for cancer diagnosis based on both pixels and entity graph based classification of tumor in neoplastic cells. In future work, this approach can be used for other cancer types, cells and tissues. We will also explore multiple instance learning (MIL) algorithms to investigate the tumor microenvironment [29]. It also can also be implemented using other cancer related imaging modalities apart from digital pathology.

Acknowledgement. The work described in this paper was supported by a grant from the College of Professional and Continuing Education, an affiliate of The Hong Kong Polytechnic University under project SEHS-2021-229(I).

References

1. Xie, J., et al.: Successful treatment of "accelerated" chronic lymphocytic leukemia with single agent ibrutinib: a report of two cases. Leukemia Res. Rep. **15**, 100247 (2021)
2. Bruzzi, J., et al.: Detection of Richter's transformation of chronic lymphocytic Leukemia by PET/CT. J. Nucl. Med. **47**, 1267–1273 (2006). https://jnm.snmjournals.org/content/47/8/1267
3. Elnair, R., Ellithi, M., Kallam, A., Bleeker, J., Bociek, G.: Survival analysis of CLL/SLL patients with Richter's transformation to DLBCL: an analysis of the SEER database. J. Clin. Oncol. **38**, 1–10 (2020)
4. Bezdek, J., Ehrlich, R., Full, W.: FCM: The fuzzy c-means clustering algorithm. Comput. Geosci. **10**, 191–203 (1984)
5. Ban, Z., Liu, J., Cao, L.: Superpixel segmentation using Gaussian mixture model. IEEE Trans. Image Process. **27**, 4105–4117 (2018)

6. Pope, P., Kolouri, S., Rostami, M., Martin, C., Hoffmann, H.: Explainability methods for graph convolutional neural networks. In: 2019 IEEE/CVF Conference On Computer Vision And Pattern Recognition (CVPR), pp. 10764–10773 (2019)
7. Dwivedi, V., Joshi, C., Luu, A., Laurent, T., Bengio, Y., Bresson, X.: Benchmarking Graph Neural Networks (2020). arxiv:2003.00982
8. Selvaraju, R., Cogswell, M., Das, A., Vedantam, R., Parikh, D., Batra, D.: Grad-CAM: visual explanations from deep networks via gradient-based localization. In: 2017 IEEE International Conference On Computer Vision (ICCV), pp. 618–626 (2017)
9. Chen, P., Aminu, M., El Hussein, S., Khoury, J., Wu, J.: Cell SpatialGraph: integrate hierarchical phenotyping and graph modeling to characterize spatial architecture in tumor microenvironment on digital pathology. Softw. Impacts. **10**, 100156 (2021)
10. Pati, P., Jaume, G., Foncubierta-Rodríguez, M., et al.: Hierarchical graph representations in digital pathology. Med. Image Anal. **75**, 102264 (2022)
11. Failmezger, H., Muralidhar, S., Rullan, A., Andrea, C., Sahai, E., Yuan, Y.: Topological tumor graphs: a graph-based spatial model to infer stromal recruitment for immunosuppression in melanoma histologystromal recruitment and immunosuppression in melanoma. Can. Res. **80**, 1199–1209 (2020)
12. Heindl, A., Nawaz, S., Yuan, Y.: Mapping spatial heterogeneity in the tumor microenvironment: a new era for digital pathology. Lab. Investig. **95**, 377–384 (2015). https://www.nature.com/articles/labinvest2014155#citeas
13. Ben Cheikh, B., Bor-Angelier, C., Racoceanu, D.: Graph-based approach for spatial heterogeneity analysis in tumor microenvironment. Diagn. Pathol. **1**, 1–3 (2016)
14. Xu, H., Cong, F., Hwang, T.: Machine learning and artificial intelligence-driven spatial analysis of the tumor immune microenvironment in pathology slides. Eur. Urol. Focus **7**, 706–709 (2021)
15. AbdulJabbar, K., Raza, S.-A., Rosenthal, R., et al.: Geospatial immune variability illuminates differential evolution of lung adenocarcinoma. Nat. Med. **26**, 1054–1062 (2020)
16. Heindl, A., et al.: Microenvironmental niche divergence shapes BRCA1-dysregulated ovarian cancer morphological plasticity. Nat. Commun. **9**, 3917 (2018)
17. Zormpas-Petridis, K., Failmezger, H., Raza, S., Roxanis, I., Jamin, Y., Yuan, Y.: Superpixel-based conditional random fields (SuperCRF): incorporating global and local context for enhanced deep learning in melanoma histopathology. Front. Oncol. **9**, 1045 (2019). https://www.frontiersin.org/article/10.3389/fonc.2019.01045
18. Hagos, Y., Narayanan, P., Akarca, A., Marafioti, T., Yuan, Y.: ConCORDe-Net: Cell Count Regularized Convolutional Neural Network for Cell Detection in Multiplex Immunohistochemistry Images. (2019). arxiv.org:1908.00907
19. Raza, S., et al.: Deconvolving convolutional neural network for cell detection. In: 2019 IEEE 16th International Symposium On Biomedical Imaging (ISBI 2019), pp. 891–894 (2019)
20. Narayanan, P., Raza, S., Dodson, A., Gusterson, B., Dowsett, M., Yuan, Y.: DeepSDCS: dissecting cancer proliferation heterogeneity in Ki67 digital whole slide images (2018). arxiv:1806.10850
21. Khan, A., Yuan, Y.: Biopsy variability of lymphocytic infiltration in breast cancer subtypes and the IMMUNOSKEW score. Sci. Rep. **6**, 36231 (2016). https://www.nature.com/articles/srep36231#citeas
22. Chen, P., Aminu, M., Hussein, S., Khoury, J., Wu, J.: Hierarchical Phenotyping and Graph Modeling of Spatial Architecture in Lymphoid Neoplasms (2021). arxiv:2106.16174

23. Csikász-Nagy, A., Cavaliere, M., Sedwards, S.: Combining Game theory and graph theory to model interactions between cells in the tumor microenvironment. In: d'Onofrio, A., Cerrai, P., Gandolfi, A. (eds.) New Challenges for Cancer Systems Biomedicine. SIMAI Springer Series, pp. 3–18. Springer, Milano (2012). https://doi.org/10.1007/978-88-470-2571-4_1
24. Sanegre, S., Lucantoni, F., Burgos-Panadero, R., La Cruz-Merino, L., Noguera, R., Álvaro Naranjo, T.: Integrating the tumor microenvironment into cancer therapy. Cancers. **12**, 1677 (2020). https://www.mdpi.com/2072-6694/12/6/1677
25. Yegnanarayanan, V., Krithicaa Narayanaa, Y., Anitha, M., Ciurea, R., Marceanu, L., Balas, V.: Graph theoretical way of understanding protein-protein interaction in ovarian cancer. J. Intell. Fuzzy Syst. **43**, 1877–1886 (2022). https://doi.org/10.3233/JIFS-219289
26. Shin, D., et al.: Quantitative analysis of high-resolution microendoscopic images for diagnosis of esophageal squamous cell carcinoma. Clin. Gastroenterol. Hepatol. **13**(2), 272–279 (2015)
27. Lewis, J.S., Jr., et al.: A quantitative histomorphometric classifier (QuHbIC) identifies aggressive versus indolent p16-positive oropharyngeal squamous cell carcinoma. Am. J. Surg. Pathol. **38**(1), 128 (2014)
28. Lu, C., et al.: Feature-driven local cell graph (FLocK): new computational pathology-based descriptors for prognosis of lung cancer and HPV status of oropharyngeal cancers. Med. Image Anal. **68**, 101903 (2021)
29. Waqas, M., Tahir, M.A., Qureshi, R.: Deep Gaussian mixture model based instance relevance estimation for multiple instance learning applications. Appl. Intell. **17**, 1–6 (2022)

Handling Domain Shift for Lesion Detection via Semi-supervised Domain Adaptation

Manu Sheoran, Monika Sharma(✉), Meghal Dani, and Lovekesh Vig

TCS Research, New Delhi, India
monika.sharma1@tcs.com

Abstract. As the community progresses towards automated Universal Lesion Detection (ULD), it is vital that the techniques developed are robust and easily adaptable across a variety of datasets coming from different scanners, hospitals, and acquisition protocols. In practice, this remains a challenge due to the complexities of the different types of domain shifts. In this paper, we address the domain-shift by proposing a novel domain adaptation framework for ULD. The proposed model allows for the transfer of lesion knowledge from a large labeled source domain to detect lesions on a new target domain with minimal labeled samples. The proposed method first aligns the feature distribution of the two domains by training a detector on the source domain using a supervised loss, and a discriminator on both source and unlabeled target domains using an adversarial loss. Subsequently, a few labeled samples from the target domain along with labeled source samples are used to adapt the detector using an over-fitting aware and periodic gradient update based joint few-shot fine-tuning technique. Further, we utilize a self-supervision scheme to obtain pseudo-labels having high-confidence on the unlabeled target domain which are used to further train the detector in a semi-supervised manner and improve the detection sensitivity. We evaluate our proposed approach on domain adaptation for lesion detection from CT-scans wherein a ULD network trained on the DeepLesion dataset is adapted to 3 target domain datasets such as LiTS, KiTS and 3Dircadb. By utilizing adversarial, few-shot and incremental semi-supervised training, our method achieves comparable detection sensitivity to the previous methods for few-shot and semi-supervised methods as well as to the Oracle model trained on the labeled target domain.

1 Introduction

Universal Lesion Detection (ULD) aims to assist radiologists by automatically detecting lesions in CT-scans across different organs [1–4]. Although, existing ULD networks perform well over a trained source domain, they are still far from practically deployable for clinical applications due to their limited generalization capabilities across target datasets acquired using different scanners and protocols. This domain shift often degrades the detection performance of ULD by over 30–40% when tested on an unseen but related target domain.

A naive approach to circumvent domain-shift is to fine-tune a ULD network, trained on source domain, over sufficient labeled target domain samples. However, obtaining

Y. Zheng et al. (Eds.): ACCV 2022, LNCS 13848, pp. 102–116, 2023.
https://doi.org/10.1007/978-3-031-27066-6_8

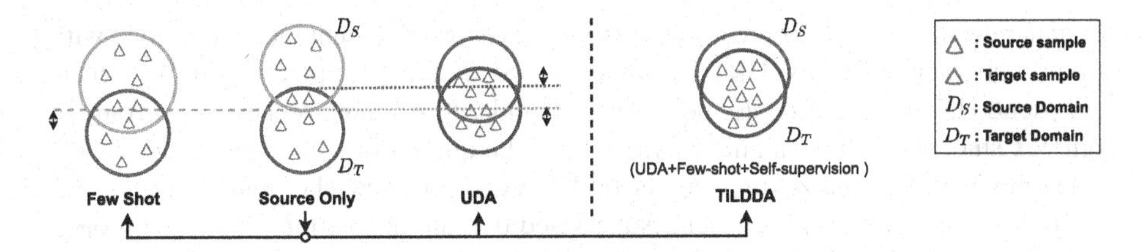

Fig. 1. Visualization of knowledge space of the detector for different adaptation methods across source and target domain. UDA stands for unsupervised domain adaptation. Here, we utilize feature alignment property of unsupervised domain adaptation (UDA) along with few-shot labeled samples from target domain to widen the knowledge space of the detector network for precise lesion detection.

requisite amount of annotations in every new domain is impractical due to the expensive and time-consuming annotation process. Simple fine-tuning may improve sensitivity on the target domain but it suffers from performance drop on the source domain which is not desirable in practical scenarios. For example, when a new CT-machine is added to a facility, then it is expected from a ULD network to maintain its detection sensitivity on new datasets along with the source domain. Therefore, domain adaptation [5–8] is the most effective and widely used technique to easily transfer knowledge from source to new unseen target domains. Widely, there are two approaches to reduce the domain-gap between source and target domain, either by image-to-image translation or by aligning the feature-space. In image-to-image translation techniques [9, 10], researchers have utilized networks such as StyleGAN [11], CycleGAN [10, 12, 13] etc. to generate source images in the style of target images and train a network on the target translated source-images. On the other hand, in feature-space alignment techniques [14–18], authors align the feature-space between source and target domain using either unsupervised adversarial training or prototype alignment. The underlying idea is to generate non-discriminatory features such that the discriminator cannot differentiate between the domains and the task-network trained on a labeled source domain can give similar performance on the new target domain. While large scale annotation of medical scans is expensive, it is often feasible to obtain a few labeled target samples for real world applications. This small amount of annotated data can often provide significant gains for domain-adaptation [19–22].

To learn from few examples of rare objects, two-stage fine-tuning approach (TFA) [23] is proposed where the detector network is first trained with abundant base images and subsequently, only the last layer of trained detectors are finetuned by jointly training few samples from base classes and few samples available for rare/novel classes. However, TFA can help to improve performance on rare classes only if the data for rare classes belong to the same distribution as that of base/source classes. Similarly, in another paper [24], a two-stage semi-supervised object detection method is proposed where detector model is first trained with labeled source data followed by training on unlabelled target data. It utilises an approach called Unbiased Teacher (UBT) where it jointly trains a student and a gradually progressing teacher model by using pseudo labeling technique. The teacher model is provided with weakly augmented inputs and the generated pseudo-labels are used for the supervision of student model provided with strongly augmented inputs. UBT is utilized to reduce the false positives in the generated pseudo labels, as these false positives can hinder the training process. We avoid the complex student-teacher training by improving the quality of generated pseudo labels by learning better initialization weights via UDA and joint few-shot finetuning.

In this paper, we propose a semi-supervised domain adaptation [25–27] approach which utilizes a combination of unsupervised feature alignment at image as well as instance level similar to Every Pixel Matters (EPM) [28], and few-shot labels from the target domain to further expand the representation-space of the ULD network for adaptation to the target domain, as visualized in Fig. 1. Subsequently, we utilize self-supervised learning where we apply the few-shot adapted ULD network on the unlabeled target dataset and obtain pseudo labels using a high confidence threshold. These pseudo labels are used to re-train the ULD network in a semi-supervised manner on the unlabeled target domain. As the combined data for joint few-shot training is dominated by the source domain [29], we train the network via a robust over-fitting aware and periodic gradient update based training scheme which iteratively performs gradient updates on source and target domain samples while accounting for the imbalance in the source and target domain data. The proposed approach can be applied to different convolution based detection backbones and the performance of feature-space alignment based unsupervised domain adaptation techniques can be enhanced and made comparable to that of the Oracle detection network by incorporating few-shot training over target domain labels and semi-supervision using generated pseudo labels on an unlabeled target dataset. To the best of our knowledge, there is very limited research on domain adaptation for lesion detection [30] and we perform transfer of knowledge from a ULD model trained on a large multi-organ dataset to organ-specific target datasets with minimal labeled samples. To summarize, our contributions in this paper are as follows:

- We propose a novel semi-supervised domain adaptation network for ULD via adversarial training, which utilizes few-shot learning for better understanding of the target domain and pseudo-labels based self-supervised learning for more accurate lesion detection on target domain. The network is named *TiLDDA*: **T**owards Universal **L**esion **D**etection via **D**omain **A**daptation.
- A simple anchor-free training scheme is used for lesion detection network which has less design parameters and can handle lesions of multiple sizes from different domains more effectively.
- We evaluate TiLDDA, trained over a source DeepLesion [31] CT dataset, and on 3 target datasets namely, KiTS [32], LiTS [33] and 3Dircadb [34]. The results show consistent improvement in detection sensitivity across all target datasets.
- Owing to the non-availability of lesion detection datasets, we generate bounding box (bbox) annotations of lesions from ground-truth pixel-level segmentation masks on above 3 target domain datasets and release bbox annotations for benchmarking and motivating further research in this area.

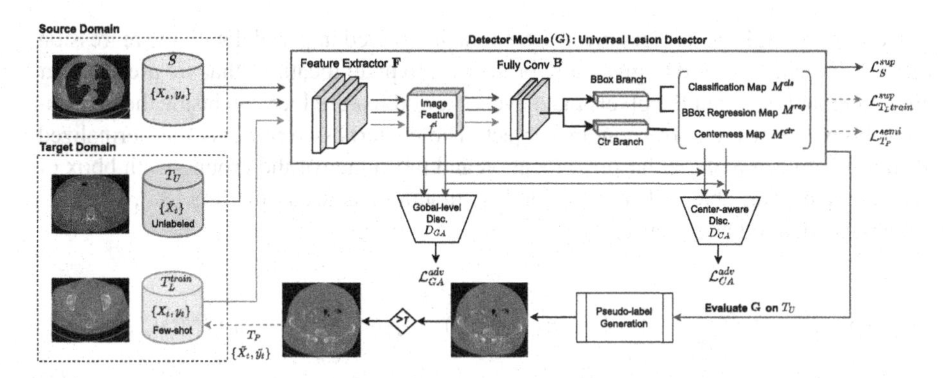

Fig. 2. Overview of our proposed TiLDDA architecture. Source dataset S and unlabeled target dataset T_U are used to train discriminators D_{GA} and D_{CA} using adversarial losses $\mathcal{L}_{GA}^{adv}$ and $\mathcal{L}_{CA}^{adv}$ for domain adaptation. Labeled source domain samples S and few labeled target domain samples T_L^{train} are used to train ULD detector in a few-shot way using supervised losses $\mathcal{L}_S^{sup}$ and $\mathcal{L}_{T_L^{train}}^{sup}$. Further, the few-shot domain adapted ULD is used to generate pseudo-labels on T_U having confidence above a threshold τ. Finally, the pseudo labels are used to re-train the detector in a semi-supervised manner using loss $\mathcal{L}_{T_P}^{semi}$.

2 Methodology

Given a labeled dataset $S = \{(X_s, y_s)\}$ from a source domain D_S, and a dataset T from a different but related target domain D_T split into: an unlabeled set $T_U = \{\tilde{X}_t\}$ and a much smaller labeled set $T_L^{train} = \{(X_t, y_t)\}$, where $T = T_U + T_L^{train}$. Both S and T share the same task, i.e., given an input CT-image X, find the bounding box (Bbox) of the lesion present y. Therefore, the aim of our proposed domain adaptation network is to learn a single set of detector model parameters G_θ such that the model trained on the source domain D_S and few labeled target domain samples T_L^{train} can work efficiently on an unseen target test-set T^{test} without degradation in lesion detection performance. The different components of our proposed domain adaptation pipeline (shown in Fig. 2) are as follows:

2.1 Universal Lesion Detection

To cater to the need of detecting multi-sized lesions across different domains, we utilize a robust anchor-free lesion detector (G) based on a fully convolutional one-stage (FCOS) [3,35] network which performs detection in a per-pixel prediction fashion rather than utilizing the pre-defined anchor-boxes. As shown in Fig. 2, for an input image X, we first extract the feature maps (f^i) at i^{th} feature pyramid network (FPN) level using a convolutional feature-extractor F. Next, using a fully-connected detection head B, each pixel location (x, y) of f^i is classified with probability $(p_{x,y})$ as foreground (with class label ${c^*}_{x,y} = 1$) or background (with class label ${c^*}_{x,y} = 0$) and then, for each positive pixel location, a $4D$ vector $u_{x,y}$ is regressed against the corresponding ground-truth bbox annotation ${u^*}_{x,y}$. To further decrease the low-quality bbox detections, a single centerness layer (Ctr) branch is added in parallel with the regression (Bbox) branch. It is used to give more preference to pixel locations that are present near the center and filter out the pixels that have a skewed feature location inside the ground-truth bbox (y) of the corresponding object. The centerness represents the normalized distance between a particular pixel location and the center of the ground-truth bbox of the corresponding object. The detection loss function, as used in FCOS [35], for ULD baseline is defined as follows:

$$\mathcal{L}^{det}(p_{x,y}, u_{x,y}) = \frac{1}{N_{pos}} \sum_{x,y} \mathcal{L}^{cls}(p_{x,y}, {c^*}_{x,y}) + \frac{\lambda}{N_{pos}} \sum_{x,y} \mathbb{1}_{c^*_{x,y}>0} \mathcal{L}^{reg}(u_{x,y}, {u^*}_{x,y}) \tag{1}$$

$$centerness = \sqrt{\frac{min(l^*, r^*)}{max(l^*, r^*)} \times \frac{min(t^*, b^*)}{max(t^*, b^*)}} \tag{2}$$

Here, $\mathcal{L}^{cls}$ and $\mathcal{L}^{reg}$ are the classification focal loss and regression IoU loss for location (x, y), N_{pos} is the no. of positive samples, λ is the balance weight, $\mathbb{1}_{c^*_{x,y}>0}$ is an indicator function for every positive location, c^* and u^* are ground-truth labels for classification and regression, respectively. For given regression targets l^*, t^*, r^* & b^* of a location, the term centerness (as defined in Eq. 2) is trained with binary cross entropy (BCE) loss $\mathcal{L}^{ctr}$ and added to the loss function defined in Eq. 1 for the refined results. Finally, the ULD network is trained using a supervised loss $\mathcal{L}^{sup}$ function as defined in Eq. 3.

$$\mathcal{L}^{sup}(X, y) = \mathcal{L}^{det} + \mathcal{L}^{ctr} \tag{3}$$

2.2 Feature Alignment via Adversarial Learning

Here, as inspired by EPM network [28], we utilize unsupervised domain adaptation (UDA) to align the feature distribution for both the domains, source D_S and target D_T, which would result in an increase in the detection sensitivity on the target domain test dataset T^{test} in an unsupervised manner. First, the detector network G is trained on S using a supervised loss-function $\mathcal{L}_S^{sup}$, as defined in Eq. 3. To train the discriminators, we first extract source $({f^i}_s)$ and target $({f^i}_t)$ feature maps by applying feature extractor F on S and T_U samples, and perform global feature alignment via a global discriminator D_{GA} which is optimized by minimizing a binary cross-entropy loss $\mathcal{L}_{GA}^{adv}$. This is a domain-prediction loss that aims to identify whether the pixels on i^{th} feature map (f^i) belong to the source/target domain. For a location (x, y) on f^i, $\mathcal{L}_{GA}^{adv}$ can be defined as below:

$$\mathcal{L}_{GA}^{adv}(X_s, \tilde{X}_t) = -\sum_{x,y} z \log\left(D_{GA}(f_s^i)^{(x,y)}\right) + (1-z) \log\left(1 - D_{GA}(f_t^i)^{(x,y)}\right) \tag{4}$$

We set the domain label z of source and target as 1 and 0, respectively. Next, the detection head B predicts pixel-wise objectness maps M^{obj} and centerness maps M^{cls}

Algorithm 1: Proposed Joint Few-shot Learning

Data: Source dataset S and few-shot labeled target dataset T_L^{train}, detector model G_θ, and Hyper-parameters: α, β, and κ.
$n(S), n(T_L^{train}) \leftarrow$ Total source and labeled target samples
for iterations $= 1, 2, 3, ...$ **do**
  Train-source: Gradients $\nabla_\theta = G_\theta^{'}(S;\theta)$;
  Updated parameters: $\theta^{'} \leftarrow \theta - \alpha\nabla_\theta$;
  $\eta = \frac{n(S)}{n(T_L^{train})*\kappa}$;
  if (iterations $\textbf{mod}\ \eta) = 0$ **do**
    Train-target: Gradients $\nabla_{\theta^{'}} = G_{\theta^{'}}^{'}(T_L^{train};\theta^{'})$;
    Updated parameters: $\theta \leftarrow \theta^{'} - \beta\nabla_{\theta^{'}}$;
  else
    $\theta \leftarrow \theta^{'}$;

which are combined to generate a centre-aware map M^{CA} [28]. The extracted feature maps f^i along with M^{CA} are utilized to train another center-aware discriminator D_{CA} with domain-prediction loss $\mathcal{L}_{CA}^{adv}$, as given in Eq. 5 in order to perform center-aware alignment at the pixel level.

$$\begin{aligned}\mathcal{L}_{CA}^{adv}(X_s,\tilde{X}_t) = -\sum_{x,y} z\log\left(D_{CA}(M_s^{CA}\odot f_s^i)^{(x,y)}\right)\\ +(1-z)\log\left(1-D_{CA}(M_t^{CA}\odot f_t^i)^{(x,y)}\right)\end{aligned} \quad (5)$$

We apply the gradient reversal layer (GRL) [36] before each discriminator for adversarial learning, which reverses the sign of the gradient while optimizing the detector. The loss for the discriminators is minimized via Eq. 4 and Eq. 5, while the detector is optimized by maximizing these loss functions, in order to deceive the discriminator. Hence, the overall loss function for UDA using δ and γ as balancing weights, can be expressed as follows:

$$\mathcal{L}^{UDA}(S,T_U) = \mathcal{L}_S^{sup}(X_s,y_s) + \delta\mathcal{L}_{GA}^{adv}(X_s,\tilde{X}_t) + \gamma\mathcal{L}_{CA}^{adv}(X_s,\tilde{X}_t) \quad (6)$$

2.3 Proposed Joint Few-Shot Learning (FSL)

Different from standard few-shot learning, where the tasks for target domain are different from the source domain and hence, we either train the network on available source samples first and then use the trained model weights as initialization or, in case of no source domain, we can use weights from already trained models such as ImageNet weights to fine-tune the task network on few target samples separately. Here in this paper, we are trying to solve the domain shift problem where we have the same task, i.e. lesion detection from CT-scans, for both the source (D_S) and target (D_T) domains.

Therefore, we can jointly fine-tune the ULD baseline G on a small labeled target domain dataset T_L^{train} combined with the larger labeled source dataset S. However, this setting suffers from data imbalance as the combined data is dominated by the source domain and hence, the training will be biased towards the source domain. To mitigate this issue, we propose a modified version of the few-shot training paradigm as given by Algorithm 1, which aims to regularize the ULD network and enable it to focus more on target domain samples without over-fitting on one particular domain. The idea is to train the detector G on both domains by alternatively updating their weights so as to ensure balanced updation across source and target samples. This is achieved by finding the best possible gradient direction due to the shared parameter optimization of the two losses. The loss on source train set S is computed using model parameter θ. The loss on the target train set T_L^{train} is computed using shared updated parameter $\theta^{'} = \theta - \alpha\nabla_\theta$ after each η iterations. To avoid over-fitting on target domain, we compute η such that κ epochs of target are trained when 1 epoch of source is trained. We empirically determined the optimal value of $\kappa = 3$. The supervised loss function for the proposed FSL is defined in Eq. 7, where $\mathbb{1}_\eta$ is an indicator function that takes a value of 1 after each η iteration.

$$\mathcal{L}^{few}(S, T_L^{train}) = \mathcal{L}_S^{sup}(X_s, y_s, \theta) + \mathbb{1}_\eta \mathcal{L}_{T_L^{train}}^{sup}(X_t, y_t, \theta^{'}) \tag{7}$$

2.4 Few-Shot Domain Adaptation (FDA)

Next, we apply the adversarial learning ($\mathcal{L}_{GA}^{adv}$ and $\mathcal{L}_{CA}^{adv}$) over source and target domain for feature alignment with the proposed FSL ($\mathcal{L}^{few}$) on the combined domain. This helps in increasing the similarity between the two domains via feature-alignment and also widens the knowledge space of ULD by incorporating information from the target domain in the form of few-shot labeled samples. The loss function for FDA is defined as follows:

$$\mathcal{L}^{FDA}(S, T_U, T_L^{train}) = \mathcal{L}^{few}(X_s, X_t, y_s, y_t) + \delta\mathcal{L}_{GA}^{adv}(X_s, \tilde{X}_t) + \gamma\mathcal{L}_{CA}^{adv}(X_s, \tilde{X}_t) \tag{8}$$

2.5 Self-supervision

As unlabeled samples T_U of target domain are available in abundance, we utilize a self-supervised learning mechanism to further improve the ULD performance on T by expanding the few-shot labeled sample space for T. Here, we obtain bbox predictions $(\tilde{y}_t)$, having confidence-score above a detection threshold (τ), on unlabeled target samples $\tilde{X}_t$ by applying the few-shot adapted UDA network. Hence, we generate pseudo samples $(T_P = \{\tilde{X}_t, \tilde{y}_t\})$ to further fine-tune the FDA network in a semi-supervised manner using $(\mathcal{L}_{T_P}^{semi})$ (defined in Eq. 9) on target domain.

$$\mathcal{L}_{T_P}^{semi} = \mathcal{L}^{sup}(\tilde{X}_t, \tilde{y}_t) \tag{9}$$

Table 1. Data distribution of Target Domain Datasets T. Here, T_U^{train}, T_L^{train} and T^{test} represent the unlabeled train data, labeled few-shot train data and test-data.

KiTS (230 Patients)	Data split	No. of Patients	No. of Images	No. of Lesions
	T_U^{train}	180	3914	4083
	T_L^{train}	10	919	1305
	T^{test}	40	923	949
LiTS (130 Patients)	**Data split**	**No. of Patients**	**No. of Images**	**No. of Lesions**
	T_U^{train}	80	4270	11932
	T_L^{train}	10	847	2342
	T^{test}	40	2073	4571
3Dircadb (15 Patients)	**Data split**	**No. of Patients**	**No. of Images**	**No. of Lesions**
	T_U^{train}	4	144	430
	T_L^{train}	3	113	163
	T^{test}	8	311	676

3 Experiments and Results

3.1 Overall Training Scheme of TiLDDA

We train the ULD network G on source samples (S) and use the source domain weights for initializing our proposed TiLDAA network. For domain adaptation on T, we first train the detector G and discriminators D_{GA} & D_{CA} via the FDA training method using loss defined in Eq. 8. Subsequently, we apply the adapted detector G on unlabeled target images $\tilde{X}_t$ and generate pseudo-labels ($T_P = \{\tilde{X}_t, \tilde{y}_t\}$). Next, we re-train the ULD network using the semi-supervised loss defined in Eq. 9. Hence, the final objective loss-function of our proposed TiLDDA network using hyper-parameters $\delta, \gamma, \eta,$ & λ is as follows:

$$\mathcal{L}^{TiLDDA}(S, T_U, T_L^{train}, T_P) = \mathcal{L}_S^{sup}(S, \theta) + \delta \mathcal{L}_{GA}^{adv}(X_s, \tilde{X}_t) + \gamma \mathcal{L}_{CA}^{adv}(X_s, \tilde{X}_t) + \mathbb{1}_\eta(\mathcal{L}_{T^{train}}^{sup}(T_L^{train}, \theta') + \lambda \mathcal{L}_{T_P}^{semi}(T_P, \theta')) \quad (10)$$

3.2 Data and Evaluation Metric

We evaluate our TiLDDA network on lesion-detection from CT-scans by adapting the ULD model trained on DeepLesion [31] as source domain dataset S to 3 different target domain datasets T such as KiTS [32], LiTS [33] and 3Dircadb [34]. We provide details for different Source and Target domain datasets as follows:

– **Source Domain Database** S: DeepLesion[1] is the largest publicly available multi-organ lesion detection dataset, released by National Institutes of Health (NIH) Clinical Center. It consists of approximately $32,000$ annotated lesions from $10,594$

[1] DeepLesion: https://nihcc.app.box.com/v/DeepLesion.

CT-scans of 4,427 unique patients having 1–3 lesions bounding boxes annotated for each CT scan by radiologists.
- **Target Domain Database** T: Since, we were unable to find relevant CT datasets for lesion detection, we utilized the ground-truth segmentation masks for lesions provided in the following target datasets to generate the bounding box-annotations. To introduce domain shift properly, we have selected target datasets that are collected across different geographical locations. The details of these datasets are given as below:
  1. *KiTS*[2] This cohort includes 230 CT-scans of patients who underwent partial or radical nephrectomy for suspected renal malignancy between 2010 and 2018 at University of Minnesota Medical Center, US. The kidney-region and the kidney-tumors in this dataset are annotated by experts and segmentation masks are released publicly.
  2. *LiTS*[3] This dataset consists of 130 pre- and post-therapy CT-scans released by Technical University of Munich, Germany. The image data is also very diverse with respect to resolution and image quality. The manual segmentations of tumors present in liver region are provided in the dataset.
  3. *3Dircadb*[4] 3D Image Reconstruction for Comparison of Algorithm Database (3Dircadb) released by Research Institute against Digestive Cancer, Strasbourg Cedex, France. It consists of 15 CT-scans of patients with manual segmentation of liver tumors performed by clinical experts.

Please refer Table 1 for data-split used for training and testing. For all the experiments, we have used labeled data of 10 patients from LiTS and KiTS dataset. But due to the very small size of 3Dircadb, we utilize labeled data of 3 patients only. The idea behind using very small-sized 3Dircadb dataset is to evaluate how effectively the proposed TiLDDA network can adapt with minimal target domain samples. As part of pre-processing the CT-images, we include black-border clipping, re-sampling voxel space to $0.8 \times 0.8 \times 2$ mm^3 and HU-windowing with a range of $[-1024, 3072]$. We also perform data augmentations such as horizontal and vertical flipping, resizing and pixel translations along x- and y-axis. For evaluation, average of detection sensitivities over four false positive rates (FPs $= \{0.5, 1, 2, 4\}$) is computed and for all future references in the paper, detection sensitivity means average detection sensitivity.

3.3 Experimental Setup

The feature extractor F is composed of ResNet-101 backbone along with 5 FPN levels and the fully-convolutional block B consists of 3 branches for classification, regression and centerness computations. For robust DA, feature alignment is done across all FPN levels and the architectures of detector G and discriminators D_{GA} & D_{CA} are similar to that used in EPM [28]. We implement TiLDDA in PyTorch-1.4 and train it on a

[2] KiTS: https://kits19.grand-challenge.org/data.
[3] LiTS: https://competitions.codalab.org/competitions/17094.
[4] 3Dircadb: https://www.ircad.fr/research/3d-ircadb-01.

Table 2. Average sensitivity (%) on target datasets using different training schemes.

Training scheme	KiTS [32]	LiTS [33]	3Dircadb [34]
Source only (DeepLesion) (FCOS) [35]	34.2	36.7	37.3
Vanilla Few-shot	44.8	40.8	20.7
UBT (few-shot) [24]	36.4	40.2	18.7
UBT (few-shot + semi-supervision) [24]	44.1	47.3	27.2
TFA (joint few-shot) [23]	54.1	51.2	42.2
Proposed joint few-shot (FSL)	56.6	53.1	45.6
UDA (EPM) [28]	39.4	44.6	42.1
Fewshot DA (FDA)	58.6	53.8	47.1
Fewshot DA + Self-supervision (TiLDDA)	*71.6*	*55.2*	*49.5*
Oracle (Target only)	**77**	**57.6**	**61.1**

NVIDIA $V100$ 16GB GPU using a batch-size of 4. For all our experiments, we set the values of κ, δ, γ, λ, and τ to 3, 0.01, 0.1, 0.5, and 0.7, respectively. The weights used in GRL for adversarial training are set to 0.01 and 0.02 for D_{GA}& D_{CA}, respectively. The detector network G for FDA is initialized using weights learned via pre-training on source S. An SGD optimizer is used to train FDA network for $65,000$ iterations with a learning rate of e^{-3} and decay-factor of 10 after $32,000$ and 52000 iterations. For overall training of TiLDAA, FDA model is further fine-tuned on S and updated T samples using a learning rate of e^{-4} for $25,000$ iterations.

3.4 Result and Ablation Study

The lesion detection sensitivity on S using the ULD baseline with ResNet-101 backbone is 80% and the aim of our proposed TiLDDA network is to perform well on target domain dataset T as well while maintaining the performance on source domain. Table 2 presents the performance of different training schemes on test-split T^{test} of target domain. The upper bound of detection sensitivity on T^{test} is determined by supervised training of the ULD baseline G directly on target samples (T_U^{train}) only in a supervised manner and this setting is referred to as Oracle setting. The lower bound is computed by evaluating the target T test-set T^{test} directly using the ULD model trained on source (S) only.

It is evident from Table 2 that there is a drop of about 30% to 40% in the detection sensitivity compared to that of Oracle setting due to the domain-shift issue. To circumvent this issue, we begin by training the network using different few-shot techniques without any domain adaptation. First, we use the vanilla few-shot finetuning, where the ULD network is initialized with ImageNet weights and trained only using few labeled target samples T_L^{train}. As expected, the ULD network performs better as compared to Source only training scheme on the test set T^{test}, except for 3Dircadb target domain where training samples are low. Next, we use a semi-supervised method UBT proposed by Yen et. al [24] which utilizes pseudo labels along with few-shot finetuning, there is a further improvement (7% to 8%) in detection sensitivity but it's still limited as lesion detection knowledge from source domain is not being utilized till now by these methods. Hence, we also experimented with a two-step joint few-shot finetuning approach (TFA) [23], where we used the entire source domain data in second step

Table 3. Experiment to show that our proposed training scheme can be used with feature-space alignment based unsupervised domain adaptation methods to further enhance their detection performance. Average sensitivity (%) for LiTS test dataset using different training schemes applied on UaDAN [37] method and EPM [28] method.

Training scheme	UaDAN [37]	EPM [28]
Source only (DeepLesion)	35.6	36.7
UDA	40.1	44.6
FDA	50.7	53.8
TiLDDA	*52.8*	*55.2*
Oracle	**56.5**	**57.6**

of fine-tuning resulting in a steep increase in the detection sensitivity value, especially for the 3Dircadb target domain. This confirms that training target samples with source samples can help to improve the detection sensitivity for the target dataset. However, as mentioned in Sect. 2.3 simple joint few-shot training suffers from data-imbalance issue, hence we train the network with our proposed joint few-shot training scheme (FSL) described in Algorithm 1 and demonstrate an increase in sensitivity (2% to 3%). Further, utilizing source domain data alone for handling domain shift issue is not enough, hence we apply a UDA method which utilizes adversarial training to align cross-domain features, similar to EPM [28], and observe that even without using any data from target domain, there is a small but significant improvement (5% to 7%) in sensitivity as compared to Source only training scheme. Subsequent to this, we combine the UDA and proposed joint few-shot method for few-shot adaptation (FDA) to train the ULD network and obtain an enhanced performance. At last, we generate psuedo labels (T_P) using the FDA model and further, fine-tune it via TiLDDA model. It can be seen clearly that we obtain a remarkable improvement (12% to 35%) in lesion detection as compared to source only training using our proposed TiLDDA network with very few labeled target samples.

Next, in order to support our claim that our proposed training scheme can be used with different convolution-based detection backbones and the performance of feature-space alignment based unsupervised DA methods can be improved, we utilize a UDA method proposed in [37] and apply the proposed incremental training schema (joint few-shot + pseudo label based self-supervision) on LiTS as target dataset. We observe a similar trend for improvement of the lesion detection sensitivity as obtained with EPM baseline used for TiLDAA, as shown in Table 3.

Further, we present the ablation-study in Table 4 on the number of few-shot labeled samples (T_L^{train}) of different target domain datasets and hyper-parameter κ used in Algorithm 1. We observe that 10 is the optimal number of few-shot labeled samples to obtain best performance. But due to the very small size of 3Dircadb dataset, we utilize labeled data of 3 patients only from 3Dircadb. As the combined data in few-shot learning is dominated by source samples, so we train the network on target samples for more number of epochs as compared to source domain using different values of κ and

Table 4. Average sensitivity (%) for different number of labeled target samples ($n(T_L^{train})$) and hyper-parameter κ for different target domains using FDA training scheme.

Target domain	No. of patients	$n(T_L^{train})$	κ	Sensitivity (%)
LiTS	1	81	1	46.3
	5	428	1	50.3
			1	51.4
	10	**847**	**3**	**53.8**
			5	53.3
KiTS			1	56.1
	10	**919**	**3**	58.6
			5	57.1
3Dircadb			1	45.4
	3	**113**	**3**	**47.1**
			5	46.2

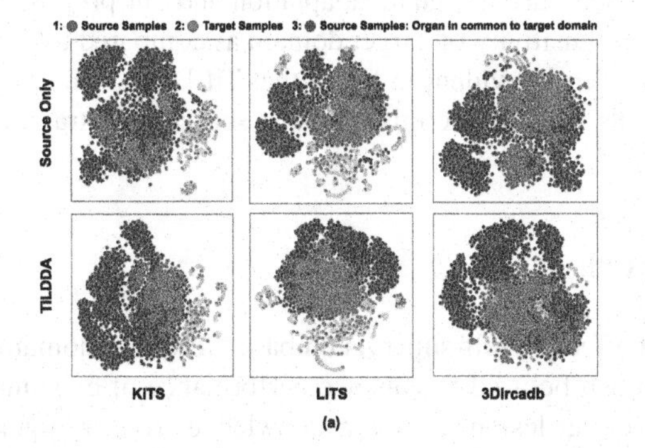

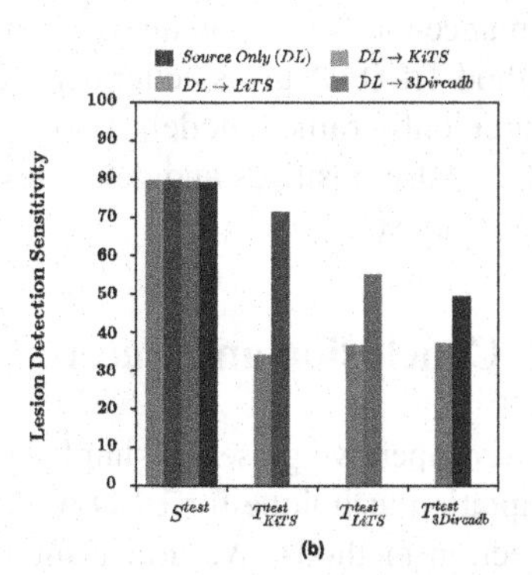

Fig. 3. (a) The t-SNE visualization of source and target sample distributions before and after using TiLDDA.(b) Effect of domain-adaptation on lesion-detection sensitivity of test-set of S and T domain datasets. Here, DL refers to DeepLesion dataset.

found that a value of 3 is optimal that avoids the model from over-fitting over target domain.

Additionally, we present a qualitative comparison using t-SNE [38] plots in Fig. 3(a) to visualize the distribution from test-split of source D_S and target D_T domain samples using Source-Only and TiLDDA training schemes. We extract embeddings of the test-samples using feature extractor F. The labels 1, 2, and 3 correspond to samples from source, target and samples of source domain organ in common to the target domain, respectively. The common organ of DeepLesion and, LiTS and 3Dircadb is liver. However, the common organ of DeepLesion and KiTS is kidney. We can clearly observe that after adaptation, the embeddings of target domain organs are now aligned better with the common organ of source domain resulting in an enhanced detection sensitivity for the target domain. It validates our claim that the detection knowledge from source

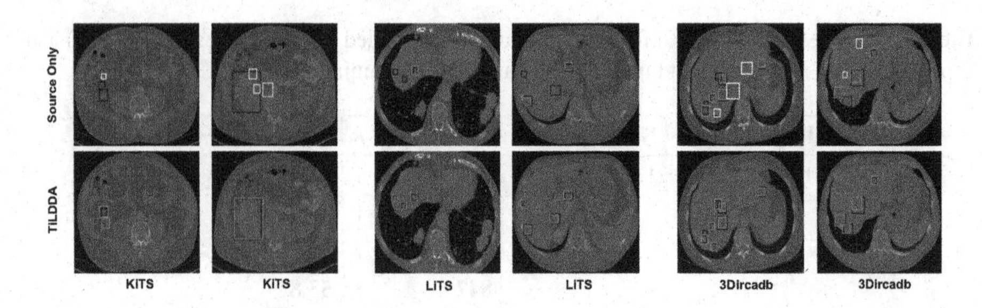

Fig. 4. Qualitative comparison of Lesion Detection before and after using TiLDDA. Here green, magenta, and red color boxes represent ground-truth, true-positive (TP), and false-positive (FP) lesion detection, respectively. (Color figure online)

domain can be transferred to the target domain to improve the lesion detection performance. Further, we present the comparison of detection-sensitivity on test-set of S and T datasets before and after applying TiLDAA in Fig. 3(b). It can be seen that the performance on source domain is maintained during domain adaptation and our proposed method TiLDAA gives better detection sensitivity on target domain as compared to the Source only trained model. Also, it is clearly evident in Fig. 4 that TiLDDA is able to reduce false positives and detect lesions which were missed using source-only trained lesion detector.

4 Conclusion and Future Work

In this paper, we present a simple but effective self-supervision based few-shot domain adaptation technique for ULD which can be used to enhance performance of existing detection methods. We utilize multi-organ lesion detection knowledge from a larger universal lesion detection source domain dataset to efficiently detect lesions on three organ-specific target domains, and achieve comparable performance to the Oracle training scheme by utilizing only a few labeled target samples. We first adversarially align the representation space of the two domains via unsupervised domain adaptation and with a few labeled target samples, further fine-tune the detector in a semi-supervised way using the self-generated pseudo labels. We experimentally show the efficacy of our method by reducing the performance drop on unseen target domains compared to an Oracle model trained on a fully labeled target dataset. In the current setup, both source and target domains have a common task of detecting lesions from CT images across a common set of organs. Going forward, we would like to propose a network that can adapt to out-of-distribution organs and work across cross-modality domains.

References

1. Yan, K., et al.: MULAN: multitask universal lesion analysis network for joint lesion detection, tagging, and segmentation. In: Shen, D., et al. (eds.) MICCAI 2019. LNCS, vol. 11769, pp. 194–202. Springer, Cham (2019). https://doi.org/10.1007/978-3-030-32226-7_22

2. Yan, K., et al.: Universal Lesion Detection by learning from multiple heterogeneously labeled datasets. arXiv preprint arXiv:2005.13753 (2020)
3. Sheoran, M., Dani, M., Sharma, M., Vig, L.: An efficient anchor-free universal lesion detection in CT-scans. In: 2022 IEEE 19th International Symposium on Biomedical Imaging (ISBI), pp. 1–4. IEEE (2022)
4. Sheoran, M., Dani, M., Sharma, M., Vig, L.: DKMA-ULD: domain knowledge augmented multi-head attention based robust universal lesion detection. arXiv preprint arXiv:2203.06886 (2022)
5. Gopalan, R., Li, R., Chellappa, R.: Domain adaptation for object recognition: an unsupervised approach. In: 2011 International Conference on Computer Vision, pp. 999–1006. IEEE (2011)
6. Long, M., Zhu, H., Wang, J., Jordan, M.I.: Unsupervised domain adaptation with residual transfer networks. In: Advances in Neural Information Processing Systems, vol. 29 (2016)
7. Pan, S.J., Yang, Q.: A survey on transfer learning. IEEE Trans. Knowl. Data Eng. **22**, 1345–1359 (2010)
8. Saito, K., Ushiku, Y., Harada, T., Saenko, K.: Strong-weak distribution alignment for adaptive object detection. In: Proceedings of the IEEE/CVF Conference on Computer Vision and Pattern Recognition, pp. 6956–6965 (2019)
9. Saxena, S., Teli, M.N.: Comparison and analysis of image-to-image generative adversarial networks: a survey. CoRR abs/2112.12625 (2021)
10. Zhu, J.Y., Park, T., Isola, P., Efros, A.A.: Unpaired image-to-image translation using cycle-consistent adversarial networks. In: Proceedings of the IEEE International Conference on Computer Vision, pp. 2223–2232 (2017)
11. Rojtberg, P., Pollabauer, T., Kuijper, A.: Style-transfer GANs for bridging the domain gap in synthetic pose estimator training. In: 2020 IEEE International Conference on Artificial Intelligence and Virtual Reality (AIVR), pp. 188–195 (2020)
12. Yang, J., Dvornek, N.C., Zhang, F., Chapiro, J., Lin, M.D., Duncan, J.S.: Unsupervised domain adaptation via disentangled representations: application to cross-modality liver segmentation. In: Shen, D., et al. (eds.) MICCAI 2019. LNCS, vol. 11765, pp. 255–263. Springer, Cham (2019). https://doi.org/10.1007/978-3-030-32245-8_29
13. Dou, Q., et al.: PnP-AdaNet: plug-and-play adversarial domain adaptation network at unpaired cross-modality cardiac segmentation. IEEE Access **7**, 99065–99076 (2019)
14. Li, H., Pan, S.J., Wang, S., Kot, A.C.: Domain generalization with adversarial feature learning. In: 2018 IEEE/CVF Conference on Computer Vision and Pattern Recognition, pp. 5400–5409 (2018)
15. Lee, S.M., Kim, D., Kim, N., Jeong, S.G.: Drop to adapt: learning discriminative features for unsupervised domain adaptation. In: 2019 IEEE/CVF International Conference on Computer Vision (ICCV), pp. 91–100 (2019)
16. Tanwisuth, K., Fan, X., Zheng, H., Zhang, S., Zhang, H., Chen, B., Zhou, M.: A prototype-oriented framework for unsupervised domain adaptation. CoRR abs/2110.12024 (2021)
17. Kamnitsas, K., et al.: Unsupervised domain adaptation in brain lesion segmentation with adversarial networks. In: Niethammer, M., et al. (eds.) IPMI 2017. LNCS, vol. 10265, pp. 597–609. Springer, Cham (2017). https://doi.org/10.1007/978-3-319-59050-9_47
18. Shin, S.Y., Lee, S., Summers, R.M.: Unsupervised domain adaptation for small bowel segmentation using disentangled representation. In: de Bruijne, M., et al. (eds.) MICCAI 2021. LNCS, vol. 12903, pp. 282–292. Springer, Cham (2021). https://doi.org/10.1007/978-3-030-87199-4_27
19. Zhao, A., et al.: Domain-adaptive few-shot learning. In: 2021 IEEE Winter Conference on Applications of Computer Vision (WACV), pp. 1389–1398 (2021)
20. Teshima, T., Sato, I., Sugiyama, M.: Few-shot domain adaptation by causal mechanism transfer. CoRR abs/2002.03497 (2020)

21. Wang, T., Zhang, X., Yuan, L., Feng, J.: Few-shot adaptive faster R-CNN. In: 2019 IEEE/CVF Conference on Computer Vision and Pattern Recognition (CVPR), pp. 7166–7175 (2019)
22. Li, S., et al.: Few-shot domain adaptation with polymorphic transformers. In: de Bruijne, M., et al. (eds.) MICCAI 2021. LNCS, vol. 12902, pp. 330–340. Springer, Cham (2021). https://doi.org/10.1007/978-3-030-87196-3_31
23. Wang, X., Huang, T.E., Darrell, T., Gonzalez, J.E., Yu, F.: Frustratingly simple few-shot object detection. arXiv preprint arXiv:2003.06957 (2020)
24. Liu, Y.C., et al.: Unbiased teacher for semi-supervised object detection. arXiv preprint arXiv:2102.09480 (2021)
25. Pan, F., Shin, I., Rameau, F., Lee, S., Kweon, I.S.: Unsupervised intra-domain adaptation for semantic segmentation through self-supervision. In: Proceedings of the IEEE/CVF Conference on Computer Vision and Pattern Recognition, pp. 3764–3773 (2020)
26. Li, J., Li, G., Shi, Y., Yu, Y.: Cross-domain adaptive clustering for semi-supervised domain adaptation. In: Proceedings of the IEEE/CVF Conference on Computer Vision and Pattern Recognition, pp. 2505–2514 (2021)
27. RoyChowdhury, A., et al.: Automatic adaptation of object detectors to new domains using self-training. In: Proceedings of the IEEE/CVF Conference on Computer Vision and Pattern Recognition, pp. 780–790 (2019)
28. Hsu, C.C., Tsai, Y.H., Lin, Y.Y., Yang, M.H.: Every pixel matters: center-aware feature alignment for domain adaptive object detector. arXiv abs/2008.08574 (2020)
29. Li, Z., Hoiem, D.: Learning without forgetting. IEEE Trans. Pattern Anal. Mach. Intell. **40**, 2935–2947 (2017)
30. Wang, J., He, Y., Fang, W., Chen, Y., Li, W., Shi, G.: Unsupervised domain adaptation model for lesion detection in retinal oct images. Phys. Med. Biol. **66**, 215006 (2021)
31. Yan, K., et al.: DeepLesion: automated mining of large-scale lesion annotations and universal lesion detection with deep learning. J. Med. Imaging (2018)
32. Heller, N., et al.: The KiTS19 challenge data: 300 kidney tumor cases with clinical context, CT semantic segmentations, and surgical outcomes. arXiv preprint arXiv:1904.00445 (2019)
33. Bilic, P., et al.: The liver tumor segmentation benchmark (LITS). arXiv preprint arXiv:1901.04056 (2019)
34. Huang, Q., Sun, J., Ding, H., Wang, X., Wang, G.: Robust liver vessel extraction using 3D U-Net with variant dice loss function. Comput. Biol. Med. **101**, 153–162 (2018)
35. Tian, Z., et al.: FCOS: fully convolutional one-stage object detection. In: ICCV, pp. 9627–9636 (2019)
36. Ganin, Y., Lempitsky, V.: Unsupervised domain adaptation by backpropagation. In: International Conference on Machine Learning, pp. 1180–1189. PMLR (2015)
37. Guan, D., Huang, J., Xiao, A., Lu, S., Cao, Y.: Uncertainty-aware unsupervised domain adaptation in object detection. IEEE Trans. Multimedia **24**, 2502–2514 (2021)
38. Van der Maaten, L., Hinton, G.: Visualizing data using t-SNE. J. Mach. Learn. Res. **9** (2008)

Photorealistic Facial Wrinkles Removal

Marcelo Sanchez[1,2(✉)], Gil Triginer[1], Coloma Ballester[2], Lara Raad[3], and Eduard Ramon[1]

[1] Crisalix S.A, Lausanne, Switzerland
marcelosanchezortega@gmail.com
[2] Universitat Pompeu Fabra, Barcelona, Spain
[3] Université Gustave Eiffel, Paris, France

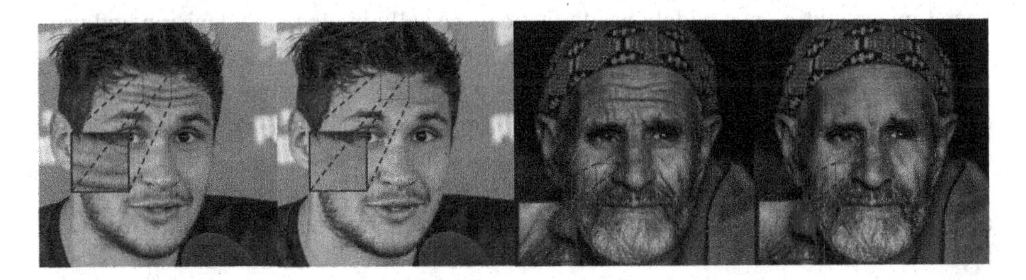

Fig. 1. Obtained results by our proposed wrinkle cleaning pipeline. Our method is able to obtain photorealistic results, learning to synthesize the skin distribution and obtain extremely realistic inpainting. Our model solves both wrinkle detection and wrinkle cleaning. For each of the two examples, the original image is shown on the left and the resulting image without wrinkles on the right, including a zoom-in of wrinkle regions.

Abstract. Editing and retouching facial attributes is a complex task that usually requires human artists to obtain photo-realistic results. Its applications are numerous and can be found in several contexts such as cosmetics or digital media retouching, to name a few. Recently, advancements in conditional generative modeling have shown astonishing results at modifying facial attributes in a realistic manner. However, current methods are still prone to artifacts, and focus on modifying global attributes like age and gender, or local mid-sized attributes like glasses or moustaches. In this work, we revisit a two-stage approach for retouching facial wrinkles and obtain results with unprecedented realism. First, a state of the art wrinkle segmentation network is used to detect the wrinkles within the facial region. Then, an inpainting module is used to remove the detected wrinkles, filling them in with a texture that is statistically consistent with the surrounding skin. To achieve this, we introduce a novel loss term that reuses the wrinkle segmentation network to penalize those regions that still contain wrinkles after the inpainting. We evaluate our method qualitatively and quantitatively, showing state of the art results for the task of wrinkle removal. Moreover, we introduce the first high-resolution dataset, named *FFHQ-Wrinkles*, to evaluate wrinkle detection methods.

E. Ramon—This work was done prior to joining Amazon.

Y. Zheng et al. (Eds.): ACCV 2022, LNCS 13848, pp. 117–133, 2023.
https://doi.org/10.1007/978-3-031-27066-6_9

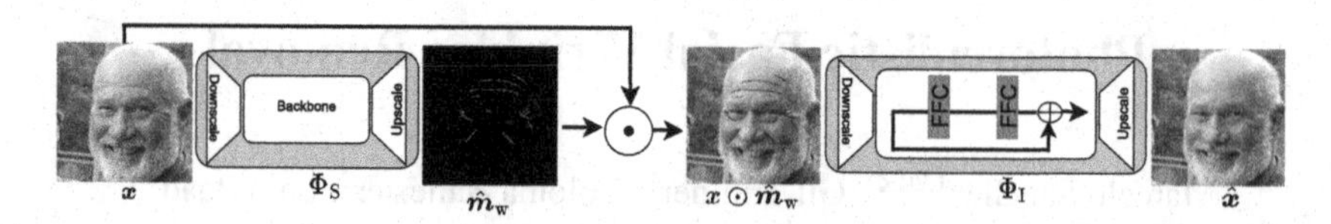

Fig. 2. Proposed pipeline for wrinkle removal. Input image x is first forwarded through ϕ_S obtaining $\hat{m}_w$. This segmentation map $\hat{m}_w$ is masked with x and passed through the inpainting module ϕ_I, obtain the wrinkle-free color image $\hat{x}$.

1 Introduction

Facial image editing is a widely used practice, specially in cinema, television and professional photography. Many techniques require human artists to spend much time on manual editing using complex software solutions [1] in order to obtain realistic modifications of the original image. Removing skin imperfections, and specifically softening or removing wrinkles, is one of the most common tasks.

In order to ease this process, there have been general efforts on its automation [2], and more concretely, in the automatic removal of facial wrinkles [3]. In [3] a two-stage approach is proposed in which the wrinkles are first detected using Gabor filters, and then removed using a texture synthesis method. While the results from this approach are interpretable, the edited images lack photorealism and hardly maintain the statistics of the skin of the person (Fig. 1).

Generative models based on deep learning, which learn the distribution of a particular category of data, have shown impressive results at generating novel faces and manipulating their content [4–10]. More concretely, Generative Adversarial Networks (GAN) [11] can learn to generate highly realistic faces. As shown in [12], it is possible to inpaint facial regions using GANs by optimizing the latent space to match a masked facial image, which in turn can be used to remove skin imperfections. However, this optimization process is slow.

Conditional generative models and in general image-to-image translation methods [8, 13–15] provide a more flexible and fast framework than vanilla generative models, and allow conditioning the generation of new samples on input images and optionally on local semantic information. These methods have been successfully applied to the task of automatic professional face retouching [16] in combination with a large scale dataset with ground truth annotations. Recent work [17] introduces an image-to-image translation model for the task of image inpainting in which fast Fourier convolutions (FFC) [18] are exploited, obtaining photorealistic inpainted images with a model learnt in an unsupervised fashion and that excels at modelling repetitive patterns, which is a desirable property for skin synthesis and wrinkle removal.

In this work, we propose a modern take on wrinkle removal by improving the standard pipeline proposed in [3] using data-driven techniques. First, we propose to replace the wrinkle detection stage using a state-of-the-art image segmentation neural network [19] in order to gain robustness and accuracy. In addition, we propose to replace the patch-based wrinkle cleaning block by an image inpainting network based on fast Fourier convolutions [17] to maintain the overall distribution of the person's skin in the modified area. In summary, our contributions are as follows:

- A modern take on wrinkle removal using state-of-the-art segmentation and inpainting methods in order to obtain gains in accuracy and photorealism.
- A novel loss that leverages a segmentation network to supervise the inpainting training process, leading to improved results.
- The first publicly available dataset for evaluating wrinkle segmentation methods.

Our paper is organized as follows. In Sect. 2, we review the state of the art for facial image editing and wrinkle removal. Next, in Sect. 3 we describe our two-stage solution for automatic wrinkle removal. In Sect. 4, we provide quantitative and qualitative results, and show that our method obtains state-of-the-art results. Finally, in Sect. 6 we end up with our conclusions.

2 Related Work

Wrinkle Removal. To the best of our knowledge, and also supported by a recent literature survey [20], few works exist for wrinkle removal. An initial solution was proposed by Bastanfard et al. [21] using face anthropometrics theory and image inpainting to erase the wrinkles. The current state-of-the-art for wrinkles removal is [3], which follows a two-stage approach. The first stage detects wrinkles using an algorithm that combines Gabor features and texture orientation filters. Once the detection is obtained, they proposed an exemplar-based texture synthesis to inpaint the wrinkle regions. Beyond the novelty of the method, it only succeeds in half megapixels images due to memory constrains and the images obtained lack photorealism, introducing artifacts near the inpainting zone.

Image-to-Image Translation. Image-to-Image translation appeared with [13] as a solution for solving the mapping of an image from one domain to another. Isola et al. used a GAN-based scheme with a conditional setup. Several methods [8,13–15] are able to translate domains that require to modify mid-size features such as glasses or mustache, while preserving the global structure of the face. Even though some of these methods need paired images in order to address the image translation problem [13,16,22], which is sometimes unfeasible in wrinkle removal, other pipelines can work with unpaired images [14,23]. However, these methods tend to attend to global regions and are not able to modify small features such as wrinkles or other skin marks.

Image Inpainting. Image Inpainting aims to fill in missing or corrupted regions of an image so that the reconstructed image appears natural to the human eye. Accordingly, filled regions should contain meaningful details and preserve the original semantic structure. Image Inpainting is closely related to wrinkle cleaning in the sense that it allows modifying the desired (sometimes small-size) regions while preserving the uncorrupted, i.e. wrinkle-free, regions. Initially, deep learning based inpainting methods used vanilla convolutional neural networks (CNN). More complex methods regularize the structure of the filled regions with edges or segmentation maps [24–27].

Many of these methods are based on CNNs with limited receptive fields, while it is well known that a large receptive field allows the network to better understand the global structure of an image. Recent work [17] uses Fourier convolutions that introduce global context by optimizing on the frequency domain of the image. This allows the network to incorporate the texture patterns of the entire image and thus obtain a better inpainting result. Also this non-local receptive field allows the inpainting model to generalize to high resolution images.

3 Method

Given a color image of a face $\boldsymbol{x} \in \mathbb{R}^{H\times W\times 3}$, our goal is to remove all the wrinkles present in $\boldsymbol{x}$ while preserving photorealism. Ideally, we should only modify the facial areas with wrinkles and, at the same time, the modified regions should preserve the local statistics of the skin of the person. We propose to solve both wrinkle removal and wrinkle cleaning via a two-stage model based on image segmentation and inpainting techniques. First we estimate a segmentation map of all the wrinkles in the image using a state-of-the-art CNN segmentation model. Once this is obtained, we propose an inpainting model based on LAMA [17] to fill in the wrinkles regions with photorealistic clear skin. We illustrate our pipeline in Fig. 2.

3.1 Wrinkle Segmentation

In this first step, a segmentation model Φ_S detects the wrinkles in $\boldsymbol{x}$, predicting a segmentation map $\hat{\boldsymbol{m}}_\text{w} \in \mathbb{R}^{H\times W}$. The segmentation model Φ_S is a state-of-the-art fully convolutional neural network with nested skip pathways, namely Unet++ [19].

Wrinkle segmentation has an inherent problem of class imbalance, since usually the facial wrinkles occupy smaller regions on the image compared with the clear skin. We manage to tackle the class imbalance by using a region-based loss instead of a classical distribution-based one such as cross entropy. In particular, we choose to minimize the Dice loss, defined as

$$\mathcal{L}_\text{Dice}(\boldsymbol{m}, \hat{\boldsymbol{m}}) = \frac{2\sum_{i=1}^{H\times W} m_i \hat{m}_i}{\sum_{i=1}^{H\times W} m_i^2 + \sum_{i=1}^{H\times W} \hat{m}_i^2} \tag{1}$$

over a set of ground truth wrinkle annotations $\boldsymbol{m}_w$ and predicted segmentation maps $\hat{\boldsymbol{m}}_w = \Phi_\text{S}(\boldsymbol{x})$. For clarity, in (1) we have omitted the subindices w and the subindices i indicate iteration over image pixels.

3.2 Wrinkle Cleaning

Once the wrinkles have been detected using the module Φ_S, we aim at cleaning them using an inpainting neural network Φ_I. The resulting inpainted image $\hat{\boldsymbol{x}}$ must be photorealistic and the generated skin must be statistically similar to the skin of the input image subject (Fig. 3).

The regions to be inpainted are defined on a binary mask $\boldsymbol{m}_\text{I} \in \mathbb{R}^{H\times W}$. We mask the input image $\boldsymbol{x} \in \mathbb{R}^{H\times W\times 3}$ and obtain the masked image $\boldsymbol{x} \odot \boldsymbol{m}_\text{I}$. In order to add redundancy, we follow the approach from [17] and concatenate m_I to the masked image $x \odot m_\text{I}$, obtaining a 4-channel input $x' = \texttt{stack}(x \odot m_\text{I}, m_\text{I})$. Therefore, our wrinkle-free image is $\hat{x} = \Phi_\text{I}(x')$ where $\hat{x} \in \mathbb{R}^{H \times W \times 3}$ and Φ_I corresponds to our inpainting module. In the following, we describe the training of Φ_I, which is illustrated in Fig. 5.

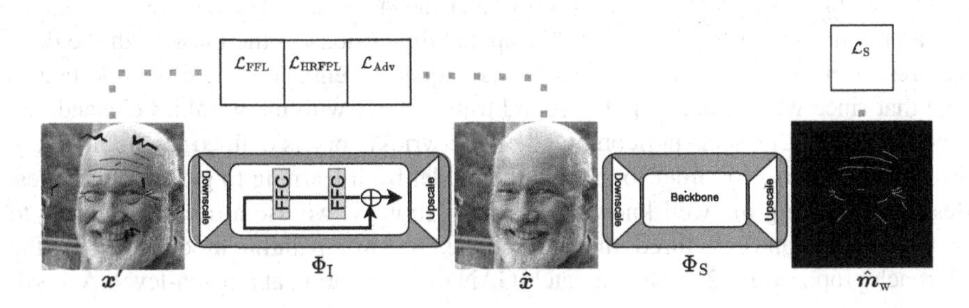

Fig. 3. Scheme of the proposed inpainting training pipeline. Image x' is forwarded through the inpainting module. The resulting inpainted image is used in multi-component loss.

Mask Generation During Training. In order to train Φ_I, we generate pairs of images and masks. The inpainting masks m_I are generated as the pixelwise logical union of two binary masks: the first marking pixels with wrinkles, m_w, and the second, m_g, marking wrinkle-shaped regions randomly generated using polygon chains [17]. The latter is added in order to provide more diverse training data and wrinkle-free skin regions. The mask generation policy plays a big role in inpainting pipelines [17,28] and influences to a high extent the final performance of the model. The training mask policy should generate masks as similar as possible to the the inpainting regions at test time. A huge difference between training and testing masks can lead to a poor inpainting generalization.

We leverage the LAMA architecture [17] in our inpainting module. The underlying components of this network are fast Fourier convolutions (FFC) [18]. While CNNs have a limited local receptive field, and use kernels that operate at a fixed spatial scale, FFCs capture global information of the input image and reuse the same kernels at different frequencies (scales). These are useful properties in wrinkle inpainting, due to the repetitive and multi-scale nature of skin texture.

Training Loss. The inpainting problem is ill-posed, as multiple image restorations can be appealing to the human eye. Because of this, training inpainting models is normally a challenging problem [29], sometimes needing multiple losses in order to obtain good results. Simple supervised losses such as the squared Euclidean norm L_2 force the generator to inpaint the corrupted regions precisely. However, in some cases the visible parts do not provide enough information to recover the corrupted part successfully, leading to blurry results. More complex losses like the perceptual loss [30] evaluate distance in the feature space of a pre-trained network $\varphi(\cdot)$, encouraging reconstructions less equal to the ground truth in favor of global perceptual structure. In order to obtain photorealistic results, we also make use of multiple loss terms, which we define next.

First, we use a perceptual loss based on a high receptive field network $\varphi_{\text{HRF}}(\cdot)$ trained on an image segmentation task [17]. Namely, we use:

$$\mathcal{L}_{\text{HRFPL}}(\boldsymbol{x}, \hat{\boldsymbol{x}}) = \Upsilon([\varphi_{\text{HRF}}(\boldsymbol{x}) - \varphi_{\text{HRF}}(\hat{\boldsymbol{x}})]^2 \odot (1 - \texttt{resize}(\boldsymbol{m}_w))), \tag{2}$$

where $[\cdot - \cdot]^2$ is the squared element-wise difference operator and Υ is the two stage mean operation, inter-layer mean of intra-layer means. Recall that the use of `resize` operator is needed in order to match the spatial dimensions of the mask with the deep features from φ_{HRF}. All channel features are equally weighted to 1. It is important to note that since we can not acquire ground truth images with the wrinkles cleaned, the supervised loss $\mathcal{L}_{\text{HRFPL}}$ is not computed on the wrinkle pixels defined by the non-zero pixel-values in $\boldsymbol{m}_{\text{w}}$, in order to prevent the model from learning to generate wrinkles. Besides, following the well-known adversarial strategy, we use a discriminator D to ensure coherent and structured images generated by our inpainting module. We use the approach proposed in [28] using a patch-GAN discriminating at a patch-level. Accordingly, the corresponding loss terms are defined as:

$$\begin{aligned}
\mathcal{L}_{\mathcal{D}} &= -\mathbb{E}_{\boldsymbol{x}\sim\mathbb{P}}[\log D(\boldsymbol{x})] - \mathbb{E}_{\hat{\boldsymbol{x}}\sim\mathbb{P}_{\phi_I}}[\log D(\hat{\boldsymbol{x}}) \odot \boldsymbol{m}] \\
&\quad - \mathbb{E}_{\hat{\boldsymbol{x}}\sim\mathbb{P}_{\phi_I}}[\log(1 - D(\hat{\boldsymbol{x}})) \odot (1 - \boldsymbol{m})], \\
\mathcal{L}_{\phi_{\mathcal{I}}} &= -\mathbb{E}_{\hat{\boldsymbol{x}}\sim\mathbb{P}_{\phi_I}}[\log D(\hat{\boldsymbol{x}})], \\
\mathcal{L}_{\text{Adv}} &= \mathcal{L}_{\mathcal{D}} + \mathcal{L}_{\phi_{\mathcal{I}}},
\end{aligned} \tag{3}$$

where $\boldsymbol{x}$ denotes a sample from the dataset distribution $\mathbb{P}$, and $\hat{\boldsymbol{x}}$ a sample from the inpainted images distribution $\mathbb{P}_{\phi_I}$.

The losses presented so far are optimized in the spatial domain. In order to enforce similarity in the spectral domain and encourage high frequencies, we add the Focal Frequency Loss (FFL) defined in [31]. This frequency domain loss enforces Φ_I to recover the Fourier modulus and phase of the input image. Concurrently, the authors of [32] have used the same strategy. The FFL is computed as:

$$\mathcal{L}_{\text{FFL}} = \frac{1}{MN} \sum_{u=0}^{M-1} \sum_{v=0}^{N-1} w(u, v) |F_{\boldsymbol{x}}(u, v) - F_{\hat{\boldsymbol{x}}}(u, v)|^2, \tag{4}$$

where $F_{\boldsymbol{x}}$ denotes the 2D discrete Fourier transform operator of the input image $\boldsymbol{x}$ and $F_{\hat{\boldsymbol{x}}}$ denotes the 2D discrete Fourier transform of the inpainted image $\hat{\boldsymbol{x}}$. The matrix $w(u, v)$ correspond to the weight of the spatial frequency at coordinates (u, v) and are defined as $w(u, v) = |F_{\boldsymbol{x}}(u, v) - F_{\hat{\boldsymbol{x}}}(u, v)|$. The gradient through the matrix w is locked, only serving as a weight for each frequency.

Up to this point, none of the presented losses are specific to the problem of wrinkle removal. We propose a novel *wrinkle loss* $\mathcal{L}_{\mathcal{S}}$ for inpainting that creates a connection between the segmentation and the inpainting module. Inspired by the fact that our final goal is to generate images without wrinkles, we make use of our wrinkle segmentation module $\boldsymbol{\Phi}_{\text{S}}$ to detect wrinkles in the inpainted image $\hat{\boldsymbol{x}}$ and to generate a supervision signal that discourages the presence of wrinkles in the inpainted image. More concretely, we first use the previously trained segmentation module $\boldsymbol{\Phi}_{\text{S}}$ to segment $\hat{\boldsymbol{x}}$ and obtain a wrinkle map. Then, we enforce that the generated wrinkle map does not contain any wrinkle pixels. The loss term $\mathcal{L}_{\mathcal{S}}$ is defined as follows:

$$\mathcal{L}_{\mathrm{S}}(x) = \Psi(\Phi_{\mathrm{S}}(\Phi_{\mathrm{I}}(\boldsymbol{x}))), \tag{5}$$

where Ψ is the spatial mean operator.

In addition to the introduced losses, we also make use of the gradient penalty loss [33] defined as $\mathcal{R}_1 = \mathbb{E}_{\boldsymbol{x}\sim\mathbb{P}}||\nabla D(\boldsymbol{x})||^2$, and the feature matching loss $\mathcal{L}_{\mathrm{DiscPL}}$ which is similar to $\mathcal{L}_{\mathrm{HRFPL}}$ but the features are obtained from the discriminator D, which is known to stabilize training [28]. The final loss function for our wrinkle inpainting model can be written as:

$$\mathcal{L} = \lambda_0\mathcal{L}_{\mathrm{Adv}} + \lambda_1\mathcal{L}_{\mathrm{HRFPL}} + \lambda_2\mathcal{L}_{\mathrm{DiscPL}} + \lambda_3\mathcal{R}_1 + \lambda_4\mathcal{L}_{\mathrm{FFL}} + \lambda_5\mathcal{L}_{\mathrm{S}}, \tag{6}$$

where $\lambda_i > 0,\ i \in \{0,\ldots,5\}$ are fixed hyperparameters (detailed in Sect. 4).

Even though our pipeline solves wrinkle segmentation and wrinkle cleaning, both solutions are not learnt simultaneously. The two problems require different learning strategies and have different time of convergence, applying simultaneous learning would damage the global accuracy of the system.

3.3 Inference

The segmentation and inpainting module are used sequentially to obtain the output image $\hat{\boldsymbol{x}}$. First, we forward the image through the segmentation model Φ_{S} to obtain the wrinkle map. With the segmentation map we then mask our input image and generate the inpainted image through Φ_{I}. A schematic representation of the inference process is displayed in Fig. 5.

However, the inpainting results are always limited by the performance of the segmentation module. The inpainting is always conditioned by the segmentation map $\hat{m}_{\mathrm{w}}$, wrinkles that are not detected by Φ_{S} will remain in $\hat{\boldsymbol{x}}$. This problem can be observed in Fig. 6

4 Experimental Results

In this section we provide quantitative and qualitative results showing the effectiveness of our method. Next, we introduce the datasets and the metrics that will be used along the section.

4.1 Dataset and Metrics

We introduce a novel facial dataset to evaluate wrinkle segmentation methods. Based on FFHQ [9], our new dataset, which we dub *FFHQ-Wrinkles*, is composed by 100 images of different subjects paired with corresponding manually annotated wrinkle segmentation masks. In order to decide which samples to annotate, we have used FFHQ-Aging dataset [34] age metadata to easily identify potential cases. Even though older people are more prone to develop wrinkles, we also added younger people samples with facial

Fig. 4. Samples from FFHQ-Wrinkles dataset.

expression like the ones showed in Fig. 4 to obtain a more diverse population. We have also ensured to collect samples from different ethnicities and skin tones. All the metrics scores have been computed on the *FFHQ-Wrinkles* dataset.

For training the different methods, we use a proprietary dataset containing 5k samples of RGB frontal face pictures with its corresponding wrinkle annotations. This dataset contains images from subjects diverse in age, gender and ethnicity. Pictures were taken under uncontrolled lightning condition with a native resolution of 1024×1024.

In order to assess the different wrinkle removal methods, we adopt the common metrics reported in inpainting literature. We use LPIPS (Learned Perceptual Image Patch Similarity) [35] and FID (Fréchet inception distance) [36] metrics. However, both LPIPS and FID require a target distribution image to be computed. Since we do not have ground truth images with wrinkles removed, we propose the following: first, we synthetically create wrinkle-shaped masks that lie on the skin of the subjects, but purposefully avoid overlapping with wrinkles. Then, we evaluate our inpainting method using these masks. Since our model should recover the clear skin, the inpainted result can be directly compared with the original image.

4.2 Implementation Details

The entire pipeline is implemented using Pytorch [22] framework. The segmentation module is based on the implementation from [37] Unet++ [19] with a ResNeXt-50 [38] backbone pre-trained on Imagenet classification task. We train Φ_S for 200 epochs using Adam optimizer [39] with a initial learning rate of 0.001 scaled at 100 epochs by 0.5. Images are downsampled to 512×512, other image transformations such as VerticalFlip, HorizontalFlip, RandomShift from Albumentations [40] are also applied. We threshold with 0.5 the probability segmentation map in order to set if a given pixel is part of a wrinkle.

The inpainting module is based on the official LAMA implementation [17], the generator is composed by 9 blocks. We use Adam optimizer [39] for both generator and discriminator with a learning rate of 0.0001. We train using 256×256 crops with different spatial image transformation such us Rotation and VerticalFlip. We use a batch size of 16 and train for 300 epochs. In all the experiments, the hyperparameters search is conducted on separated validation dataset. All methods have been trained with our proprietary dataset and following the recommendations suggested by each author. Figures and metrics reported are computed with our new *FFHQ-Wrinkles* dataset.

4.3 Comparisons to the Baselines

We compare the proposed approach on wrinkle cleaning with state-of-the-art methods in Fig. 8. Batool et al. [3] method contrary to ours needs human intervention to obtain correct wrinkle segmentation. This intervention is done by selecting the potential facial regions containing wrinkles in order to reduce the search space. Without human intervention it tends to segment as wrinkles high gradient sections such us hair or clothes. In order to provide a fair comparison and favour their results, we compare our method with two variants of [3], which are displayed in Fig. 8. The first variant consists in using the segmentation map obtained directly from their detection module (with no user intervention). The corresponding results with the segmentation map obtained by [3] are shown in the right-most column. For each experiment, the inpainting result is shown in the first row and the segmentation result in the second row. The second variant consists in manually filtering that segmentation mask obtained automatically by [3] by applying a mask marking the region of valid wrinkles (i.e., avoiding hair, glasses, ...). As we do not have access to the authors' inpainting module, we used a patch based inpainting approach [41]. similar to the one used in [3] which is based on the texture synthesis method [42].

Our model largely outperforms method [3] and is robust to external elements of the face such us hair, clothes or earrings. Also is able to fill the wrinkle regions with a more natural and smooth skin, leading to skin patches indistinguishable to the human eye (Table 1).

Table 1. Quantitative comparison with previous state-of-the-art wrinkle cleaning method on FFHQ-Wrinkles dataset.

Configuration	IoU ↑	LPIPS ↓	FID ↓
Ours	0.561	0.231	7.24
Batool et al. [3]	0.192	3.43	14.11

5 Ablations

In order to verify the effectiveness and importance of the proposed modules and principal loss terms, we perform a qualitative and quantitative ablation study on FFHQ-Wrinkles dataset.

Segmentation Module. The performance of the overall system is heavily influenced by the accuracy of the wrinkle segmentation module. Our method is heavily influenced by the quality of the detected masks. As it can be observed in Fig. 6, the wrinkles that are not detected are still present in the final image. While this is a weak point of our method, it can be solved either by improving the segmentation model or by correcting the predicted segmentation mask at test time (Table 2).

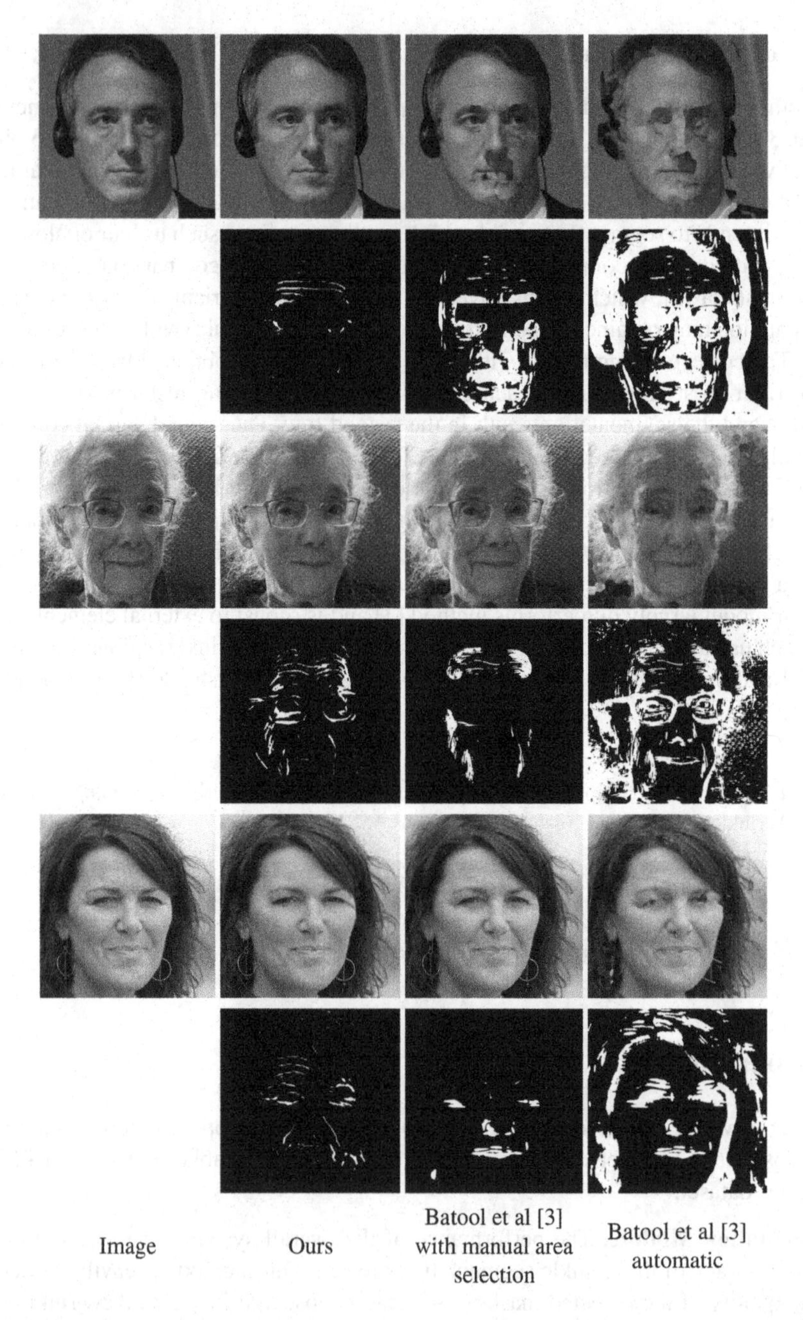

Fig. 5. Qualitative comparison to the state-of-the-art methods on wrinkle cleaning on FFHQ-Wrinkles dataset. Our methods outperforms current state-of-the-art algorithms and is robust to external elements of the face such us hair, clothes or earrings.

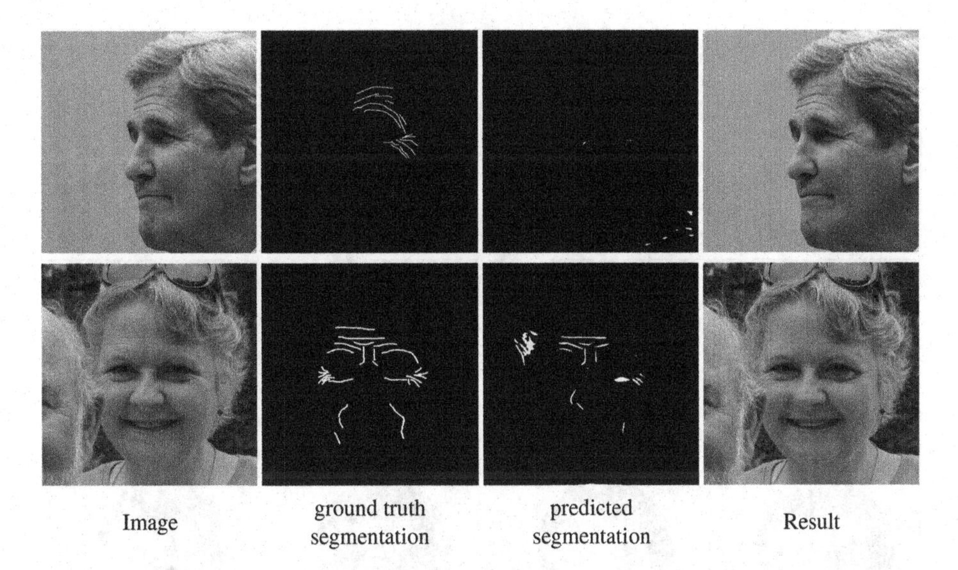

Fig. 6. Qualitative results of our wrinkle cleaning pipeline. This examples display the limitations of our method and the importance of a correct segmentation in the wrinkle cleaning pipeline.

As reported in Table 3, the Unet++ [19] model provides us with the highest intersection over union (IoU) score. We associate this improvement to the nested pathways of Unet++ that reduces the semantic gap between the encoder and decoder, making possible to detect fine-grained details such as small wrinkles. This is reflected in Fig. 8, where the row associated with Unet++ is also the best performing one.

Qualitatively, vanilla Unet tends to miss subtle wrinkles (Fig. 8). DeepLabV3 and Unet++ are better at detecting small wrinkles compared with Unet. However, the latter is able to segment more precisely the entire wrinkle instead of segmenting small pieces independently along the wrinkle.

Inpainting Module. In terms of inpainting accuracy, we report the LPIPS and the FID metrics in Table 3. Our method obtains the lowest errors for both metrics, suggesting that the dual loss term forces the network not only to generate a wrinkle-free image, but also to generate a skin texture that is statistically consistent with the surrounding skin.

Finally, in Fig. 7, we show qualitative results for different examples taken from the proposed *FFHQ-Wrinkles* dataset. Our results are compared to those obtained with different loss configurations. In general, our method is more precise at removing the wrinkles on the whole face due to the dual loss term $\mathcal{L}_S$ which strengthens the network to generate wrinkle-free images. Moreover, these results show that the generated skin and non-skin textures are more consistent with the surrounding textures. Looking closely the other loss configurations display the well-known checkboard effect [17,32], a pattern that clearly indicates that the model is not able to inpaint properly the corrupted regions. This is consistent with the quantitative results obtain in Tables 3.

Fig. 7. Ablation study on different inpainting configurations. Demonstrates that our configuration is able to obtain better photorealistic results, learning to synthesize the skin distribution and obtain extremely realistic inpainting. We are able to generate the high frequency pattern of human skin. Surprisingly, our method is able to inpaint diverse face features such as facial or head hair.

Table 2. Comparison of different inpainting and segmentation methods on FFHQ-Wrinkles dataset. This figure displays metrics of segmentation and inpainting of the different ablations conducted. All the methods use the same wrinkle masks for inpainting. The arrow next to each metric indicates if larger values are better (↑) or smaller values are better (↓).

Inpainting		
Configuration	LPIPS ↓	FID ↓
w/ $\mathcal{L}_s$ and w/ $\mathcal{L}_{FFL}$	34.625	11.321
w/ $\mathcal{L}_{FFL}$	31.112	8.931
Ours	**0.231**	**7.24**
Segmentation		
Configuration	Iou ↑	
Unet++ [19]	**0.561**	
DeepLabV3 [43]	0.492	
Unet [44]	0.401	

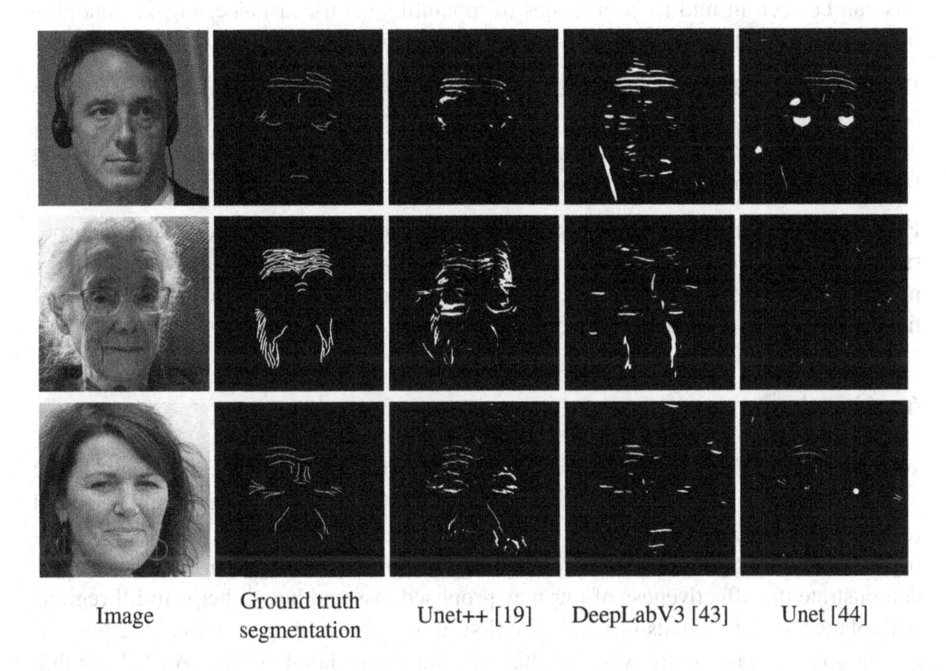

Fig. 8. Qualitative results of ablation studies for the wrinkle segmentation module. The last 3 columns correspond to the different segmentation networks evaluated for the task. All methods share the same ResNeXt-50 backbone [38].

Table 3. Quantitative evaluation of inpainting and segmentation modules on the FFHQ-Wrinkles dataset. We report the LPIPS metric (the lower the better). The different segmentation models and inpainting modules are listed on rows and columns respectively. Our setup consistently outperforms other configurations.

	w/ $\mathcal{L}_s$ and w/ $\mathcal{L}_{FFL}$	w/ $\mathcal{L}_{FFL}$	Ours
Unet [44]	42.11	35.98	32.786
Unet++ [19]	34.625	31.112	**29.547**
DeepLabV3 [43]	35.45	34.58	30.786

Focal Frequency Loss. $\mathcal{L}_{FFL}$ forces the model to learn the whole spectrum of frequencies which substantially improves FID scores of the inpainted images Table 3). As shown in Fig. 2 qualitatively, the model is able to properly inpaint with a natural skin texture.

Training the model without optimizing $\mathcal{L}_{FFL}$ degrades the quality of the inpainting, this can be seen in mid to large zones of inpainting of, for instance, Fig. 2, obtaining blurry regions that do not capture the high frequencies of the skin.

Dual Loss. $\mathcal{L}_s$ add small but subtle changes in the inpainting results. The intuition behind is that our inpainting module is able to use the segmentation loss as a wrinkle expert playing the role of tatic wrinkle discriminator which is able to detect inpainting that do not have the appearance of clean skin.

Fast Fourier Convolutions. FFCs allow to capture global information and incorporate texture patterns from the entire image. Also the inductive bias of FFC allow the model to generalize to higher resolutions. This is a key component because due to GPU limitations, is not possible to train at full resolution.

6 Conclusions

We have presented a pipeline for photorealistic wrinkle removal. This pipeline is based on state-of-the-art methods for segmentation and inpainting, achieving unprecedented results. The proposed two-stage pipeline pushes the state of the art in wrinkle cleaning however, struggles with non-frontal images. Our experiments on FFHQ-Wrinkles demonstrate the effectiveness of our new proposed loss and how it helps to fill regions with plausible skin. We also provide the first public dataset that for wrinkle segmentation in order to ease future work on this topic and provide a baseline. We believe that solving wrinkle detection and wrinkle cleaning independently is a key feature that adds interpretability to the system. However, exploring ways to jointly fuse this pipeline and train jointly is an exciting and interesting future work.

References

1. Adobe Inc.: (Adobe photoshop)
2. visage-lab: (Visage lab face retouch)

3. Batool, N., Chellappa, R.: Detection and inpainting of facial wrinkles using texture orientation fields and Markov random field modeling. IEEE Trans. Image Process. **23**, 3773–3788 (2014)
4. Alaluf, Y., Patashnik, O., Cohen-Or, D.: Only a matter of style: age transformation using a style-based regression model. ACM Trans. Graph. (TOG) **40**, 1–12 (2021)
5. Song, X., Shao, M., Zuo, W., Li, C.: Face attribute editing based on generative adversarial networks. SIViP **14**(6), 1217–1225 (2020). https://doi.org/10.1007/s11760-020-01660-0
6. Lample, G., Zeghidour, N., Usunier, N., Bordes, A., Denoyer, L., Ranzato, M.: Fader networks: Manipulating images by sliding attributes. In: Advances in Neural Information Processing Systems, vol. 30 (2017)
7. He, Z., Zuo, W., Kan, M., Shan, S., Chen, X.: AttGAN: facial attribute editing by only changing what you want. IEEE Trans. Image Process. **28**, 5464–5478 (2019)
8. Ding, C., Kang, W., Zhu, J., Du, S.: InjectionGAN: unified generative adversarial networks for arbitrary image attribute editing. IEEE Access **8**, 117726–117735 (2020)
9. Karras, T., Laine, S., Aila, T.: A style-based generator architecture for generative adversarial networks. In: Proceedings of the IEEE/CVF Conference on Computer Vision and Pattern Recognition, pp. 4401–4410 (2019)
10. Wu, P.W., Lin, Y.J., Chang, C.H., Chang, E.Y., Liao, S.W.: RelGAN: multi-domain image-to-image translation via relative attributes. In: Proceedings of the IEEE/CVF International Conference on Computer Vision, pp. 5914–5922 (2019)
11. Goodfellow, I., et al.: Generative adversarial nets. In: Advances In Neural Information Processing Systems, vol. 27 (2014)
12. Abdal, R., Qin, Y., Wonka, P.: Image2styleGAN++: how to edit the embedded images? In: Proceedings of the IEEE/CVF Conference on Computer Vision and Pattern Recognition, pp. 8296–8305 (2020)
13. Isola, P., Zhu, J.Y., Zhou, T., Efros, A.A.: Image-to-image translation with conditional adversarial networks. In: Proceedings of the IEEE Conference on Computer Vision and Pattern Recognition, pp. 1125–1134 (2017)
14. Zhu, J.Y., Park, T., Isola, P., Efros, A.A.: Unpaired image-to-image translation using cycle-consistent adversarial networks. In: Proceedings of the IEEE International Conference on Computer Vision, pp. 2223–2232 (2017)
15. Liu, M., et al.: STGAN: a unified selective transfer network for arbitrary image attribute editing. In: Proceedings of the IEEE/CVF Conference on Computer Vision and Pattern Recognition, pp. 3673–3682 (2019)
16. Shafaei, A., Little, J.J., Schmidt, M.: AutoreTouch: automatic professional face retouching. In: Proceedings of the IEEE/CVF Winter Conference on Applications of Computer Vision, pp. 990–998 (2021)
17. Suvorov, R., et al.: Resolution-robust large mask inpainting with Fourier convolutions. In: Proceedings of the IEEE/CVF Winter Conference on Applications of Computer Vision, pp. 2149–2159 (2022)
18. Chi, L., Jiang, B., Mu, Y.: Fast Fourier convolution. Adv. Neural. Inf. Process. Syst. **33**, 4479–4488 (2020)
19. Zhou, Z., Siddiquee, M.M.R., Tajbakhsh, N., Liang, J.: UNet++: redesigning skip connections to exploit multiscale features in image segmentation. IEEE Trans. Med. Imaging **39**, 1856–1867 (2019)
20. Yap, M.H., Batool, N., Ng, C.C., Rogers, M., Walker, K.: A survey on facial wrinkles detection and inpainting: datasets, methods, and challenges. IEEE Trans. Emerg. Top. Comput. Intell. **5**, 505–519 (2021)
21. Bastanfard, A., Bastanfard, O., Takahashi, H., Nakajima, M.: Toward anthropometrics simulation of face rejuvenation and skin cosmetic. Comput. Anim. Virtual Worlds **15**, 347–352 (2004)

22. Paszke, A., et al.: Pytorch: an imperative style, high-performance deep learning library. In: Advances in Neural Information Processing Systems, vol. 32, pp. 8024–8035. Curran Associates, Inc. (2019)
23. Park, T., Efros, A.A., Zhang, R., Zhu, J.-Y.: Contrastive learning for unpaired image-to-image translation. In: Vedaldi, A., Bischof, H., Brox, T., Frahm, J.-M. (eds.) ECCV 2020. LNCS, vol. 12354, pp. 319–345. Springer, Cham (2020). https://doi.org/10.1007/978-3-030-58545-7_19
24. Nazeri, K., Ng, E., Joseph, T., Qureshi, F., Ebrahimi, M.: EdgeConnect: structure guided image inpainting using edge prediction. In: Proceedings of the IEEE/CVF International Conference on Computer Vision Workshops (2019)
25. Liao, L., Xiao, J., Wang, Z., Lin, C.-W., Satoh, S.: Guidance and evaluation: semantic-aware image inpainting for mixed scenes. In: Vedaldi, A., Bischof, H., Brox, T., Frahm, J.-M. (eds.) ECCV 2020. LNCS, vol. 12372, pp. 683–700. Springer, Cham (2020). https://doi.org/10.1007/978-3-030-58583-9_41
26. Yang, J., Qi, Z., Shi, Y.: Learning to incorporate structure knowledge for image inpainting. In: Proceedings of the AAAI Conference on Artificial Intelligence, vol. 34, pp. 12605–12612 (2020)
27. Liao, L., Xiao, J., Wang, Z., Lin, C.W., Satoh, S.: Image inpainting guided by coherence priors of semantics and textures. In: Proceedings of the IEEE/CVF Conference on Computer Vision and Pattern Recognition, pp. 6539–6548 (2021)
28. Wang, T.C., Liu, M.Y., Zhu, J.Y., Tao, A., Kautz, J., Catanzaro, B.: High-resolution image synthesis and semantic manipulation with conditional GANs. In: Proceedings of the IEEE Conference on Computer Vision and Pattern Recognition, pp. 8798–8807 (2018)
29. Suthar, R., Patel, M.K.R.: A survey on various image inpainting techniques to restore image. Int. J. Eng. Res. Appl. **4**, 85–88 (2014)
30. Johnson, J., Alahi, A., Fei-Fei, L.: Perceptual losses for real-time style transfer and super-resolution. In: Leibe, B., Matas, J., Sebe, N., Welling, M. (eds.) ECCV 2016. LNCS, vol. 9906, pp. 694–711. Springer, Cham (2016). https://doi.org/10.1007/978-3-319-46475-6_43
31. Jiang, L., Dai, B., Wu, W., Loy, C.C.: Focal frequency loss for image reconstruction and synthesis. In: ICCV (2021)
32. Lu, Z., Jiang, J., Huang, J., Wu, G., Liu, X.: Glama: joint spatial and frequency loss for general image inpainting. In: Proceedings of the IEEE/CVF Conference on Computer Vision and Pattern Recognition, pp. 1301–1310 (2022)
33. Ross, A., Doshi-Velez, F.: Improving the adversarial robustness and interpretability of deep neural networks by regularizing their input gradients. In: Proceedings of the AAAI Conference on Artificial Intelligence, vol. 32 (2018)
34. Or-El, R., Sengupta, S., Fried, O., Shechtman, E., Kemelmacher-Shlizerman, I.: Lifespan age transformation synthesis. In: Vedaldi, A., Bischof, H., Brox, T., Frahm, J.-M. (eds.) ECCV 2020. LNCS, vol. 12351, pp. 739–755. Springer, Cham (2020). https://doi.org/10.1007/978-3-030-58539-6_44
35. Zhang, R., Isola, P., Efros, A.A., Shechtman, E., Wang, O.: The unreasonable effectiveness of deep features as a perceptual metric. In: Proceedings of the IEEE Conference on Computer Vision and Pattern Recognition, pp. 586–595 (2018)
36. Heusel, M., Ramsauer, H., Unterthiner, T., Nessler, B., Hochreiter, S.: GANs trained by a two time-scale update rule converge to a local Nash equilibrium. In: Advances in Neural Information Processing Systems, vol. 30 (2017)
37. Yakubovskiy, P.: Segmentation models PyTorch (2020). https://github.com/qubvel/segmentation_models.pytorch
38. Xie, S., Girshick, R., Dollár, P., Tu, Z., He, K.: Aggregated residual transformations for deep neural networks. In: Proceedings of the IEEE Conference on Computer Vision and Pattern Recognition, pp. 1492–1500 (2017)

39. Kingma, D.P., Ba, J.: Adam: A method for stochastic optimization. arXiv preprint arXiv:1412.6980 (2014)
40. Buslaev, A., Parinov, A., Khvedchenya, E., Iglovikov, V., Kalinin, A.: Albumentations: fast and flexible image augmentations. ArXiv e-prints (2018)
41. Newson, A., Almansa, A., Gousseau, Y., Pérez, P.: Non-local patch-based image inpainting. Image Process. Line **7**, 373–385 (2017). https://doi.org/10.5201/ipol.2017.189
42. Efros, A.A., Freeman, W.T.: Image quilting for texture synthesis and transfer. In: Proceedings of the 28th Annual Conference on Computer Graphics and Interactive Techniques, pp. 341–346 (2001)
43. Chen, L.C., Papandreou, G., Schroff, F., Adam, H.: Rethinking atrous convolution for semantic image segmentation. arXiv preprint arXiv:1706.05587 (2017)
44. Ronneberger, O., Fischer, P., Brox, T.: U-Net: convolutional networks for biomedical image segmentation. In: Navab, N., Hornegger, J., Wells, W.M., Frangi, A.F. (eds.) MICCAI 2015. LNCS, vol. 9351, pp. 234–241. Springer, Cham (2015). https://doi.org/10.1007/978-3-319-24574-4_28

Improving Segmentation of Breast Arterial Calcifications from Digital Mammography: Good Annotation is All You Need

Kaier Wang(✉), Melissa Hill, Seymour Knowles-Barley, Aristarkh Tikhonov, Lester Litchfield, and James Christopher Bare

Volpara Health Technologies, Wellington, New Zealand
kyle.wang@volparahealth.com

Abstract. Breast arterial calcifications (BACs) are frequently observed on screening mammography as calcified tracks along the course of an artery. These build-ups of calcium within the arterial wall may be associated with cardiovascular diseases (CVD). Accurate segmentation of BACs is a critical step in its quantification for the risk assessment of CVD but is challenging due to severely imbalanced positive/negative pixels and annotation quality, which is highly dependent on annotator's experience. In this study, we collected 6,573 raw tomosynthesis images where 95% had BACs in the initial pixel-wise annotation (performed by a third-party annotation company). The data were split with stratified sampling to 80% train, 10% validation and 10% test. Then we evaluated the performance of the deep learning models deeplabV3+ and Unet in segmenting BACs with varying training strategies such as different loss functions, encoders, image size and pre-processing methods. During the evaluation, large numbers of false positive labels were found in the annotations that significantly hindered the segmentation performance. Manual re-annotation of all images would be impossible owing to the required resources. Thus, we developed an automatic label correction algorithm based on BACs' morphology and physical properties. The algorithm was applied to training and validation labels to remove false positives. In comparison, we also manually re-annotated the test labels. The deep learning model re-trained on the algorithm-corrected labels resulted in a 29% improvement in the dice similarity score against the re-annotated test labels, suggesting that our label auto-correction algorithm is effective and that good annotations are important. Finally, we examined the drawbacks of an area-based segmentation metric, and proposed a length-based metric to assess the structural similarity between annotated and predicted BACs for improved clinical relevance.

Keywords: Breast arterial calcification · Deep learning · Segmentation

Supplementary Information The online version contains supplementary material available at https://doi.org/10.1007/978-3-031-27066-6_10.

Y. Zheng et al. (Eds.): ACCV 2022, LNCS 13848, pp. 134–150, 2023.
https://doi.org/10.1007/978-3-031-27066-6_10

1 Introduction

Mammography is a diagnostic imaging technique that is used to detect breast cancer and other early breast abnormalities. During mammography, each patient normally has one mediolateral oblique (MLO) and one craniocaudal (CC) projections for the left and right breasts. Breast arterial calcifications (BACs) noted on mammograms are calcium deposited in the walls of arteries in the breast [5], appearing in various structures and patterns [7]. The presence and progression of BACs have shown to be associated with coronary artery disease and cardiovascular disease (CVD) in recent clinical studies [16,20]. Prevalence of BACs in screening mammograms has been estimated at 12.7% [12]. Breast radiologists may note BACs as an incidental finding, but doing so is subjective and time-consuming, thus automation may be beneficial. Computer aided detection is not new, and several authors have reported promising BACs segmentation results using either classical computer vision algorithms [7,8,10,35] or deep learning models [3,11,28].

Deep learning models have gained increasing popularity in various medical domains [22,23] for their outstanding performance in different tasks such as lesion detection [21], tumor segmentation [27] and disease classification [30] et al. The performance of a deep learning model is typically influenced by hyper-parameters chosen during model training. These include parameters related to experimental factors such as epochs, batch size and input image size; parameters related to training strategies such as loss function, learning rate, optimiser; parameters related to model architecture such as number of layers and choice of encoder. Kaur et al. [17] and Thambawita et al. [31] demonstrated how a model's performance could be improved by properly configuring the hyper-parameters.

Apart from hyper-parameters, the effect of annotation quality on object segmentation has received little attention. In medical image segmentation especially in the scope of BACs segmentation, it is hard or impossible to conduct sufficiently sophisticated annotation due to cost and required domain expertise. Yu et al. [37] indicated that medical imaging data paired with noisy annotation is prevalent. Their experimental results revealed that the model trained with noisy labels performed worse than the model trained using the reference standard.

To address the challenge of noisy labels, in this study we propose a label correction algorithm to automatically remove false positives from manual BACs annotations. The effect of the corrected labels is evaluated along with other hyper-parameters such as image size, image normalisation methods and model architectures. Finally, we analyse the drawbacks of area-based segmentation metric, and propose a length-based metric to evaluate the structural similarity between annotated and predicted BACs for better clinical relevance.

2 Materials

2.1 Dataset

The de-identified image data were collected at a single health institution in the United States. There are 6,573 raw tomosynthesis images acquired from two x-ray systems: 5,931 from GE Pristina system and 642 from GE Senographe Essential. GE Pristina images have a resolution of 2,850 × 2,394 and 0.1 mm per pixel; and GE Senographe Essential images have a resolution of 3,062 × 2,394 and 0.1 mm per pixel.

Digital breast tomosynthesis is a clinical imaging technology in which an x-ray beam sweeps in an arc across the breast, producing tomographic images for a better visibility of malignant structures [2]. In this study, we use the central projection image from those collected in each scan for simplicity as the source is normal to the detector for this image.

To the best of our knowledge, it may be one of the largest BACs datasets, and the first one reported to use tomosynthesis images for training and evaluating deep learning models. In comparison, other reported datasets are summarised in Table 1.

Table 1. Literature reported BACs datasets.

Literature	No. of images	BACs+ %	Modality
[34]	840	60	2D mammography
[11]	661	NA	2D mammography
[28]	5,972	14.93	2D mammography
[3]	826	50	2D mammography

2.2 Annotation

The annotation task was performed using Darwin.v7labs[1] by a third-party annotation company where the data were first split to batches then assigned to multiple annotators. Cross check or consensus reading were not performed on the annotation results, so each image was only reviewed by a single annotator.

All annotators have prior annotation experience in general domain, but little specific medical knowledge especially in radiology. An introduction to BACs and annotation guidance were supplied to each annotator. Briefly, the task is to use the brush tool to label the BACs pixels. The first 50 annotated samples were reviewed by an in-house imaging scientist with 3 years experience in mammographic image analysis. Upon the acceptance of these samples, the annotators completed the remaining images.

3 Methods

3.1 Image Pre-processing

Digital mammography generally creates two types of images: *raw* and *processed*. Raw images are the images as acquired, with some technical adjustment such as pixel calibration and inhomogeneity correction; processed images are manipulated

[1] https://www.v7labs.com/.

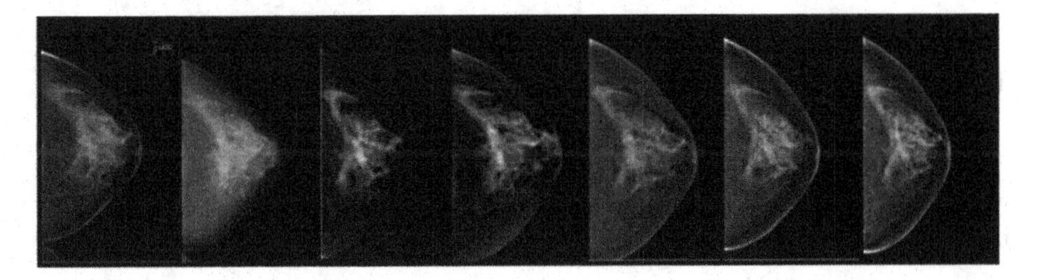

Fig. 1. Seven consecutive years of mammograms for the same patient showing as processed images by three different mammography systems (from left to right): Hologic, GE, Hologic, Hologic, Siemens, Siemens, Siemens. Presentation of the processed images varies significantly based on the mammography system characteristics and image processing.

Fig. 2. Demonstration of a compressed breast during mammography (left) and its mammographic segmentation in craniocaudal (middle) and mediolateral oblique (right) views. The compressed region (contact area) is labelled in white, pectoral muscle in light gray and periphery in dark gray. The directly exposed area to x-ray beam is labelled in black as background.

from the raw images by manufacturer specific algorithm to enhance the image contrast to better fit human eye response in detecting tissue lesions. Manufacturers' preferences in image processing may result in the same breast imaged at two x-ray systems having distinctive appearances, as shown in Fig. 1. Furthermore, [36] reported a deep learning model trained from processed images from one manufacturer cannot be transferred to an unseen external dataset, possibly due to image inconsistency. In contrast, raw images record the original information of the detector response, making it ideal for further image processing to achieve a consistent contrast and visual appearance across different mammography systems. Here, we presented three different normalisation methods in ascending complexity: simple gamma correction, self-adaptive contrast adjustment and Volpara® density map. All three methods depend on a segmentation map (see Fig. 2) labelling pectoral muscle, fully compressed and peripheral regions, which were produced by VolparaDensity™ software. The segmentation accuracy of the software was validated by Woutjan et al. [4].

Simple Gamma Correction. Given a raw image I^{raw}, a logarithm transform is applied on each pixel as in Eq. (1) [24] that brightens the intensities of the breast object:

$$I^{\ln} = \ln(I^{\text{raw}} + 1.0) . \tag{1}$$

Then, a gamma correction Eq. (2) is applied to the log-transformed image

$$I^{\text{gamma}} = [(I^{\ln} - I^{\ln}_{\min})/(I^{\ln}_{\max} - I^{\ln}_{\min})]^{1/2} , \tag{2}$$

where $I^{\ln}_{\min}$ and $I^{\ln}_{\max}$ are the minimum and maximum pixel values respectively in the breast region of $I^{\ln}$.

Self-adaptive Contrast Adjustment. The algorithm brings a given raw mammogram Y_0 to a target mean intensity $\overline{Y}$ in the breast region, by iteratively applying gamma correction. At each step, the gamma value is computed as Eq. (3a), and the image is gamma corrected as Eq. (3b):

$$\gamma_{i+1} = \gamma_i \times \ln(\overline{Y} - \overline{Y_i}) \tag{3a}$$

$$Y_{i+1} = Y_i^{\gamma_{i+1}} \tag{3b}$$

where $\gamma_0 = 1$, and $\overline{Y_i}$ is the mean pixel intensity in the breast region after gamma transformation in the previous step. After a set number of iterations, or after $\overline{Y_i}$ stops converging to the target $\overline{Y}$, the process is terminated. See [19] for detailed implementation.

Volpara® Density Map. The Volpara® algorithm finds an area of the breast within a region in contact with the compression paddle that corresponds to entirely fatty tissues, referred as P^{fat}, then using it as a reference level to compute the thickness of the dense tissue h^{dt} at each pixel location (x, y) based on Eq. (4) [14,18]

$$h^{\text{dt}}(x, y) = \frac{\ln(P(x, y)/P^{\text{fat}})}{\mu^{\text{fat}} - \mu^{\text{dt}}} , \tag{4}$$

where the pixel value $P(x, y)$ is linearly related to the energy imparted to the x-ray detector. μ^{fat} and μ^{dt} are the effective x-ray attenuation coefficients for fat and dense tissues respectively at a particular peak potential (kilovoltage peak, or kVp) applied to the x-ray tube [13]. Equation (4) converts a raw mammographic image to a density map where the pixel value corresponds to the dense tissue thickness. The volumetric breast density is then computed by integrating over the entire breast area in the density map. The VolparaDensity™ algorithm has shown strong correlation with the ground truth reading (magnetic resonance imaging data) [33] and its density measurements are consistent across various mammography systems [9].

Figure 3 shows the normalisation results of the above three methods on raw images acquired from two x-ray systems. Despite the raw images are displayed in the same intensity range, GE Pristina tomosynthesis is clearly brighter than the GE Senographe Essential image, and the dense tissues are hardly visible in both images. The simple gamma correction and self-adaptive contrast adjustment stretch the contrast between fat and dense tissue in minor and moderate levels respectively, while the Volpara® density map is a complete nonlinear transform revealing the volumetric tissue properties.

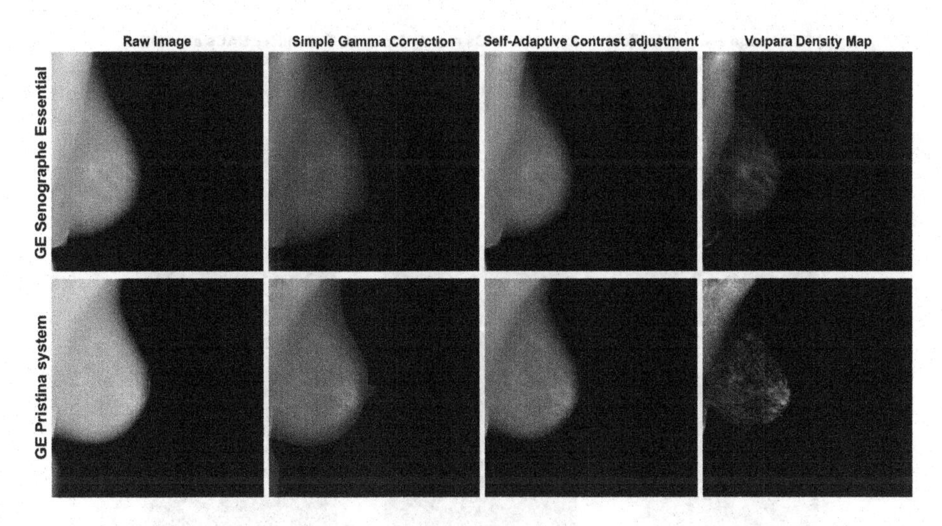

Fig. 3. Examples of normalising raw GE tomosynthesis in three different methods: simple gamma correction, self-adaptive contrast adjustment and Volpara® density map. The 16-bit raw images (first column) are displayed in the same range of [62299, 65415] for better visibility.

3.2 Other Training Variables

Apart from different image normalisation methods, we also investigated other training variables as below:

- Image size: 1024×1024 and 1600×1600[2].
- Loss function: $1/2\times$(dice loss + MSE Loss) [29].
- Model architecture: see Table 2

Table 2. Deep learning models [15] in this study.

Model name	Architecture	Encoder	Parameters, M
Unet^{R}	Unet	Resnet34	24.43
DeepLabV3+^{R}	DeepLabV3+	Resnet34	22.43
DeepLabV3+^{M}	DeepLabV3+	Mobilenetv3_large_100	4.70

[2] In this study, we used a single Tesla T4 GPU, which can accommodate a maximum of batch size 3 and 1600×1600 input image size in the training phase. The experiment of patch-based implementation on full resolution images is reported in the Supplementary Material. We found its performance is not comparable with full image implementation, and its slow inference is not practical in a busy clinical environment (GPU is normally not available on a Picture Archiving and Communication System).

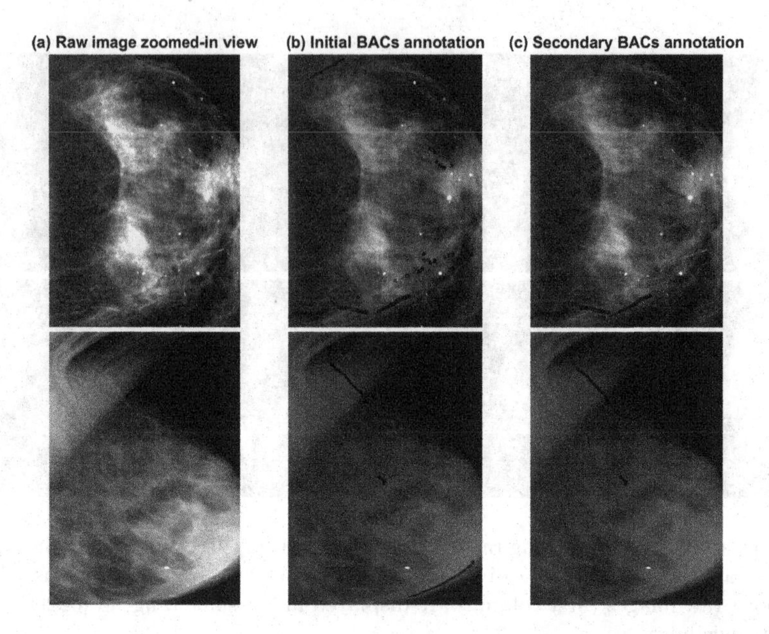

Fig. 4. Zoomed mammogram image sections for comparison between (b) third-party initial BACs annotations and our (c) in-house secondary manual annotations on the same (a) raw image (contrast enhanced for visualisation). In the top panel, micro calcifications and linear dense tissues are initially labelled as BACs; in the bottom panel, the edge of dense tissues is a false positive annotation.

3.3 Label Correction Algorithm

During evaluating the model performance, we discovered large amount of false positives in the BACs annotations. Mostly seen are mislabelling micro-calcifications and dense tissues as BACs, as shown in Fig. 4.

Instead of manually re-annotating all images, we developed a correction algorithm that automatically removes these false positive labels based on BACs' morphology. In [13], Highnam et al. derived Eq. (5) to calculate the calcification thickness (millimeter) from a Volpara® density map:

$$h^{\text{calc}}(x,y) = \frac{(\mu^{\text{dt}} - \mu^{\text{fat}})(h^{\text{dt}}(x,y) - h^{\text{dt}}_{\text{bkg}}(x,y))}{\mu^{\text{calc}}} \,. \tag{5}$$

μ^{calc} is the effective x-ray attenuation coefficient for calcification. Using the values of the linear attenuation coefficients at 18 keV: $\mu^{\text{dt}} = 1.028$ and $\mu^{\text{calc}} = 26.1$. $h^{\text{dt}}_{\text{bkg}}$ is the background tissue thickness, and can be estimated from morphological opening operation as demonstrated in Fig. 5. Thus, a Volpara® density map describing dense tissue thickness can be converted to a calcification thickness map. Then, overlay the annotation on the calcification thickness map, the labels either too short (stand-alone micro calcifications) or having insufficient mean

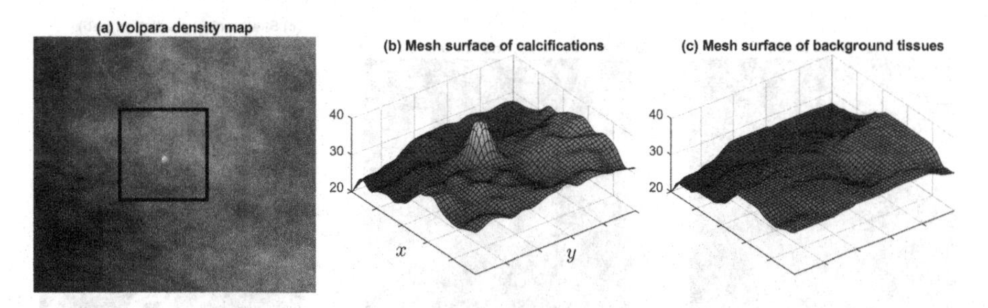

Fig. 5. Demonstration of calcifications and the corresponding background tissues in Volpara® density map. (a) Zoomed-in view of a Volpara® density map where calcifications region is highlighted. (b) Mesh surface of the tissue density (thickness in millimeters) in the highlighted region of (a). (c) Background tissues resulting from morphological opening operation of (b).

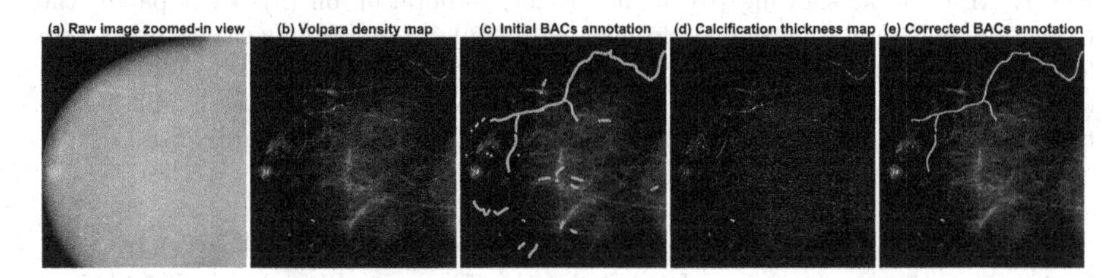

Fig. 6. Demonstration of correcting BACs annotations. First, Volpara® density algorithm converts (a) raw image (zoomed image section) to (b) density map where the pixel value represents dense tissue thickness. Then, Eq. (5) converts (b) density map to (d) calcification thickness map where the pixel value is the potential calcification thickness. The automatic label correction algorithm examines (c) initial BACs annotation. After removing the labels with insufficient length or mean calcification thickness (based on (d) calcification thickness map), the algorithm yields (e) corrected BACs annotation.

calcification thickness (dense tissues) are removed. An example of correcting BACs annotation is shown in Fig. 6.

3.4 Length-Based Dice Score

Typical semantic segmentation evaluation metrics, such as recall, precision, accuracy, F1-score/dice score, examine the overlapping area between prediction and annotation. However, BACs are small in size, and slight differences in their segmentation region may result in strong negative effects on the standard metric like dice score, despite the segmentation still capturing sufficient clinical relevant calcifications. Furthermore, the quantification of BACs is clinically measured by their length rather than area size [1, 32]. Therefore, we introduced a length-based dice score to better focus on the linear trace of the BACs.

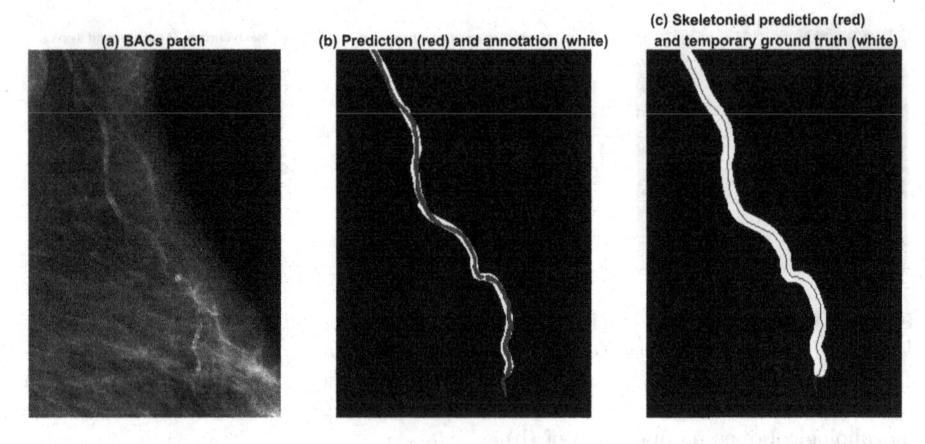

Fig. 7. An example showing (b) prediction and annotation on (a) BACs patch (the prediction and annotation were performed on the full image. We only show a patch here for better visibility). A standard area based dice similarity score from (b) is 0.7081. In (c), the temporary ground truth is derived from skeletonisation followed by dilation on the annotation of (b). The length based dice score reads 0.9750, according to Eq. (6).

Following the work of Wang et al. [35], we first derived the skeletons for both annotation and predicted BACs labels. Then we dilated the skeletonised annotation labels to a width of 2.1 mm, the typical width of a breast artery [34]. The dilated labels were taken as temporary ground truth. A length-based dice score is defined as Eq. (6)

$$\mathrm{Dice}^{\mathrm{L}} = 2 * \frac{\text{Length of predicted BACs within temporary ground truth zone}}{\text{Length of predicted BACs} + \text{Length of annotation BACs}}. \tag{6}$$

The numerator is simply the length of the skeletonised prediction within the region of the temporary ground truth, as seen in Fig. 7(c). In the denominator, the length of the predicted and annotation BACs can be calculated from their respective skeletons. $\mathrm{Dice}^{\mathrm{L}}$ ranges between 0 and 1 where 1 being a complete match of the length between prediction and ground truth BACs.

From Figs. 7(a) and (b), clearly there are under-segmented annotation at multiple locations along the BACs trace. Wang et al. also reported similar annotation defects in Fig. 4 of [35]. Figure 7(b) shows a strong visual agreement between the prediction and annotation, and the prediction seems to have a better segmentation quality than the annotation. But their subtle mismatch in area only yields a moderate dice similarity of 0.7081. In comparison, our proposed $\mathrm{Dice}^{\mathrm{L}}$ focuses on the correlation of the linear trace. As shown in Fig. 7(c), the skeleton of the prediction delineates the BACs signals in high agreement, resulting in a more clinical relevant $\mathrm{Dice}^{\mathrm{L}}$ score of 0.9750.

4 Experiments

4.1 Data Split

The annotation results indicate positive BACs in 95% of the images. The images were split with stratified sampling to 80%/10%/10% for train/validation/test, so each split has a same positive rate of 95%.

4.2 Secondary Annotation on Test Data

The test data were carefully re-annotated by an in-house imaging scientist to remove the false positive segments from the initial annotation as much as possible. After the re-annotation, 93 images are unchanged, 8 images have added BACs segments, and 399 images have false positive segments removed. There are 155 images which have both added and removed segments. The re-annotated labels, noted as ExpertGT (expert ground truth), were utilised to test the deep learning models trained on the original and algorithm corrected train and validation labels.

4.3 Training Settings

The input mammograms are 16-bit, gray scale images. The image went through a contrast normalisation via the methods in Sect. 3.1 followed by a standard pixel value redistribution to achieve a mean of 0 and standard deviation of 1. We also applied common augmentation such as blurring the image using a random-sized kernel, random affine transforms and random brightness and contrast [6].

The models were trained using Adam optimiser with a initial learning rate of $1e^{-4}$. During training, the loss on the validation dataset was monitored at the end of each epoch. The learning rate was reduced by a factor of 10 once the validation loss plateaued for 5 epochs. After a maximum of 100 epochs, or after 10 epochs of no improvement on validation loss, the training was terminated. The model with the best validation loss was saved.

The deep learning segmentation model outputs a probability map. A final binary mask is obtained by applying a cut-off threshold to the probability map. In this study, such cut-off threshold was determined by a parameter sweep from 0.05 to 0.95 at a step of 0.05 on the validation dataset to achieve a highest dice similarity score.

4.4 Results

Table 3 and Fig. 8 present the BACs segmentation results from various combinations of models, image size, normalisation methods and annotation labels. For ease of interpretation, we categorised the results into two groups as illustrated in Table 3. The bottom group comprises five runs from rt0 to rt4 (r stands for run, and t is short for third-party annotated labels), examining the impact of the image size (rt3 vs. rt4) and normalisation (rt0, rt1, rt2 and rt4) on the segmentation performance. Clearly, the model trained on larger image size yields a better dice score since higher image resolutions preserve more subtle texture information [31]. Meanwhile, simple gamma correction outperforms auto gamma correction (i.e. self-adaptive contrast adjustment) and Volpara® density map in both Dice^{TPA} and $\text{Dice}^{\text{ExpertGT}}$ scores (i.e. Dice scores from the respective third-party annotated, TPA in short, and ExpertGT labels of the test data). Comparing to the reference run rt0 (no image normalisation), image normalisation shows positive influence on improving the segmentation performance.

The top group (ra0 to ra3 where 'a' stands for algorithm corrected labels) in Table 3 investigates the impact of annotations. As mentioned in Sect. 3.3, we developed an automatic algorithm to correct false positives in TPA labels. Here, the models were trained on the corrected labels then evaluated on the original TPA and our re-annotated test labels (ExpertGT in the table). As the ExpertGT labels were corrected from TPA labels, the two kinds of labels would fall into distinct distributions. As a result, the models trained on TPA labels did not perform well on the ExpertGT labels, and vice versa. In contrast, the models trained on the algorithm corrected labels have significantly improved $\text{Dice}^{\text{ExpertGT}}$ scores (compare $\text{Dice}^{\text{ExpertGT}}$ before and after rt4 in Fig. 8). For example, ra1's Dice (0.4485) is 29% higher than rt4's Dice score (0.3477) on the ExpertGT labels of the test data, suggesting the effectiveness of our label correction algorithm.

The top group in Table 3 also probes different model structures. The runs ra2 and ra3 correspond to Unet and DeepLabV3+ architectures respectively, and they both use the same encoder structure of Resnet34, which has 5 times more parameters than $\text{DeepLabV3+}^{\text{M}}$ (see Table 2 for details). Performance-wise, $\text{DeepLabV3+}^{\text{R}}$ slightly falls behind Unet^{R}, and they both surpass $\text{DeepLabV3+}^{\text{M}}$ (ra1) in relatively large margins of 4.88% and 8.12% in $\text{Dice}^{\text{ExpertGT}}$ separately.

Shifting our attention from Dice to Dice^{L} in the last column of Table 3, the length-based Dice^{L} score provides a better intuition of how the predicted BACs clinically correlate to the ExpertGT annotation. As Dice^{L} mitigates the slight difference in BACs segmentation region or width, its score reads higher than the Dice metric. Generally, the Dice^{L} and Dice results are aligned. They both show the advantages of using larger image size, larger model and better quality annotation labels. Among these benefits, quality annotation labels perhaps play the most important role in improving segmentation performance. Figure 8 shows models rt0 to rt4 perform similarly as measured by either Dice or Dice^{L}. Models ra0 to ra3 trained on the corrected labels show consistently higher scores. The major difference between ra0 and rt0 – rt4 is whether or not the corrected labels were used during training.

Figure 9 shows the examples of BACs segmentation results from the top three performing models ra1, ra2 and ra3. The overall performance for BACs segmentation is visually very close to the annotation, and the Dice^{L} score better matches with our perception than the Dice score. In the example at the third row, we can see a surgical scar similar to BACs in appearance. The models ra1

Table 3. Comparison of BACs segmentation performance in terms of the area based Dice and length based Dice^L scores on various training configurations. Note the Dice score was computed against both third-party annotated (TPA) and the re-annotated (ExpertGT) test labels, while the Dice^L score was computed against the ExpertGT only. The highest Dice and Dice^L scores are highlighted.

Run ID	Model Name	Image		Label	Test Dice		Test Dice^L
		Normalisation	Size	Train/Val	TPA	ExpertGT	ExpertGT
ra3	DeepLabV3+^{R}	Simple gamma	1600	Algorithm	0.3061	0.4704	**0.6261**
ra2	Unet^R	Simple gamma	1600	Algorithm	0.3122	**0.4849**	0.6121
ra1	DeepLabV3+^{M}	Simple gamma	1600	Algorithm	0.3072	0.4485	0.5960
ra0	DeepLabV3+^{M}	Raw image	1600	Algorithm	0.3042	0.3993	0.5456
rt4	DeepLabV3+^{M}	Simple gamma	1600	TPA	0.3771	0.3477	0.5088
rt3	DeepLabV3+^{M}	Simple gamma	1024	TPA	0.3500	0.3287	0.4962
rt2	DeepLabV3+^{M}	Auto gamma	1600	TPA	0.3610	0.3447	0.5053
rt1	DeepLabV3+^{M}	Density map	1600	TPA	0.3680	0.3442	0.5084
rt0	DeepLabV3+^{M}	Raw image	1600	TPA	0.3590	0.3355	0.5066

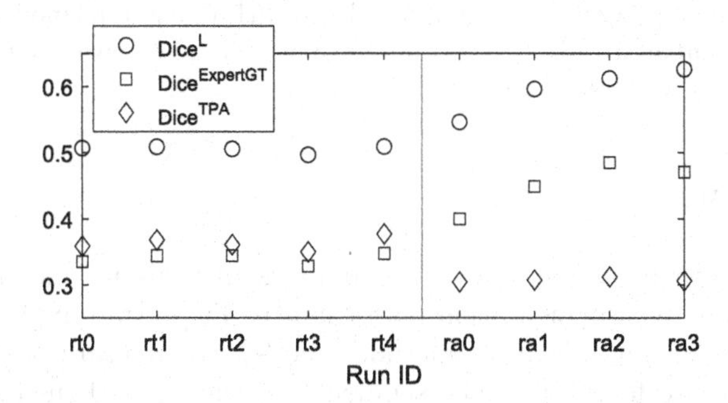

Fig. 8. Visualisation of the Dice and Dice^L scores in Table 3. The vertical line between rt4 and ra0, as the horizontal mid-line in Table 3, divides the data into two groups.

and ra2 incorrectly label the scar as BACs while ra3 does not have such false positive prediction. Notably, ra1 has a comparable performance with ra2 and ra3 despite significantly fewer parameters.

4.5 Additional Results

Molloi et al. [25] and Wang et al. [34] reported a symmetrical presence of BACs between the two views (CC vs MLO) and between the two breasts (left vs right). Indeed as shown in Fig. 10, by using segmentation result as a binary prediction, we find a strong correlation between the left and right breasts in a weighted F1-score of: 0.6711 for the annotation, and 0.8811 for our method. Further, a similar correlation is found between the CC and MLO views of: 0.8613 for the annotation, and 0.9139 for our method.

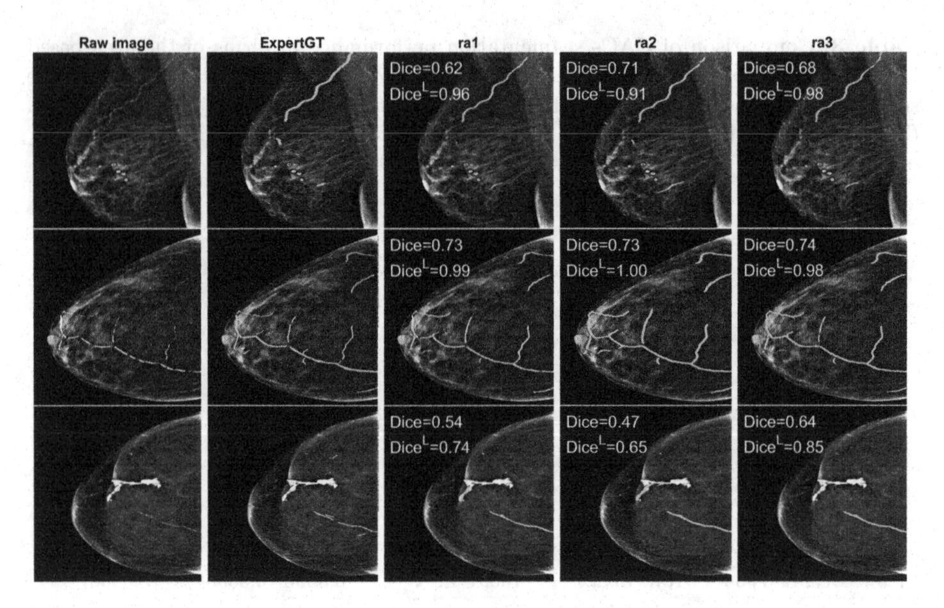

Fig. 9. Examples of BACs segmentation results for ra1, ra2 and ra3 models (see Table 3 for details) as compared to ExpertGT annotations. The raw images were zoomed and contrast enhanced for better visibility.

5 Discussion

This paper presents a comprehensive analysis of training factors and their impacts on the BACs segmentation performance. These training factors include input image size, normalisation method, model architecture and annotation quality. Firstly, we found that the segmentation accuracy benefits from using larger image size for which more subtle features are preserved, relative to down-sampled images. In the experiments of image normalisation, although the model trained on normalised images outperforms the model trained on raw images, their performance does not vary significantly where the simplest gamma correction shows a modest advantage over other more complicated methods. Further, we compared the performance of three model structures: DeepLabV3+^{M}, DeepLabV3+^{R} and Unet^{R}. DeepLabV3+^{M} has 5 times fewer parameters than Unet^{R} and DeepLabV3+^{R} but their performances are very comparable, and their segmentation results are visually close. Lastly, we revealed the importance of high quality annotation. Among other training factors, annotation seems to be the most critical factor determining segmentation performance. A good (here, algorithm corrected) annotation alone shows the highest increase in the BACs segmentation performance, relative to the other hyper-parameter settings tested. However, for hand-crafted annotation it is practically difficult to achieve a pixel-level perfection, nor consistency between readers, and delineation of fine BACs structures in the presence of imaging artefacts is particularly challenging. Expert annotation is expensive and consensus reads from many experts even more so.

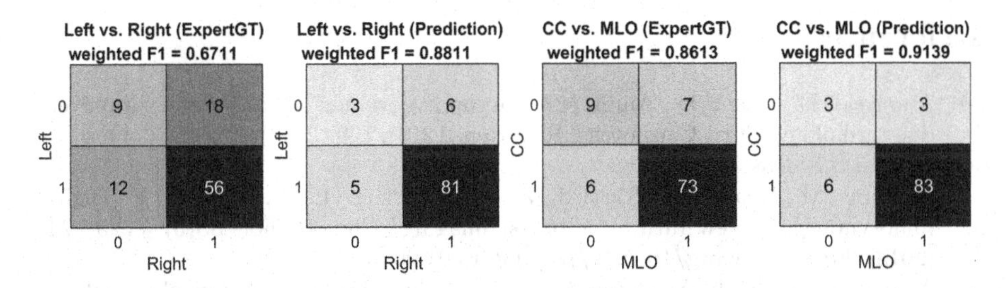

Fig. 10. Confusion matrices for BACs presence between left and right breasts, and between CC and MLO views on test data. Out of 655 test images, there are 380 images belonging to 95 patients with all 4 views (LCC, LMLO, RCC and RMLO) available. The binary prediction is achieved by examining if a segmentation mask is empty. The presented results are from ra3.

Pixel-perfect consistency is difficult even for experts. Crowd-sourced non-expert annotation is inexpensive, but imperfect. To bridge this annotation quality gap, we developed a label correction algorithm that automatically removes false positive labels from BACs annotations. The algorithm demonstrated its effectiveness by allowing a significantly improved segmentation performance.

In addition to the investigation of training factors, we also developed what may be a more clinically relevant metric to improve the evaluation of BACs segmentation. Subtle differences in artery segmentation may have a significant detrimental effect on standard evaluation metrics like Dice score, but may still be able to capture clinically important calcifications with acceptable results. To quantify such clinical relevance, we rectified Dice calculation from its focus on area similarity to trace similarity. The new metric, namely Dice^{L}, has shown to provide a more intuitive measure for the structural correlation between BACs prediction and annotation.

A limitation of this work is the lack of ground truth that is independent of human annotation, e.g. this could come from either physical or digital (simulated) phantom images [26]. We are interested to carry out a validation study where BACs segmentation can be compared to known calcification size and location.

Another important limitation is the lack of negative control images to more thoroughly train and test the model and to comprehensively validate the false positive rate. E.g., Fig. 10 is of limited value without more cases that have no BACs.

In summary, deep learning models have demonstrated promising performance on BACs segmentation. An optimal result can be achieved with higher input image resolution, appropriate image contrast adjustment and larger deep learning model. The annotation quality is found to be a key factor determining the segmentation performance. In general, a model trained with noisy labels is inferior to that trained with good annotation. We recommend other researchers conduct a comprehensive quality control over the annotation process. A thorough review would be required if the annotations were made by non-expert readers.

References

1. Abouzeid, C., Bhatt, D., Amin, N.: The top five women's health issues in preventive cardiology. Curr. Cardiovasc. Risk Rep. **12**(2), 1–9 (2018). https://doi.org/10.1007/s12170-018-0568-7
2. Alakhras, M., Bourne, R., Rickard, M., Ng, K., Pietrzyk, M., Brennan, P.: Digital tomosynthesis: a new future for breast imaging? Clin. Radiol. **68**(5), e225–e236 (2013). https://doi.org/10.1016/j.crad.2013.01.007
3. AlGhamdi, M., Abdel-Mottaleb, M., Collado-Mesa, F.: DU-Net: convolutional network for the detection of arterial calcifications in mammograms. IEEE Trans. Med. Imaging **39**(10), 3240–3249 (2020). https://doi.org/10.1109/TMI.2020.2989737
4. Branderhorst, W., Groot, J.E., Lier, M.G., Highnam, R.P., Heeten, G.J., Grimbergen, C.A.: Technical note: validation of two methods to determine contact area between breast and compression paddle in mammography. Med. Phys. **44**(8), 4040–4044 (2017). https://doi.org/10.1002/mp.12392
5. Bui, Q.M., Daniels, L.B.: A review of the role of breast arterial calcification for cardiovascular risk stratification in women. Circulation **139**(8), 1094–1101 (2019). https://doi.org/10.1161/CIRCULATIONAHA.118.038092
6. Buslaev, A., Iglovikov, V.I., Khvedchenya, E., Parinov, A., Druzhinin, M., Kalinin, A.A.: Albumentations: fast and flexible image augmentations. Information **11**(2), 125 (2020). https://doi.org/10.3390/info11020125
7. Cheng, J.Z., Chen, C.M., Cole, E.B., Pisano, E.D., Shen, D.: Automated delineation of calcified vessels in mammography by tracking with uncertainty and graphical linking techniques. IEEE Trans. Med. Imaging **31**(11), 2143–2155 (2012). https://doi.org/10.1109/TMI.2012.2215880
8. Cheng, J.Z., Chen, C.M., Shen, D.: Identification of breast vascular calcium deposition in digital mammography by linear structure analysis. In: 2012 9th IEEE International Symposium on Biomedical Imaging (ISBI), Barcelona, Spain, pp. 126–129. IEEE (2012). https://doi.org/10.1109/ISBI.2012.6235500
9. Damases, C.N., Brennan, P.C., McEntee, M.F.: Mammographic density measurements are not affected by mammography system. J. Med. Imaging **2**(1), 015501 (2015). https://doi.org/10.1117/1.JMI.2.1.015501
10. Ge, J., et al.: Automated detection of breast vascular calcification on full-field digital mammograms. In: Giger, M.L., Karssemeijer, N. (eds.) Medical Imaging, San Diego, CA, p. 691517 (2008). https://doi.org/10.1117/12.773096
11. Guo, X., et al.: SCU-Net: a deep learning method for segmentation and quantification of breast arterial calcifications on mammograms. Med. Phys. **48**(10), 5851–5861 (2021). https://doi.org/10.1002/mp.15017
12. Hendriks, E.J.E., de Jong, P.A., van der Graaf, Y., Mali, W.P.T.M., van der Schouw, Y.T., Beulens, J.W.J.: Breast arterial calcifications: a systematic review and meta-analysis of their determinants and their association with cardiovascular events. Atherosclerosis **239**(1), 11–20 (2015). https://doi.org/10.1016/j.atherosclerosis.2014.12.035
13. Highnam, R., Brady, J.M.: Mammographic Image Analysis. Computational Imaging and Vision, Springer, Dordrecht (1999). https://doi.org/10.1007/978-94-011-4613-5
14. Highnam, R., Brady, S.M., Yaffe, M.J., Karssemeijer, N., Harvey, J.: Robust breast composition measurement - Volpara^{TM}. In: Martí, J., Oliver, A., Freixenet, J., Martí, R. (eds.) IWDM 2010. LNCS, vol. 6136, pp. 342–349. Springer, Heidelberg (2010). https://doi.org/10.1007/978-3-642-13666-5_46

15. Iakubovskii, P.: Segmentation Models Pytorch (2020). https://github.com/qubvel/segmentation_models.pytorch
16. Iribarren, C., et al.: Breast arterial calcification: a novel cardiovascular risk enhancer among postmenopausal women. Circ. Cardiovasc. Imaging **15**(3), e013526 (2022). https://doi.org/10.1161/CIRCIMAGING.121.013526
17. Kaur, S., Aggarwal, H., Rani, R.: Hyper-parameter optimization of deep learning model for prediction of Parkinson's disease. Mach. Vis. Appl. **31**(5), 1–15 (2020). https://doi.org/10.1007/s00138-020-01078-1
18. Khan, N., Wang, K., Chan, A., Highnam, R.: Automatic BI-RADS classification of mammograms. In: Bräunl, T., McCane, B., Rivera, M., Yu, X. (eds.) PSIVT 2015. LNCS, vol. 9431, pp. 475–487. Springer, Cham (2016). https://doi.org/10.1007/978-3-319-29451-3_38
19. Knowles-Barley, S.F., Highnam, R.: Auto Gamma Correction. WIPO Patent WO2022079569 (2022)
20. Lee, S.C., Phillips, M., Bellinge, J., Stone, J., Wylie, E., Schultz, C.: Is breast arterial calcification associated with coronary artery disease?—a systematic review and meta-analysis. PLoS ONE **15**(7), e0236598 (2020). https://doi.org/10.1371/journal.pone.0236598
21. Li, Y., Gu, H., Wang, H., Qin, P., Wang, J.: BUSnet: a deep learning model of breast tumor lesion detection for ultrasound images. Front. Oncol. **12**, 848271 (2022). https://doi.org/10.3389/fonc.2022.848271
22. Liu, X., et al.: Advances in deep learning-based medical image analysis. Health Data Sci. **2021**, 1–14 (2021). https://doi.org/10.34133/2021/8786793
23. Lundervold, A.S., Lundervold, A.: An overview of deep learning in medical imaging focusing on MRI. Z. Med. Phys. **29**(2), 102–127 (2019). https://doi.org/10.1016/j.zemedi.2018.11.002
24. Marchesi, A., et al.: The effect of mammogram preprocessing on microcalcification detection with convolutional neural networks. In: 2017 IEEE 30th International Symposium on Computer-Based Medical Systems (CBMS), Thessaloniki, pp. 207–212. IEEE (2017). https://doi.org/10.1109/CBMS.2017.29
25. Molloi, S., Mehraien, T., Iribarren, C., Smith, C., Ducote, J.L., Feig, S.A.: Reproducibility of breast arterial calcium mass quantification using digital mammography. Acad. Radiol. **16**(3), 275–282 (2009). https://doi.org/10.1016/j.acra.2008.08.011
26. Molloi, S., Xu, T., Ducote, J., Iribarren, C.: Quantification of breast arterial calcification using full field digital mammography. Med. Phys. **35**(4), 1428–1439 (2008). https://doi.org/10.1118/1.2868756
27. Ranjbarzadeh, R., Bagherian Kasgari, A., Jafarzadeh Ghoushchi, S., Anari, S., Naseri, M., Bendechache, M.: Brain tumor segmentation based on deep learning and an attention mechanism using MRI multi-modalities brain images. Sci. Rep. **11**(1), 10930 (2021). https://doi.org/10.1038/s41598-021-90428-8
28. Riva, F.: Breast arterial calcifications: detection, visualization and quantification through a convolutional neural network. Thesis for Master of Sciences in Biomedical Engineering, Polytechnic University of Milan, Italy (2021)
29. Savioli, N., Montana, G., Lamata, P.: V-FCNN: volumetric fully convolution neural network for automatic atrial segmentation. In: Pop, M., Sermesant, M., Zhao, J., Li, S., McLeod, K., Young, A., Rhode, K., Mansi, T. (eds.) STACOM 2018. LNCS, vol. 11395, pp. 273–281. Springer, Cham (2019). https://doi.org/10.1007/978-3-030-12029-0_30

30. Srinivasu, P.N., SivaSai, J.G., Ijaz, M.F., Bhoi, A.K., Kim, W., Kang, J.J.: Classification of skin disease using deep learning neural networks with MobileNet V2 and LSTM. Sensors **21**(8), 2852 (2021). https://doi.org/10.3390/s21082852
31. Thambawita, V., Strümke, I., Hicks, S.A., Halvorsen, P., Parasa, S., Riegler, M.A.: Impact of image resolution on deep learning performance in endoscopy image classification: an experimental study using a large dataset of endoscopic images. Diagnostics **11**(12), 2183 (2021). https://doi.org/10.3390/diagnostics11122183
32. Van Berkel, B., Van Ongeval, C., Van Craenenbroeck, A.H., Pottel, H., De Vusser, K., Evenepoel, P.: Prevalence, progression and implications of breast artery calcification in patients with chronic kidney disease. Clin. Kidney J. **15**(2), 295–302 (2022). https://doi.org/10.1093/ckj/sfab178
33. Wang, J., Azziz, A., Fan, B., Malkov, S., Klifa, C., Newitt, D., Yitta, S., Hylton, N., Kerlikowske, K., Shepherd, J.A.: Agreement of mammographic measures of volumetric breast density to MRI. PLoS ONE **8**(12), e81653 (2013). https://doi.org/10.1371/journal.pone.0081653
34. Wang, J., Ding, H., Bidgoli, F.A., Zhou, B., Iribarren, C., Molloi, S., Baldi, P.: Detecting cardiovascular disease from mammograms with deep learning. IEEE Trans. Med. Imaging **36**(5), 1172–1181 (2017). https://doi.org/10.1109/TMI.2017.2655486
35. Wang, K., Khan, N., Highnam, R.: Automated segmentation of breast arterial calcifications from digital mammography. In: 2019 International Conference on Image and Vision Computing New Zealand (IVCNZ), Dunedin, New Zealand, pp. 1–6. IEEE (2019). https://doi.org/10.1109/IVCNZ48456.2019.8960956
36. Wang, X., Liang, G., Zhang, Y., Blanton, H., Bessinger, Z., Jacobs, N.: Inconsistent performance of deep learning models on mammogram classification. J. Am. Coll. Radiol. **17**(6), 796–803 (2020). https://doi.org/10.1016/j.jacr.2020.01.006
37. Yu, S., Chen, M., Zhang, E., Wu, J., Yu, H., Yang, Z., Ma, L., Gu, X., Lu, W.: Robustness study of noisy annotation in deep learning based medical image segmentation. Phys. Med. Biol. **65**(17), 175007 (2020). https://doi.org/10.1088/1361-6560/ab99e5

Machine Learning and Computing for Visual Semantic Analysis

Towards Scene Understanding for Autonomous Operations on Airport Aprons

Daniel Steininger(✉), Andreas Kriegler, Wolfgang Pointner, Verena Widhalm, Julia Simon, and Oliver Zendel

AIT Austrian Institute of Technology, Seibersdorf, Austria
{daniel.steininger,andreas.kriegler,wolfgang.pointner,verena.widhalm, julia.simon,oliver.zendel}@ait.ac.at

Abstract. Enhancing logistics vehicles on airport aprons with assistant and autonomous capabilities offers the potential to significantly increase safety and efficiency of operations. However, this research area is still underrepresented compared to other automotive domains, especially regarding available image data, which is essential for training and benchmarking AI-based approaches. To mitigate this gap, we introduce a novel dataset specialized on static and dynamic objects commonly encountered while navigating apron areas. We propose an efficient approach for image acquisition as well as annotation of object instances and environmental parameters. Furthermore, we derive multiple dataset variants on which we conduct baseline classification and detection experiments. The resulting models are evaluated with respect to their overall performance and robustness against specific environmental conditions. The results are quite promising for future applications and provide essential insights regarding the selection of aggregation strategies as well as current potentials and limitations of similar approaches in this research domain.

Keywords: Dataset design · Scene understanding · Classification · Object detection · Airport apron · Autonomous vehicles

1 Introduction

While many research activities in recent years were focused on increasing the autonomy of road vehicles, assistant and autonomy functions for vehicles in off-road domains such as airport environments are still in their infancy. These tasks pose similar requirements regarding safety and robustness aspects, but must be executed in a significantly different domain, which hinders a straight-forward application of existing approaches and datasets. Especially the transition from classic to learning-based computer vision approaches requires high amounts of domain-specific image data for training and testing purposes.

Supplementary Information The online version contains supplementary material available at https://doi.org/10.1007/978-3-031-27066-6_11.

Y. Zheng et al. (Eds.): ACCV 2022, LNCS 13848, pp. 153–169, 2023.
https://doi.org/10.1007/978-3-031-27066-6_11

Therefore, the aim of this work is to mitigate this data gap by creating a versatile dataset focusing on apron-specific objects and presenting an efficient approach for data acquisition, sampling and aggregation, which may serve as a precursor for automating mobile platforms in other challenging domains. Image data was acquired by mounting cameras on multiple transport vehicles, which were operated in apron and logistics areas throughout multiple seasons. The high number of captured sequences covers a wide range of data variability, including variations in environmental conditions such as time-of-day, seasonal and atmospheric effects, lighting conditions, as well as camera-related degradation effects. Furthermore, an efficient sampling and meta-annotation approach was developed to automatically extract a representative set of samples from the extensive amount of recorded image data while minimizing the manual effort required for annotation. We additionally introduce a novel data aggregation strategy and provide preliminary models for detecting and classifying apron-specific objects. To summarize, we propose the following contributions:

- We introduce a novel dataset[1] specialized on objects encountered in apron areas including efficient approaches for image acquisition, dataset design and annotation.
- We train and evaluate baseline models for classification and detection on multiple dataset variants and thoroughly quantify the models' robustness against environmental influences.

Overall, we believe that our contributions provide vital insights in a novel and highly relevant application domain as well as universal strategies for efficient dataset design and experiments setup.

2 Related Work

While detecting and classifying objects encountered on airport aprons represents a novel application domain, there are several links to prior research, especially regarding learning and dataset-design approaches as well as existing datasets containing relevant objects.

Automating apron vehicles requires consistent robustness under a wide range of challenging environmental conditions. While some works focus on specific aspects such as variations in either daytime [4], weather [23] or image degradation [14] for benchmarking models or investigating the variability of existing datasets [1,2], few of them provide a comprehensive analysis regarding the impact of multiple factors on model performance [31,37] and neither includes the classes relevant for the given domain. Capturing the required data for versatile learning experiments can either be accomplished by complex sensor systems providing high-quality multi-modal data, as demonstrated by KITTI [8] and ApolloScape [12] or more portable equipment usually facilitating a significantly higher number of recording sessions and therefore higher data variability, as shown by Mapillary [25]. Objects typically encountered on airport aprons include certain common

[1] Images and annotations are available at https://github.com/apronai/apron-dataset.

classes like aircraft or persons, which are part of established datasets, such as MS COCO [19], PASCAL-Context [24], ADE20K [39] and OpenImages [18]. However, their coverage in terms of number and variability in these datasets is rather limited. Additionally, a number of dedicated datasets is available for the category of persons [3,34,38,40]. Similarly, different types of aircraft are provided by specialized datasets offering a more fine-grained categorization [22,31], as well as top-view or satellite [21,29,33] imagery.

The majority of relevant classes for the target application, however, is highly specific to the airport domain and rarely occurs outside it. This includes objects such as specialized airport vehicles, traffic signs or containers, which are rarely captured due to safety-related access restrictions. To the best of our knowledge, the presented dataset is the first to put the focus not only on airplanes, but the entire environment of airport aprons.

3 The Apron Dataset

Automating transport vehicles on airport aprons requires a reliable perception of this highly specialized environment. Therefore, the dataset's label specification is focused mainly on multiple types of apron vehicles, but also includes other kinds of static and transient obstacles. Ensuring the necessary relevance and efficiency in creating the dataset requires a consistent strategy across all stages of dataset design, as presented in the following subsections.

3.1 Data Acquisition

To match the requirements of the intended application as closely as possible, data acquisition was conducted in a realistic environment from the transport vehicle's expected point of view. Therefore, cooperating with a commercial airport was essential to gain access to transport vehicles in regular operation. However, this critical infrastructure implies that compliance with safety and legal considerations was required, such as preserving the privacy of passengers and airport staff, as well as ensuring that the recording campaign never interferes with airport and logistics operations. All image data was recorded using Nextbase 612GW dashcams with a resolution of 3860×2160 pixels, providing sufficient image quality combined with low cost and efficient handling. They were mounted on the inside of the windshields of two container-transport vehicles. To ensure that recordings are paused when the vehicle is inactive, their power supply was coupled to the respective engines. To increase the flexibility for later applications, one of them was modified to incorporate a lens with 90° field of view instead of the built-in lens with 150°. Most of the data was captured in time-lapse mode at 5 fps to provide a data variability sufficiently representing the environment. Furthermore, the recordings were complemented by sequences at 30 fps for demonstration purposes as well as future developments such as multi-object tracking.

Over a time period of six months we recorded 1715 image sequences, covering the seasons of spring, summer and autumn. Since transport vehicles typically traverse between multiple locations in logistics and apron areas, these recordings conveniently contain all kinds of objects encountered along their routes. On the other hand, they include irrelevant sections with low scene activity during parking or moving along monotonous regions, as well as motion blur and noise, which need to be pruned as a first step during annotation.

3.2 Sequence and Image Annotation

While removing highly redundant or irrelevant data, each remaining sequence is additionally assigned a defined set of parameters to specify the environmental factors during recording time.

- *Time of Day* describes the variance between natural and artificial light sources throughout the day.
- *Lighting* specifies sunny and diffuse conditions during daytime based on the appearance of shadows and is undefined for night recordings.
- *Atmosphere* differentiates multiple weather and atmospheric effects.
- *Scene Dynamics* is a measure for the number and activity of dynamic objects in a sequence, as well as variations due to motion of the capturing vehicle.

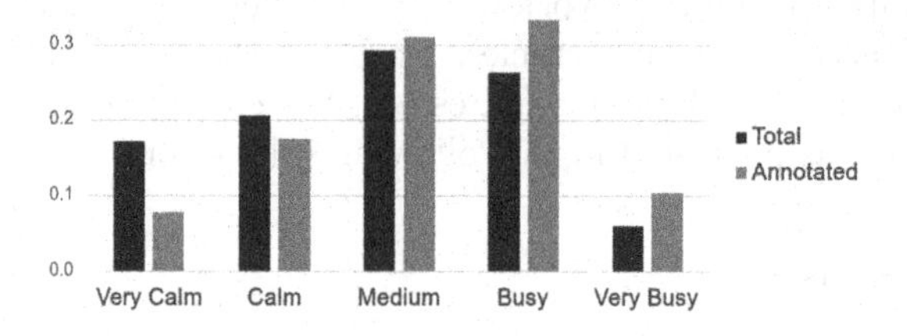

Fig. 1. Distribution of *Scene Dynamics* for total and annotated images

We eventually selected 1209 sequences representing a pool of more than 3.2 million images in total. They were sampled at varying rates proportional to the parameter *Scene Dynamics* aiming to reduce redundancies and extract a representative set of approximately 10k images. As demonstrated in Fig. 1, sequences tagged as *Busy* and *Very Busy* are oversampled by 25% and 50% respectively, whereas *Calm* and *Very Calm* scenes are undersampled analogously. Additionally, the few remaining redundant images, recorded when the vehicle was stopped with the engine running, were manually removed. The final set of 10098 images is annotated with additional per-image parameters. *Degradation* summarizes multiple factors expected to negatively influence model performance as shown in Fig. 2. While these factors typically appear simultaneously, the parameter is set to *High* if any of them significantly influences image quality. Emphasis is placed on near- and mid-range objects in a distance of up to 100 m.

The distribution of parameters for all annotated frames is displayed in Fig. 3. *Time of Day* variations are well balanced due to the airport operating hours including dawn, dusk and a significant proportion of the night. *Lighting* conditions are annotated only for 68% of the dataset since the parameter does not

Fig. 2. Representative examples for sequence and image parameters. *Time of day* differentiates between *Day* (a), *Twilight* (b) and *Night* (c), *Lighting* between *Sunny* (d) and *Diffuse* (e) and *Atmosphere* between *Clear* (f), *Rain* (g) and *Heavy Rain* (h), with rain drops significantly impacting perception. The assigned state of *Low* or *High* for *Degradation* typically depends on multiple factors such as under- (i) and overexposure (j), windshield reflections (k, l), motion blur (m) and wiper occlusions (n)

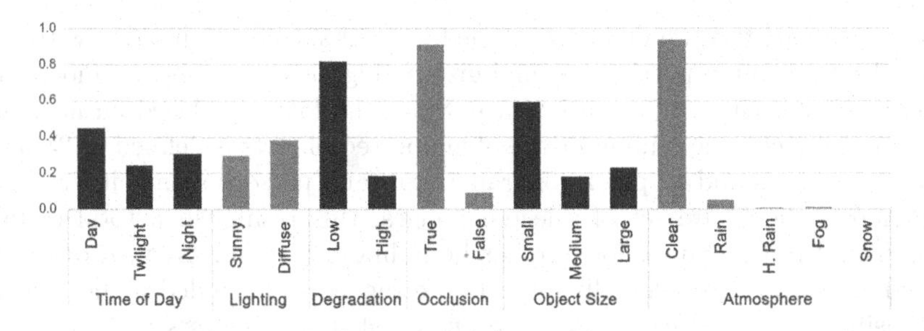

Fig. 3. Per-instance distribution of annotated meta parameters and object sizes (*Small* $< 15k$ pixels $<$ *Medium* $< 45k$ pixels $<$ *Large*)

apply to night recordings and ambiguous sequences containing both *Sunny* and *Diffuse* states. However, the remaining images are roughly evenly distributed, therefore representing a solid basis for evaluating their impact on detection and classification performance. Atmospheric effects, on the other hand, show a strong bias towards *Clear* conditions and thereby represent the environment encountered during the recording time. Nevertheless, the data includes a small number of images of *Rain* and *Heavy Rain* as well as *Fog*, which are useful for preliminary insights regarding the impact of harsh weather conditions as well as the generalization capability of trained models.

3.3 Instance Annotation

Specifying a set of object labels tailored to the target application requires an extensive analysis of sampled images to identify visually distinct categories for frequently appearing types of vehicles and obstacles. Additionally, safety-relevant classes are considered independently of their occurrence frequency. We aim at annotating a fine-grained definition of classes which can be further condensed for training specialized models on more coarse-grained dataset variants. For this purpose, we define 43 categories, which are fully listed in the supplementary material and visualized in Fig. 4.

Fig. 4. Representative examples of different object categories included in the dataset

For each category we define a detailed textual specification along with multiple selected sample patches to minimize ambiguous assignments and resolve potential corner cases at an early stage. Since the focus is placed on near- and mid-range objects, the minimum size of annotated objects is defined as 28 pixels for vehicle classes and 12 pixels for traffic signs and persons along either dimension. Across the entire set of selected images, this results in a total of more than 169k object instances localized as bounding boxes and assigned one of the defined categories. Additionally, objects are tagged as occluded if they are not fully visible due to other objects or truncated at image borders.

It is well known that object-occurrence frequency in real-world images often follows a long-tailed distribution [20,27], which is also visible in this case, as demonstrated in Fig. 5. The overabundance of a few head classes with the numerous tail classes collectively still making up a significant portion of the data [41] is challenging for learning systems. The environment on airport aprons is generally relatively structured and controlled but also crowded with an average of 16.8 objects per image. Compared to the total number of samples the object occurrence in terms of images containing a certain category, is more evenly distributed, indicating that especially head classes tend to appear in multitudes within a single image. The distribution of object sizes and its relation to the occurrence frequency is of additional interest, especially for the detection task. The mean of about 80k pixels and a surprisingly low median of 9k pixels indicate that a relatively low number of classes with exceedingly large objects stands in contrast to a large number of classes with small to medium objects.

While the method of image acquisition and the label definition affect these dataset statistics, the long-tailed distribution can be mitigated by exploiting the closed-off, controlled and repetition-driven nature of airport apron processes. Therefore, it might be easier to create accurate and robust models for this domain than it is for other automotive applications and we believe our dataset showcases a promising way to efficiently accumulate data for this purpose.

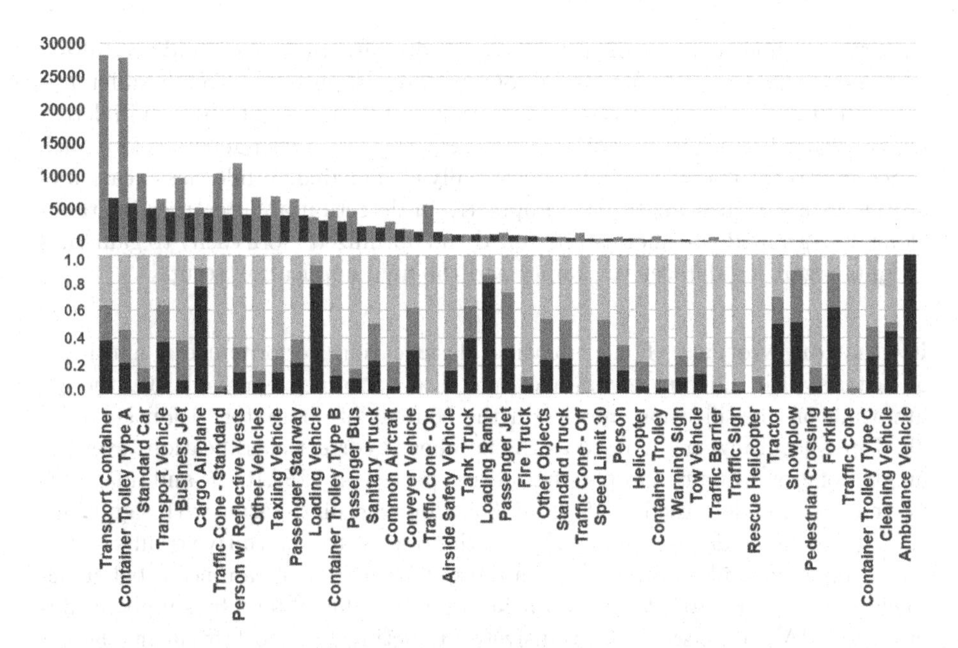

Fig. 5. Top: numbers of samples annotated for each object category (light color) and images containing them (dark color). Bottom: normalized size distribution: (Large > 45k pixels > Medium > 15k pixels > Small)

3.4 Data Aggregation

By mapping selected classes of the annotated data we define three variants to facilitate a comparative analysis of dataset balancing effects.

- *Fine* is the baseline variant including all 43 annotated labels
- *Top* limits the dataset to the 25 most frequent classes in terms of total object occurrence which contribute roughly 97% of samples
- *Coarse* uses the full set of instances but remaps them to only 23 superclasses based on semantic similarity

For all experiments described below, 68% of each dataset are used for training and 17% for validation, while the remaining 15% of samples are withheld during the experiments and exclusively reserved for testing purposes. This split is applied on a per-sequence basis to reduce the effects of over-fitting.

4 Fine-Grained Classification

Fine-grained image classification focuses on correctly identifying differences between hard-to-distinguish object (sub-)classes and predicting the specific variants accordingly. Taking a look at Fig. 4 and the corresponding categories shown in Fig. 5, many visually similar yet distinct classes can be observed in the *Apron* dataset: multiple types of container trolleys, aircraft, traffic signs and specialized cars and trucks are all commonly encountered on aprons. To correctly differentiate such visually similar classes, feature representations need to be rich in detail. On the other hand, the overall object variety requires a well generalized model, which makes the classification task especially challenging. While classification of object instances also takes place implicitly in detection architectures, a standalone, fine-grained classifier can be tuned and optimized more easily to gain vital insights regarding dataset variability and the final application setup.

Evaluation Metrics. For evaluating classification performance a multitude of metrics with different advantages and drawbacks has emerged [30], though literature on classification metrics in the context of computer vision is sparse [9]. Top-1 accuracy (α), defined as the number of correct classifications over the number of ground-truth samples, has been reported on CIFAR [17] and ImageNet [5] and is still widely used [6,15,28,35,36]. However, on datasets with significant class-imbalance or long-tailed characteristics α leads to unintuitive results, since, for example, a model evaluated on a dataset with 90% of samples s belonging to class A and only 10% to class B achieves an α value of 90% by simply always predicting 'A'. This score of exclusively predicting the most frequent class is defined as the null accuracy α_0, where s_i is the number of samples of class i and s is the total number of samples:

$$\alpha_0 = \frac{max\{s_1, \ldots, s_n\}}{s} \tag{1}$$

To evaluate the significance of α, it should be compared to α_0 as well as to the random accuracy α_r, which represents the score if predictions are equally distributed over the number of classes (n).

$$\alpha_r = \frac{1}{n} \tag{2}$$

Moreover, α is prone to be even more biased in the case of top-x accuracy with $x > 1$ where a sample counts as correct if the true label is within the x most-confident model predictions. Therefore, we use the less biased metric of per-class average recall ($\bar{r}$) for selecting the best performing models of each experiment and evaluating all fine-grained classification models. This metric represents the average of class-wise Top-1 accuracies, as used by e.g. [19], rendering each class equally important independent of the number of samples assigned to it.

$$\bar{r} = \frac{1}{n}\sum_{i=1}^{n} \frac{TP_i}{TP_i + FN_i} \tag{3}$$

Additionally, we employ the metric of per-class average precision ($\bar{p}$) calculated analogously as the average ratio between true positives and the total number of predictions for each class:

$$\bar{p} = \frac{1}{n}\sum_{i=1}^{n}\frac{TP_i}{TP_i + FP_i} \tag{4}$$

In formulas above, TP_i and FP_i denote the number true and false positive classifications for class i, respectively, while FN_i is the corresponding number of false negatives. To improve clarity when comparing results, we furthermore report the f_1 score as the harmonic mean of both measures.

$$f_1 = 2 \times \frac{\bar{p} \times \bar{r}}{\bar{p} + \bar{r}} \tag{5}$$

Setup and Optimization. We conducted a two-fold validation for comparing multiple ResNet [11] and EfficientNet [32] architectures and tuning their hyper-parameters based on $\bar{r}$. Eventually, we chose EfficientNet-B3 since it outperforms the ResNet variants and can conveniently be adapted to specific data using a single scaling coefficient for modifying width, depth and image scale.

Object instances below 30 pixels along both dimensions or an aspect ratio exceeding 10:1 were excluded, slightly reducing the original dataset to roughly 150k samples. After evaluating multiple image-augmentation techniques, the best results were observed using a random horizontal flip before resizing to the required input size of 300 pixels. An additional application of random crop and Gaussian blur did not improve results on the validation set, indicating that the visual variability in the Apron dataset is already significant. It could be observed, however, that predictions on classes with relatively few samples were more accurate using stronger augmentations, since they benefit more from the additional variability.

All models were trained from scratch for 40 epochs using the SGD optimizer with a learning rate of 0.1, which yielded slower but stable convergence unlike Adam [16], as is often the case in PyTorch [26]. We updated the learning rate every 10 epochs, using a step ratio of 0.1 and a weight decay of 0.0005. The experiments were conducted on an NVIDIA RTX 2080 Ti using a batch size of 128. Furthermore, we used the Swish activation function, a CrossEntropy loss with a dropout rate of 0.3 and Kaiming uniform [10] parameter initialization.

4.1 Classification Results

Table 1 shows the results obtained on the respective test sets of all three dataset variants. The baseline variant *Fine* poses a significant challenge, but the corresponding model still obtains an f_1 score of 68.2%. As expected, leaving out low-frequency classes (*Top*) or merging them to superclasses (*Coarse*) significantly improves performance by up to 12.6%. Precision $\bar{p}$ tends to be only slightly higher than recall $\bar{r}$ despite the long-tailed class distribution of the dataset, indicating consistent classification performance across most classes. As expected, obtained α-scores are far above $\bar{r}$, giving a skewed and less distinctive impression of the models' performance and are therefore omitted for more detailed comparisons.

Table 1. Classification results on the *Fine* (*F*), *Top* (*T*) and *Coarse* (*C*) dataset variants as average recall, average precision and f_1 score, as well as top-1, null and random accuracy. The last two columns specify the numbers of samples in the test sets and the entire dataset variants, respectively

	$\bar{r}$	$\bar{p}$	f_1	α	α_0	α_r	s_T	s
F	0.624	0.752	0.682	0.866	0.184	0.023	22.6k	150.6k
T	0.779	0.810	0.794	0.880	0.193	0.040	21.9k	145.9k
C	0.819	0.798	0.808	0.881	0.213	0.044	22.6k	150.6k
∅	**0.741**	**0.787**	**0.762**	**0.876**	–	–	–	–

4.2 Robustness Analysis

To gain more detailed insights regarding the impact of environmental effects on model performance we filter the test sets by each of the parameters defined in Sect. 3.2 as well as object size and occlusion. The corresponding evaluations on each resulting set are presented in Tables 2 and 3. Distributions of test sets are similar to those of the overall dataset presented in Fig. 3. Note that the samples do not cover the entire test set for the *Lighting* and *Atmosphere* parameters. In the former case this results from the parameter not applying to *Night* settings, while in the latter case all underrepresented conditions were omitted.

Table 2. Impact of environmental conditions on classification performance as deviation from overall f_1 scores (Table 1) for each model on the corresponding test set

	Time of day			Lighting		Degradation		Atmosphere	
	Day	Twilight	Night	Sunny	Diffuse	Low	High	Clear	Rain
F	−0.020	−0.025	−0.005	0.015	−0.008	−0.005	−0.007	0.005	−0.101
T	0.003	0.006	−0.013	0.006	0.006	0.005	−0.028	0.000	−0.008
C	0.006	−0.012	−0.011	0.018	−0.007	0.008	−0.011	0.002	−0.039
∅	**−0.004**	**−0.011**	**−0.009**	**0.013**	**−0.003**	**0.003**	**−0.015**	**0.003**	**−0.049**

Overall, as visible in Table 2, both the positive and negative deviations are relatively small across all dataset variants and parameters, indicating that most conditions are sufficiently covered in the dataset. Since the recording vehicles accumulated image data over a long period of time, they encountered a wide variety of conditions expected during long-term autonomous operation. The least deviation is reported between the different times of day. Since all three values are extremely small and recall and precision values are averaged across classes, the offsets of underrepresented classes can even sufficiently distort results for all deviations to be negative in this case. The trends are more clearly visible for the lighting and degradation parameters, where sunny conditions and low image degradation appear to be the least challenging for all models. The strongest negative impact is visible for light rain which reduces the f_1 score by more than

5% compared to clear atmosphere. It is possible that the model learns more ambiguous and blurred filters for rainy samples and as such the predictions are spread more equally across all classes. On the other hand, for samples with clear sight, the learned filters might be very class-specific and overfit on the classes with many samples, since this phenomena is most noticeable for *Fine*.

Table 3. Impact of object size (*Small* < 15*k* pixels < *Medium* < 45*k* pixels < *Large*) and occlusion on classification performance as deviation from overall f_1 scores (Table 1) for each model on the corresponding test set

	Object size			Occlusion	
	Small	Medium	Large	True	False
F	−0.066	0.001	−0.042	0.003	0.004
T	−0.070	0.001	0.013	−0.001	0.033
C	−0.071	0.012	0.033	−0.004	0.025
∅	**−0.069**	**0.005**	**0.001**	**−0.001**	**0.021**

As expected, Table 3 shows that all models perform significantly better for medium-sized and large objects than for those smaller than 15k pixels. The models seem to be suitable for fine-grained classification tasks where objects are relatively close and largely depicted and therefore most relevant for the intended application scenarios, while tiny and distant objects are more prone to errors. Furthermore, the difference in scores depending on occlusion of objects is relatively low, since more than 90% of all objects in the entire dataset are occluded, providing a rich set of representative training examples.

5 Detection

Based on the insights and promising results gained during our classification experiments, the next step towards the real-world application of autonomous vehicle operation is to analyze entire scenes by localizing and simultaneously classifying objects using an end-to-end detection approach.

Evaluation Metrics. We evaluate the results based on the established average-precision (AP) metric defined as the area under the precision-recall curve for each class. But instead of using a single IoU threshold to distinguish between correct and incorrect detections as traditionally used in detection challenges such as Pascal VOC [7], we average the results over 10 IoU thresholds ranging from 0.5 to 0.95, as suggested by the COCO challenge [19]. The reported overall values are subsequently averaged over all available classes of the respective dataset variant.

Setup and Optimization We conducted the experiments using an existing implementation of YOLOv5 (release 6.1) [13] and compared the results of the small, medium and large architectures on the defined dataset variants. All images were scaled to 1280 pixels along each dimension and subjected to standard data augmentation. We trained on a system with two NVIDIA RTX 3090, using the SGD optimizer with a linear learning rate of 0.01 and a batch size of 16. All models were initialized with the pre-trained weights provided by the authors of [13]. For each combination of architecture and dataset variant we selected the best-performing model out of 50 training epochs based on the validation results.

5.1 Detection Results

As visible in Table 4, performance increases with model complexity. Analogous to the original experiments on the COCO dataset [13] the gain is more significant between the small and medium than between the medium and large architecture.

Fig. 6. Representative detection results of the *Coarse* model on the corresponding test set (green: correct detection (*TP*), orange: correct localization, but incorrect class, red: incorrect detection (*FP*), blue: undetected ground-truth object (*FN*)) (Color figure online)

For each architecture, the dataset variant with the highest granularity (*Fine*) serves as a baseline, since it is the most challenging. As expected, limiting the

Table 4. Detection *APs* on the *Fine* (*F*), *Top* (*T*) and *Coarse* (*C*) datasets variants by model architecture on the corresponding test sets. For comparing the expected computational complexity, the numbers of parameters and floating point operations (FLOPs) for each model are given as stated in [13]

	YOLOv5s6	YOLOv5m6	YOLOv5l6
Parameters	12.6M	35.7M	76.8M
FLOPs	16.8B	50.0B	111.4B
F	0.364	0.415	0.435
T	0.441	0.480	0.501
C	0.473	0.514	0.526
⌀	**0.426**	**0.470**	**0.487**

label categories to the 25 most frequent ones (*Top*) and thereby reducing the total number of samples improves the score. However, the highest robustness is achieved by the variant combining semantically similar object categories and thereby using all available samples (*Coarse*), which results in a similar number of classes but higher variability. The evaluation therefore indicates that limiting the number of classes by remapping provides a superior alternative to simply omitting under-represented classes regarding accuracy as well as flexibility.

The high performance indicated by the scores using the challenging COCO metric is also noticeable in the qualitative results presented in Fig. 6. The model is well capable of handling occlusions in crowded scenes as well as varying environmental conditions, including different times of day, as well as moderate image degradation, as visible in the upper two rows. However, effects such as strong motion blur and significant underexposure decrease detection performance, as visible in the lower right visualization. Furthermore, even small objects can robustly be localized in most cases. However, they are more prone to being assigned wrong categories, as discussed in Sect. 4.2 and depicted in the lower left image.

5.2 Robustness Analysis

To quantify the influence of environmental effects on detection performance, we evaluate all models based on the filtered test sets analogously to Sect. 4.2, as shown in Table 5. As an indicator for the significance of each parameter the number of corresponding samples in the test set for the *Fine* and *Coarse* dataset variants are specified, with the number for *Top* being marginally smaller. As described for the classification results in Sect. 4.2, the total number of samples for each parameter does not necessarily cover the entire dataset.

The slight impacts that can be observed are strongest for changes in lighting conditions as well as image degradation. Since operating areas are well lit at night, the results for the *Time-of-day* parameter confirm that a single model is suitable for operating 24 h a day. Furthermore, light rain can be handled well with only a slight decrease in performance, while more challenging weather effects require more training and test data.

Table 5. Average impact of environmental conditions as absolute deviation from overall detection *APs* (Table 4) on each test set across all three selected model architectures

	Time of day			Lighting		Degradation		Atmosphere	
	Day	Twilight	Night	Sunny	Diffuse	Low	High	Clear	Rain
s_T	6.9k	4.5k	5.6k	5.8k	5.4k	13.6k	3.4k	16.1k	0.6k
F	0.039	−0.004	−0.001	0.037	0.027	0.002	0.002	0.002	0.058
T	0.003	0.031	0.005	0.000	0.000	0.002	−0.021	0.000	−0.019
C	0.016	−0.006	−0.003	0.025	−0.010	0.009	−0.030	0.003	−0.041
∅	**0.019**	**0.007**	**0.000**	**0.021**	**0.006**	**0.004**	**−0.016**	**0.002**	**−0.001**

6 Conclusion

In this work, we demonstrated the process of creating an extensive dataset for the novel application domain of autonomous operation on airport aprons. We introduced efficient concepts for image acquisition and annotation before training and evaluating models for classification and detection based on multiple variants of the dataset. Additionally, we enriched the analysis with annotations of environmental conditions and quantified their impact on model performance.

The results show that our models are already capable of robustly detecting and classifying most relevant near and mid-range objects, rendering them a promising foundation for the further development of assisted and autonomous vehicle operation in this application domain. We achieved our aim of training robust models covering variable conditions at the specific airport used for recording the dataset. While we are aware that the resulting models do not seamlessly generalize to different locations and novel object classes, our dataset and the presented insights represent a valuable basis for significantly reducing the effort and required data to specialize on other airport environments.

We plan to evaluate the resources required for specializing our models and dataset to novel locations by recording additional training and test data at other airports to gain further insights on the re-usability of our concepts and data and their combination with additional approaches. Especially combining the results with multi-object tracking facilitating the propagation of object instances over time holds the potential to further increase detection robustness and facilitate embedded real-time processing on a mobile vehicle.

Acknowledgement. We would like to thank the Federal Ministry for Climate Action, Environment, Energy, Mobility, Innovation and Technology, and the Austrian Research Promotion Agency (FFG) for co-financing the "ICT of the Future" research project AUTILITY (FFG No. 867556). Additionally, we want to thank our project partner Linz Airport, Quantigo AI and our annotation team consisting of Vanessa Klugsberger, Gulnar Bakytzhan and Marlene Glawischnig.

References

1. Asudeh, A., Jin, Z., Jagadish, H.: Assessing and remedying coverage for a given dataset. In: 2019 IEEE 35th International Conference on Data Engineering (ICDE), pp. 554–565. IEEE (2019)
2. Auer, S., Demter, J., Martin, M., Lehmann, J.: LODStats – an extensible framework for high-performance dataset analytics. In: ten Teije, A., et al. (eds.) EKAW 2012. LNCS (LNAI), vol. 7603, pp. 353–362. Springer, Heidelberg (2012). https://doi.org/10.1007/978-3-642-33876-2_31
3. Braun, M., Krebs, S., Flohr, F.B., Gavrila, D.M.: Eurocity persons: a novel benchmark for person detection in traffic scenes. IEEE Trans. Pattern Anal. Mach. Intell. **41**(8), 1844–1861 (2019)
4. Dai, D., Van Gool, L.: Dark model adaptation: semantic image segmentation from daytime to nighttime. In: 2018 21st International Conference on Intelligent Transportation Systems (ITSC), pp. 3819–3824. IEEE (2018)
5. Deng, J., Dong, W., Socher, R., Li, L.J., Li, K., Fei-Fei, L.: ImageNet: a large-scale hierarchical image database. In: 2009 IEEE Conference on Computer Vision and Pattern Recognition, pp. 248–255. IEEE (2009)
6. Dosovitskiy, A., et al.: An image is worth 16 × 16 words: transformers for image recognition at scale. arXiv preprint arXiv:2010.11929 (2020)
7. Everingham, M., Van Gool, L., Williams, C.K., Winn, J., Zisserman, A.: The pascal visual object classes (VOC) challenge. Int. J. Comput. Vision **88**(2), 303–338 (2010)
8. Geiger, A., Lenz, P., Stiller, C., Urtasun, R.: Vision meets robotics: the kitti dataset. Int. J. Robot. Res. **32**(11), 1231–1237 (2013)
9. Grandini, M., Bagli, E., Visani, G.: Metrics for multi-class classification: an overview. arXiv preprint arXiv:2008.05756 (2020)
10. He, K., Zhang, X., Ren, S., Sun, J.: Delving deep into rectifiers: surpassing human-level performance on ImageNet classification. In: Proceedings of the IEEE International Conference on Computer Vision, pp. 1026–1034 (2015)
11. He, K., Zhang, X., Ren, S., Sun, J.: Deep residual learning for image recognition. In: Proceedings of the IEEE Conference on Computer Vision and Pattern Recognition, pp. 770–778 (2016)
12. Huang, X., Wang, P., Cheng, X., Zhou, D., Geng, Q., Yang, R.: The ApolloScape open dataset for autonomous driving and its application. IEEE Trans. Pattern Anal. Mach. Intell. **42**(10), 2702–2719 (2019)
13. Jocher, G., Nishimura, K., Mineeva, T., Vilariño, R.: YOLOv5. Code repository (2020). https://github.com/ultralytics/yolov5
14. Kamann, C., Rother, C.: Benchmarking the robustness of semantic segmentation models with respect to common corruptions. Int. J. Comput. Vision **129**(2), 462–483 (2021)
15. Khan, A., Sohail, A., Zahoora, U., Qureshi, A.S.: A survey of the recent architectures of deep convolutional neural networks. Artif. Intell. Rev. **53**(8), 5455–5516 (2020). https://doi.org/10.1007/s10462-020-09825-6
16. Kingma, D.P., Ba, J.: Adam: a method for stochastic optimization. arXiv preprint arXiv:1412.6980 (2014)
17. Krizhevsky, A., Hinton, G., et al.: Learning multiple layers of features from tiny images. University of Toronto, Technical report (2009)
18. Kuznetsova, A., et al.: The open images dataset V4. Int. J. Comput. Vision **128**, 1956–1981 (2020)
19. Lin, T.-Y., et al.: Microsoft COCO: common objects in context. In: Fleet, D., Pajdla, T., Schiele, B., Tuytelaars, T. (eds.) ECCV 2014. LNCS, vol. 8693, pp. 740–755. Springer, Cham (2014). https://doi.org/10.1007/978-3-319-10602-1_48

20. Liu, Z., Miao, Z., Zhan, X., Wang, J., Gong, B., Yu, S.X.: Large-scale long-tailed recognition in an open world. In: Proceedings of the IEEE/CVF Conference on Computer Vision and Pattern Recognition, pp. 2537–2546 (2019)
21. Maggiori, E., Tarabalka, Y., Charpiat, G., Alliez, P.: Can semantic labeling methods generalize to any city? The Inria aerial image labeling benchmark. In: IEEE International Geoscience and Remote Sensing Symposium (IGARSS). IEEE (2017)
22. Maji, S., Rahtu, E., Kannala, J., Blaschko, M., Vedaldi, A.: Fine-grained visual classification of aircraft. arXiv preprint arXiv:1306.5151 (2013)
23. Michaelis, C., et al.: Benchmarking robustness in object detection: autonomous driving when winter is coming. arXiv preprint arXiv:1907.07484 (2019)
24. Mottaghi, R., et al.: The role of context for object detection and semantic segmentation in the wild. In: Proceedings of the IEEE Conference on Computer Vision and Pattern Recognition, pp. 891–898 (2014)
25. Neuhold, G., Ollmann, T., Rota Bulo, S., Kontschieder, P.: The Mapillary Vistas dataset for semantic understanding of street scenes. In: Proceedings of the IEEE International Conference on Computer Vision, pp. 4990–4999 (2017)
26. Paszke, A., et al.: PyTorch: an imperative style, high-performance deep learning library. In: Advances in Neural Information Processing Systems, pp. 8026–8037 (2019)
27. Salakhutdinov, R., Torralba, A., Tenenbaum, J.: Learning to share visual appearance for multiclass object detection. In: CVPR 2011, pp. 1481–1488. IEEE (2011)
28. Shen, Z., Savvides, M.: Meal V2: boosting vanilla ResNet-50 to 80%+ top-1 accuracy on ImageNet without tricks. arXiv preprint arXiv:2009.08453 (2020)
29. Shermeyer, J., Hossler, T., Etten, A.V., Hogan, D., Lewis, R., Kim, D.: RarePlanes: synthetic data takes flight (2020)
30. Sokolova, M., Lapalme, G.: A systematic analysis of performance measures for classification tasks. Inf. Process. Manage. **45**(4), 427–437 (2009)
31. Steininger, D., Widhalm, V., Simon, J., Kriegler, A., Sulzbachner, C.: The aircraft context dataset: Understanding and optimizing data variability in aerial domains. In: Proceedings of the IEEE/CVF International Conference on Computer Vision, pp. 3823–3832 (2021)
32. Tan, M., Le, Q.V.: EfficientNet: rethinking model scaling for convolutional neural networks. arXiv preprint arXiv:1905.11946 (2019)
33. Xia, G.S., et al.: DOTA: a large-scale dataset for object detection in aerial images. In: Proceedings of the IEEE Conference on Computer Vision and Pattern Recognition, pp. 3974–3983 (2018)
34. Xiao, T., Li, S., Wang, B., Lin, L., Wang, X.: Joint detection and identification feature learning for person search. In: Proceedings of the IEEE Conference on Computer Vision and Pattern Recognition, pp. 3415–3424 (2017)
35. Xie, C., Tan, M., Gong, B., Wang, J., Yuille, A.L., Le, Q.V.: Adversarial examples improve image recognition. In: Proceedings of the IEEE/CVF Conference on Computer Vision and Pattern Recognition, pp. 819–828 (2020)
36. Xie, Q., Luong, M.T., Hovy, E., Le, Q.V.: Self-training with noisy student improves ImageNet classification. In: Proceedings of the IEEE/CVF Conference on Computer Vision and Pattern Recognition, pp. 10687–10698 (2020)
37. Zendel, O., Honauer, K., Murschitz, M., Steininger, D., Domínguez, G.F.: WildDash - creating hazard-aware benchmarks. In: Ferrari, V., Hebert, M., Sminchisescu, C., Weiss, Y. (eds.) ECCV 2018. LNCS, vol. 11210, pp. 407–421. Springer, Cham (2018). https://doi.org/10.1007/978-3-030-01231-1_25

38. Zheng, L., Zhang, H., Sun, S., Chandraker, M., Yang, Y., Tian, Q.: Person re-identification in the wild. In: Proceedings of the IEEE Conference on Computer Vision and Pattern Recognition, pp. 1367–1376 (2017)
39. Zhou, B., Zhao, H., Puig, X., Fidler, S., Barriuso, A., Torralba, A.: Scene parsing through ADE20K dataset. In: Proceedings of the IEEE Conference on Computer Vision and Pattern Recognition, pp. 633–641 (2017)
40. Zhu, P., et al.: Detection and tracking meet drones challenge. IEEE Trans. Pattern Anal. Mach. Intell. **44**(11), 7380–7399 (2021)
41. Zhu, X., Anguelov, D., Ramanan, D.: Capturing long-tail distributions of object subcategories. In: Proceedings of the IEEE Conference on Computer Vision and Pattern Recognition, pp. 915–922 (2014)

Lightweight Hyperspectral Image Reconstruction Network with Deep Feature Hallucination

Kazuhiro Yamawaki and Xian-Hua Han(✉)

Yamaguchi University, Yamaguchi 753-8511, Japan
hanxhua@yamaguchi-u.ac.jp

Abstract. Hyperspectral image reconstruction from a compressive snapshot is an dispensable step in the advanced hyperspectral imaging systems to solve the low spatial and/or temporal resolution issue. Most existing methods extensively exploit various hand-crafted priors to regularize the ill-posed hyperspectral reconstruction problem, and are incapable of handling wide spectral variety, often resulting in poor reconstruction quality. In recent year, deep convolution neural network (CNN) has became the dominated paradigm for hyperspectral image reconstruction, and demonstrated superior performance with complicated and deep network architectures. However, the current impressive CNNs usually yield large model size and high computational cost, which limit the wide applicability in the real imaging systems. This study proposes a novel lightweight hyperspectral reconstruction network via effective deep feature hallucination, and aims to construct a practical model with small size and high efficiency for real imaging systems. Specifically, we exploit a deep feature hallucination module (DFHM) for duplicating more features with cheap operations as the main component, and stack multiple of them to compose the lightweight architecture. In detail, the DFHM consists of spectral hallucination block for synthesizing more channel of features and spatial context aggregation block for exploiting various scales of contexts, and then enhance the spectral and spatial modeling capability with more cheap operation than the vanilla convolution layer. Experimental results on two benchmark hyperspectral datasets demonstrate that our proposed method has great superiority over the state-of-the-art CNN models in reconstruction performance as well as model size.

Keywords: Hyperspectral image reconstruction · Lightweight network · Feature hallucination

1 Introduction

Hyperspectral imaging (HSI) systems is able of capturing the detailed spectral distribution with decades or hundreds of bands at each spatial location of a scene. The abundant spectral signature in HSI possesses the deterministic attributes about the lighting and imaged object/material, which greatly benefits the characterization of the captured scene in wide fields, including remote sensing [4,14], vision inspection [23,24], medical diagnosis [3,20] and digital forensics [10]. To capture a full 3D HSI, the conventional hyperspectral sensors have to employ multiple exposures to scan the target

Y. Zheng et al. (Eds.): ACCV 2022, LNCS 13848, pp. 170–184, 2023.
https://doi.org/10.1007/978-3-031-27066-6_12

scene [5,6,9,26], and require long imaging time failing in dynamic scene capturing. To enable the HS image measurement for moving objects, various snapshot hyperspectral imaging systems [12,18] have been exploited by mapping different spectral bands (either single narrow band or multiplexed ones) to different positions and then collecting them by one or more detectors, which unavoidably cause low-resolution in spatial domain. Motivated by the compressive sensing theory, a promising imaging modality: coded aperture snapshot spectral imaging (CASSI) [1,13,29] has attracted increasing attention. With the elaborated optical design, CASSIs encode the 3D HSI into a 2D compressive snapshot measurement, and require a reconstruction phase to recover the underlying 3D cubic data on-line.

Although extensive studies [2,17,25,27,32,39,43,44] have been exploited, to faithfully reconstruct the desirable HSI from its compressive measurement is still the bottleneck in the CASSIs. Due to the ill-posed nature of the reconstruction problem, traditional model-based methods widely incorporate various hand-crafted priors of the underlying HSIs, such as the total variation [2,33,34,40], sparsity [2,11,30] and low-rankness [19,42], and demonstrate some improvements in term of the reconstruction performance. However, the hand-crafted priors are empirically designed, and usually deficient to model the diverse attributes of the real-world spectral data.

Recently, deep convolutional neural network (DCNN) [8,22,31,35,36] has popularly been investigated for HSI reconstruction by leveraging its powerful modeling capability and automatically learning of the inherent priors in the latent HSI using the previously collected external dataset. Compared with the model-based methods, these deep learning-based paradigms have prospectively achieved superior performance, and been proven to provide fast reconstruction in test phase. However, the current researches mainly focus on designing more complicated and deeper network architecture for pursuing performance gain, and thus cause a large-scale reconstruction model. However, the large-scale model would restrict wide applicability for being implanted in real HSI systems. More recently, incorporating the deep learned priors with iterative optimization procedure has been investigated to increase the flexibility of deep reconstruction model, and the formulated deep unrolling based optimization methods e.g., LISTA [15] ADMMNet [21,37] and ISTA-Net [41] have manifested acceptable performance for the conventional compressive sensing problem but still have insufficient spectral recovery capability for the HSI reconstruction scenario.

To this end, this study aims to exploit a practical deep reconstruction model with small size and high efficiency for being easily embedded in the real imaging systems, and proposes a novel lightweight hyperspectral reconstruction network (LWHRN) via hallucinating/duplicating the effective deep feature from the already learned ones. As proven in [16], the learned feature maps in the well-trained deep models such as in the ResNet-50 using the ImageNet dataset may have abundance or even redundant information, which often guarantees the comprehensive understanding of the input data, and some features can be obtained with a more cheap transformation operation from other feature maps instead of the vanilla convolution operation. Inspired by the above insight, we specifically exploit a deep feature hallucination module (DFHM) for synthesizing more features with cheap operations as the main components of our LWHRN model, and stack multiple of them to gradually reconstruct the residual component un-recovered in the

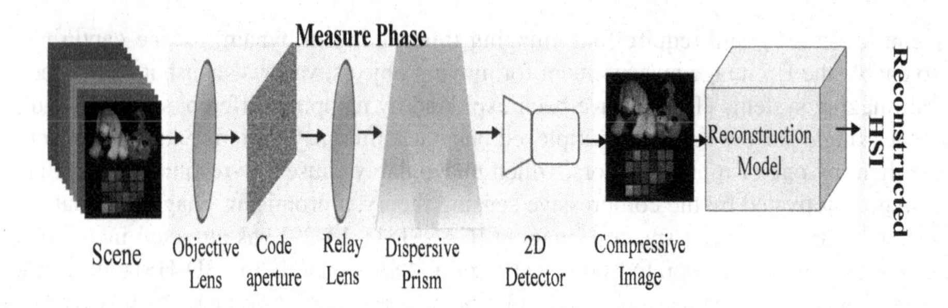

Fig. 1. The schematic concept of the CASSI system.

previous phase. Concretely, the DFHM consists of spectral hallucination block (SHB) for synthesizing more channel of features and spatial context aggregation block (SCAB) for exploiting various scales of contexts, where both SHB and SCAB are implemented using depth-wise convolution layer instead of the vanilla convolution layer, which can be expected to enhance the spectral and spatial modeling capability with the more cheap operation. Experimental results on two benchmark hyperspectral datasets demonstrate that our proposed LWHRN manifests great superiority over the state-of-the-art CNN models in reconstruction performance as well as model size.

In summary, the main contributions are three-fold:

1. We present a novel lightweight hyperspectral reconstruction network from a single snapshot measurement, which employs multiple reconstruction modules to gradually recover the residual HS components by alternatively incorporating spectral and spatial context learning.
2. We exploit a deep feature hallucination module (DFHM) as the main component of the multi-stage reconstruction module, which consists of a spectral hallucination block (SHB) for synthesizing more channel of features and a spatial context aggregation block (SCAB) for exploiting various scales of contexts using more cheap depth-wise convolution than the vanilla convolution.
3. We conduct extensive experiments on two benchmark HSI datasets, and demonstrate superior results over the SoTA reconstruction models in term of the reconstruction performance, model size and computational cost.

2 Related Work

Recently, the hyperspectral reconstruction in the computational spectral imaging have attracted extensive attention, and different kinds of methods, which are mainly divided into optimization-based methods and deep learning-based methods, have been proposed for improving reconstruction performance. In this section, we briefly survey the related work.

2.1 Optimization-Based Methods

The HSI reconstruction from a snapshot measurement is inherently a inverse problem, and can be intuitively formulated as the minimization problem of the reconstruction error of the observed snapshot. Since the number of the unknown variables in the latent HSI is much larger than the known variables in the observed snapshot image, this inverse problem has severe ill-posed nature, and would cause quite un-stable solution via directly optimization. Existing methods have striven to incorporate various hand-crafted priors such as modeling the spatial structure and spectral characteristics of the latent HSI, into the inverse problem, and then formulate as a regularization term for robust optimization. Taking the high dimensionality of spectral signatures into accounting, different image local priors for characterizing the spectral image structure within a local region have popularly been exploited. For example, Wang et al. [33] exploited a Total Variation (TV) regularized model by imposing the first-order gradient smoothness prior for spectral image reconstruction while Yuan et al. [40] proposed to employ a generalized alternating projection to solve the TV-regularized model (GAP-TV). Further, Kittle et al. [17] explored two-step iterative shrinkage/thresholding method (TwIST) for optimization. Although the incorporation of the TV prior for the HSI reconstruction potentially benefits both boundary preservation and smooth region recovery, the reconstructed result may loss some detail structure. Motivated by the successful application in the blind compressed sensing (BCS) [25], sparse representation algorithms have been applied for HSI reconstruction from the snapshot image, which optimizes the representation coefficients with the sparsity prior constraint on the learned dictionary for local image patches [18]. Later, Yuan et al. imposed the compressibility constraint instead of sparsity prior and proposed a global-local shrinkage prior to learn the dictionary and representation coefficients [39]. Moreover, Wang et al. [30] incorporated the non-local similarity into a 3D non-local sparse representation model for boosting reconstruction performance. However, the hand-crafted image priors are not always sufficient to capture the characteristics in various spectral images, and thus cause unstable reconstruction performance. Furthermore, the suitable priors for different images would be varied, and to discover a proper prior for a specific scene is a hard task in the real scenario.

2.2 Deep Learning-Based Methods

Benefiting from the powerful modeling capability, deep learning-based methods have been widely used for image restoration tasks including HSI reconstruction. The deep learning-based HSI reconstruction methods can implicitly learn the underlying prior from the previously prepared training samples instead of manually designing priors for modeling the spatial and spectral characteristics of the latent HSI, and then construct a mapping model between the compressed snapshot image and the desirable HSI. Various deep networks have been proposed for the HSI reconstruction problem. For example, Xiong et al. [36] employed several vanilla convolution layer-based network (HSCNN) to learn a brute-force mapping between the latent HS image and its spectrally under-sampled projections, and demonstrated the feasibility for HSI reconstruction from a common RGB image or a compressive sensing (CS) measurement.

Wang et al. [35] proposed a joint coded aperture optimization and HSI reconstruction network for simultaneously learning the optimal sensing matrix and the latent HS image in an end-to-end framework while Miao et al. [22] developed a λ-net by integrating both the sensing mask and the compressed snapshot measurement for hierarchically reconstructing the HSI with a dual-stage generative model. Later Wang et al. [31] conducted multi-stage deep spatial-spectral prior (DSSP) modeling to incorporate both local coherence and dynamic characteristics for boosting the HSI reconstruction performance. Although promising performance has been achieved with the deep networks, the current researches mainly focus on designing more complicated and deeper network architecture for pursuing performance gain, and thus cause a large-scale reconstruction model. However, the large-scale model would restrict wide applicability for being implanted in real HSI systems.

In order to enhance the flexibility and interpretability of the deep reconstruction model, several works recently incorporated the deep learned priors into iterative optimization procedure, and proposed the deep unrolling based optimization methods in natural compressive sensing, e.g., LISTA [15] ADMMNet [21,37] and ISTA-Net [41]. These methods unroll the iterative optimization procedure into a serial of learnable subproblems, and aim at simultaneously learning the network parameters for modelling the deep priors and the image updating parameters according to the reconstruction formula. However, they were proposed for solving natural compressive sensing problem via elaborately modelig the latent spatial structure, and are insufficient to capture the spectral prior in the high-dimensional HSIs. In order to effectively model the prior in the spectral domain, Choi et al. [8] proposed a convolutional auto-encoder network to learn spectral prior, and then incorporated the deep image priors learned in pretraining phase into the optimization procedure as a regularizer. Wang et al. [32] further conducted both spectral and non-local (NLS) prior learning, and combined the model-based optimization method with the NLS-based regularization for robust HSI reconstruction. Although these unrolling methods have manifested acceptable performance for the conventional compressive sensing problem but still have insufficient spectral recovery capability.

3 Proposed Lightweight Hyperspectral Reconstruction Network

In this section, we first present the formulation problem of the measure and reconstruction phases in the coded aperture snapshot spectral imaging (CASSI) system, and then introduce our lightweight hyperspectral reconstruction model including the overview architecture and the proposed residual reconstruction module: deep feature hallucination module.

3.1 CASSI Observation Model

CASSI [1,13,29] encodes the 3D hyperspectral data of a scene into a 2D compressive snapshot image. we denote the intensity of the incident light for a spectral scene as $X(h, w, \lambda)$, where h and w are the spatial index ($1 \leqq h \leqq H$, $1 \leqq w \leqq W$) and λ is the spectral index ($1 \leqq \lambda \leqq \Lambda$). The incoming light can be collected by the objective lens, and then spatially modulated using a coded aperture, which creates a transmission function $T(h, w)$ for the mathematical implementation. Next the modulated scene is spectrally dispersed with a wavelength-dependent dispersion function $\psi(\lambda)$ by the disperser, and a charge-coupled device (CCD) is adopted to detect the spatial and spectral coded scene as a snapshot image. The schematic concept of this measurement phase in the CASSI is shown in Fig. 1. Mathematically, the observation procedure for measuring the 2D snapshot image can be formulated as:

$$Y(h, w) = \sum T(h - \psi(\lambda))X(h - \psi(\lambda), w, \lambda). \tag{1}$$

For simplicity, we reformulate the observation model in Eq. 1 as a matrix-vector form, which is expressed as:

$$\mathbf{Y} = \mathbf{\Phi X} \tag{2}$$

where $\mathbf{\Phi} \in \Re^{(W+\Lambda-1)H \times WH\Lambda}$ is the measurement matrix of CASSI, and is the combination operation jointly determined by $T(h, w)$ and $\psi(\lambda)$. $\mathbf{Y} \in \Re^{(W+\Lambda-1)H}$ and $\mathbf{X} \in \Re^{WH\Lambda}$ represent the vectorized expression of the compressive image and the full 3D HSI, respectively.

Give the observed compressive snapshot $\mathbf{Y}$, the goal of the HSI reconstruction in the CASSI is to recover the underlying 3D spectral image $\mathbf{X}$, which is a severe ill-posed inverse problem. The traditional model-based methods usually result in insufficient performance while the existing deep learning-based paradigms usually yield large-scale model and then restrict its wide applicability in the real HSI systems despite the promising performance. This study aims to exploit a lightweight deep reconstruction model for not only maintaining the reconstruction performance but also reducing model size and computational cost.

3.2 Overview of the Lightweight Reconstruction Model

The conceptual architecture of the proposed lightweight reconstruction model (LWHRN) is illustrated in Fig. 2(a). which includes an initial reconstruction module and multiple lightweight deep feature hallucination modules (DFHM) for hierarchically reconstructing the un-recovered residual spatial and spectral components with cheaper operation than the vanilla convolution layers. The DFHM module is composed of a spectral hallucination block (SHB) for duplicating more spectral feature maps using depth-wise convolution and a spatial context aggregation block (SCAB) for exploiting the multiple contexts in various receptive fields. In order to reduce the complexity and the model parameter, we elaborately design both SHB and SCAB with more cheap operation but maintaining the amount of the learned feature maps for guaranteeing the reconstruction performance, and construct the lightweight model for practical application in real HSI systems. Moreover, we employ the residual connection structure to learn the un-recovered component in the previous module, and gradually estimate the HS image from coarse to fine.

Concretely, given the measured snapshot image $\mathbf{Y}$, the goal is to recover the full spectral image $\mathbf{X}$ using the LWHRN model. Firstly, as shown in Fig. 2(a), an initial reconstruction module, which consist of several vanilla convolution layers, transforms

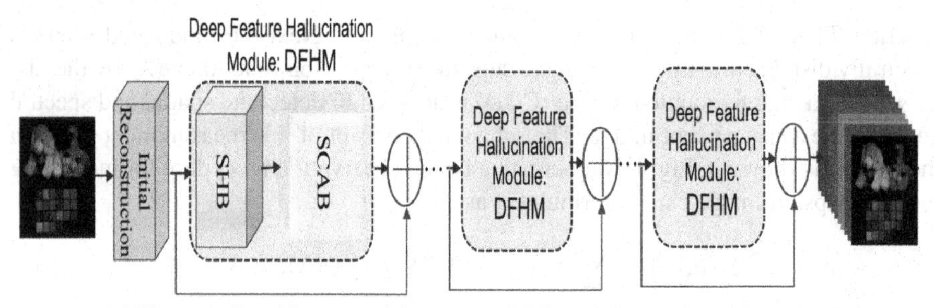

(a) The overall architecture of the proposed lightweight reconstruction model

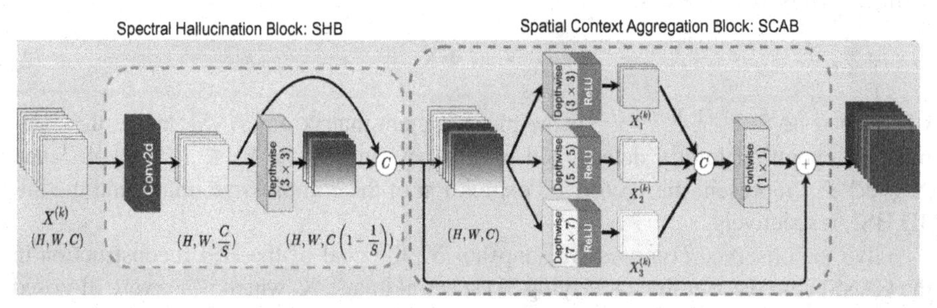

(b) The deep hallucination module

Fig. 2. The conceptual architecture of the proposed lightweight reconstruction model.

the 2D compressed image $\mathbf{Y}$ into multi-channel of feature maps, and then predict an initial HSI: $\mathbf{X}_0$ with Λ spectral bands. The initial reconstruction can be formulated as:

$$\mathbf{X}^{(0)} = f_{Ini-rec}(\mathbf{Y}), \tag{3}$$

where $f_{Ini-rec}(\cdot)$ represents the overall transformation of the initial reconstruction module. In our experiments, we simply employ 3 convolution layers with kernel size 3×3, and a RELU activation layer follows after each convolution. Then, multiple deep feature hallucination modules (DFHM) with cheap operation and residual connection are stacked to form our backbone architecture, which can hierarchically predict the residual components to reconstruct the latent HSI from coarse to fine. Let $\mathbf{X}_k$ denotes the output of the $k-th$ DFHM module, the $(k+1)-th$ DFHM module with the residual connection aims to learn a more reliable reconstruction of the latent HSI, which is expressed as

$$\mathbf{X}_{k+1} = \mathbf{X}_k + f_{DFHM}(\mathbf{X}_k), \tag{4}$$

where $f_{DFHM}(\cdot)$ denotes the transformation operators in the MFHM module. The MFHM module consists of a spectral hallucination blocks and a spatial context aggregation block, which are implemented to capture sufficient channel of feature maps based on cheap depth-wise convolution operation instead of the vanilla convolution, and is expected to reconstruct more reliable structures in both spatial and spectral directions.

Moreover, we adopt the residual connection in the MFHM module to model only the un-recovered components in the previous module as shown in Fig. 2(a). Next, we would present the detail structure of our proposed DFHM module.

3.3 The DFHM Module

In the HSI reconstruction task from a snapshot image, it require to simultaneously model detail spatial structure and abundant spectral characteristics for reconstructing more plausible HSIs. It is an extreme challenging task to reliably reconstruct the high-dimensional signal in both spectral and spatial dimensions. The existing deep models generally deepen and widen the network architecture to learn large amount of feature maps for boosting the recovering performance, which unavoidably causes large-scale model size and high computational cost. Inspired by the insight that some feature maps in the well-trained networks may be obtained by employing specific transformation operations on the already learned features, we deploy the vanilla convolution layer to learn feature maps with small number of channels (reduced spectral), and then adopt the more cheap depth-wise convolution operation transforming the previously learned ones to obtain more hallucinated spectral information, dubbed as spectral hallucination block (SHB). Moreover, with the concatenated spectral reduced and hallucinated feature maps, we further conduce the depth-wise convolution with various kernel sizes to capture multi-scale spatial context, and then aggregate them as the final feature map, dubbed as spatial context aggregation block (SCAB). Since the SCAB mainly is composed of depth-wise convolution, it also can greatly decrease the parameter compared with vanilla convolution. Finally, a point-wise convolution layer is used to estimate the un-recover residual component in the previous module. To this end, we construct the deep feature hallucination module (DFHM) with a SHB and a SCAB to reciprocally hallucinate more spectral information and spatial structure with various scale of contexts, following a point-wise convolution to achieve the output. The DFHM structure is ahown in Fig. 2(b). Next, we embody the detailed description of the SHB and SCAB.

Spectral Hallucination Block (SHB): Given the reconstructed HSI $\mathbf{X}_k \in \mathbb{R}^{H \times W \times \Lambda}$ at the $k - th$ DFHM module, the DFHM first transforms it to a feature map with C channels: $\mathbf{X}^{(k)} \in \mathbb{R}^{H \times W \times C}$, where large number channel (spectral) should have better representative capability. The SHB aims to further learn deeper representative features with the same channel number. Instead of directly learning the deeper feature with the required spectral number, the SHB first employs a pair of vanilla convolution/RELU layer to obtain a feature map with the reduced spectral channel number $\frac{C}{S}$, and then adopts a set of linear operations on the reduced spectral feature to hallucinate more spectral features. Finally, the hallucinated spectral features by linear operations and the spectral reduced feature have been stacked together as the final learned feature map of the SHB. Specifically, we use the depth-wise convolution, which is much cheaper operation than the vanilla convolution layer, to implement the linear operation. The mathematical formula of the SHB can be expressed as:

$$\mathbf{X}_{RSF}^{(k)} = f_{RSF}(\mathbf{X}^{(k)}), \tag{5}$$

$$\mathbf{X}_{SH}^{(k)} = f_{SH}(\mathbf{X}_{RSF}^{(k)}), \tag{6}$$

$$\mathbf{X}_{SHB}^{(k)} = Concat(\mathbf{X}_{RSF}^{(k)}, \mathbf{X}_{SH}^{(k)}) \tag{7}$$

where $\mathbf{X}_{RSF}^{(k)} \in \mathbb{R}^{H \times W \times \frac{C}{S}}$, $\mathbf{X}_{SH}^{(k)} \in \mathbb{R}^{H \times W \times \frac{C(S-1)}{S}}$ and $\mathbf{X}_{SHB}^{(k)} \in \mathbb{R}^{H \times W \times C}$ represent the spectral reduced feature, the hallucinated spectral feature, and the outputed feature map of the SHB, respectively. $f_{RSF}(\cdot)$ denotes the transformation of a vanilla convolution/RELU layer with the spatial kernel size 3×3 to reduce the spectral channel number from C to $\frac{C}{S}$ while f_{SH} is the transformation of a set of depth-wise convolution layers with the spatial kernel size 3×3.

It should noted if a vanilla convolution with kernel size $d \times d$ is employed to transform the feature map $\mathbf{X}^{(k)}$ into a deeper feature with the same spectral number, the number of the parameters by ignoring the bias term for simplicity would be $C \cdot d \cdot d \cdot C$. While the parameter number in the proposed SHB with the spectral reduced vanilla convolution and the spectral hallucinated depth-wise convolution is $C \cdot d \cdot d \times \frac{C}{S} + d \cdot d \cdot (C - \frac{C}{S})$. Thus, the compression ratio of the parameters with the SHB can be calculated as

$$r_p = \frac{C \cdot d \cdot d \cdot C}{C \cdot d \cdot d \times \frac{C}{S} + d \cdot d \cdot (C - \frac{C}{S})} \approx S \tag{8}$$

Similarly, we can obtain the theoretical speed-up ratio by replacing the vanilla convolution layer with the SHB as S. Therefore, the proposed SHB can not only learn the same amount of feature map but also greatly reduce the parameter number as wells as speed-up the computation.

Table 1. Performance comparisons on the CAVE and Harvard datasets (3% compressive ratio). The best performance is labeled in **bold**, and the second best is labeled in underline.

Method	CAVE			Harvard			Params (MB)	Flops(G)
	PSNR	SSIM	SAM	PSNR	SSIM	SAM		
TwIST	(–)	(–)	(–)	27.16	0.924	0.119	(–)	(–)
3DNSR	(–)	(–)	(–)	28.51	0.940	0.132	(–)	(–)
SSLR	(–)	(–)	(–)	29.68	**0.952**	**0.101**	(–)	(–)
HSCNN [36]	24.94	0.736	0.452	35.09	0.936	0.145	312	87
HyperReconNet [35]	25.18	0.825	0.332	35.94	0.938	0.160	581	152
λ-Net [22]	24.77	0.816	<u>0.314</u>	36.73	0.947	0.141	58247	12321
DeepSSPrior [31]	<u>25.48</u>	<u>0.825</u>	0.324	<u>37.10</u>	0.950	0.137	341	89
Our	**27.44**	**0.830**	**0.302**	**37.26**	<u>0.951</u>	<u>0.133</u>	197	49

Spatial Context Aggregation Block (SCAB): It is known that the reliable spectral recovery of a specific pixel would greatly depend on the around spatial context, and the required spatial range may be changed according to the physical characteristics of the pixel. The ordinary convolution networks usually carry out the context exploitation of the same spatial range for all pixels regardless to the pixel characteristic, which may yield non-optimal spectral reconstruction. To this end, we attempt to learn the feature by exploiting multiple spatial contexts under various receptive fields with the cheap depth-wise convolution operation, and adaptively aggregate them to a compact representation with a point-wise convolution, dubbed as Spatial context aggregation block (SCAB). Given the spectral hallucinated features $\mathbf{X}_{SHB}^{(k)}$ of the SHB, the SCAB firstly adopts a mixed depth-wise convolutional layer (dubbed as MixConv) [28] to adaptively exploit the spatial dependency in different sizes of local spatial regions for parameter reduction. In the detailed implementation, the spectral hallucinated feature map $\bar{\mathbf{X}}_{SHB}^{(k)} \in \mathbb{R}^{H \times W \times C}$ is partitioned into M groups: $\bar{\mathbf{X}}_{SHB}^{(k)} = [\mathbf{X}_1, \mathbf{X}_2, \cdots, \mathbf{X}_M]$ via evenly dividing the channel dimension, where $\mathbf{X}_m \in \mathbb{R}^{H \times W \times L_m}$ ($L_m = L/M$) represents the feature maps in the $m-th$ group. The MixConv layer is deployed to exploit different spatial contexts for different groups via using depth-wise convolution layers. Let's denote the parameter set of the MixConv layer as $\boldsymbol{\Theta}_{Mix}^{(k)} = \{\theta_1, \theta_2, \cdots, \theta_M\}$ in the M group of depth-wise convalution layers, where the parameters for different groups have various spatial kernel sizes for exploring spatial contexts in different local regions with $\theta_m \in \mathbb{R}^{s_m \times s_m \times L_m}$, the MixConv layer is formulated as:

$$\mathbf{X}_{Mix}^{(k)} = Concat(f_{dp}^{\theta_1}(\mathbf{X}_1), f_{dp}^{\theta_2}(\mathbf{X}_2), \cdots, f_{dp}^{\theta_M}(\mathbf{X}_M)), \tag{9}$$

where $f_{dp}^{\theta_m}(\cdot)$ represents the depth-wise convolutional layer with the weight parameter θ_m (for simplicity, we ignore the bias parameters). With the different kernel spatial sizes at different groups, the spatial correlation in various local regions is simultaneously integrated for extracting high representative features in one layer. Moreover, we employ the depth-wise convolution operations in all groups, which can greatly reduces the parameters ($\frac{1}{L_m}$) compared with a vanilla convolution layer for being easily implanted in the real imaging systems, and expect more reliable spatial structure reconstruction via concentrating on spatial context exploration. Finally, a point-wised convolution layer is employed to estimate the residual component of the $k-th$ DFHM module, and is expressed as:

$$\bar{\mathbf{X}}_k = f_{PW}(\mathbf{X}_{Mix}^{(k)}). \tag{10}$$

4 Experimental Results

To demonstrate the effectiveness of our proposed lightweight reconstruction model, we conduct comprehensive experiments on two hyperspectral datasets including the CAVE [38] dataset and the Harvard dataset [7]. The CAVE dataset consists of 32 images with spatial resolution 512 × 512 and 31 spectral channels ranging from 400 nm to 700 nm while the Harvard dataset is composed of 50 outdoor images captured under daylight conditions with the spatial resolution are 1040 × 1392 and the spectral wavelength ranging from 420 nm to 720 nm. In our experiment, we randomly select 20 images in

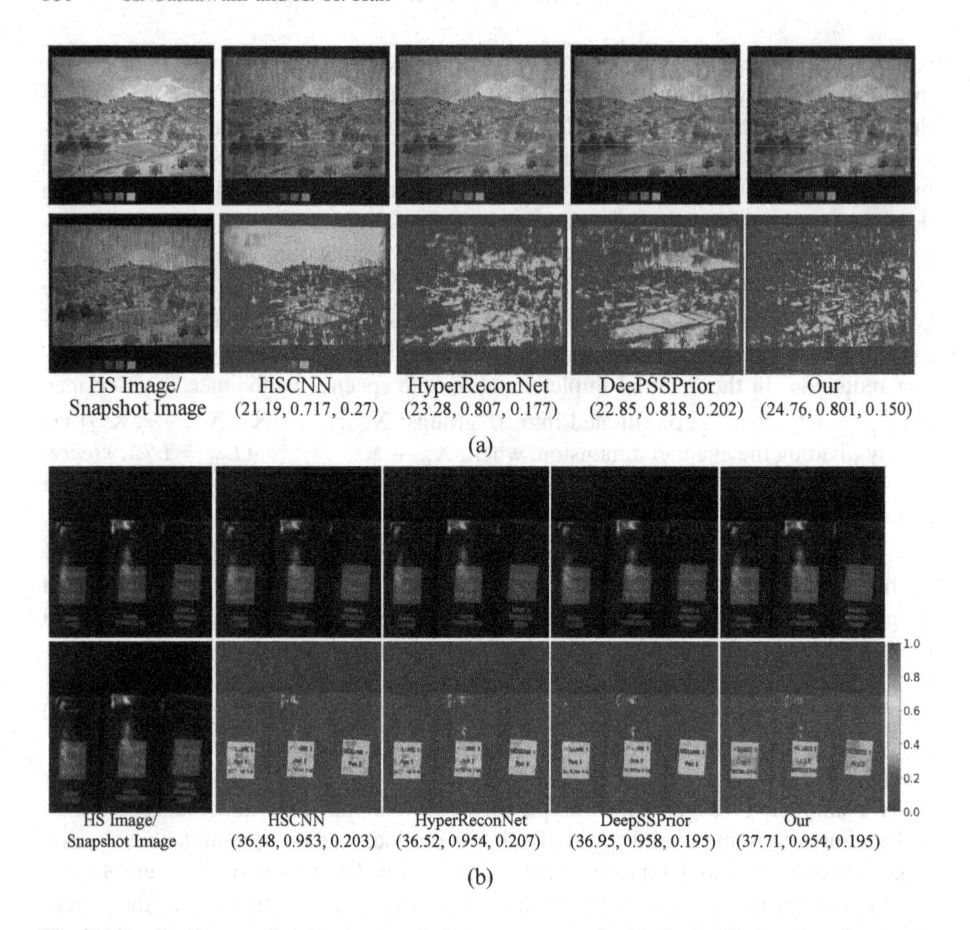

Fig. 3. Visualization results of two example images compared with the SoTA deep learning models: HSCNN [36], HyperReconNet [35], DeepSSPrior [31], and our proposed lightweight models, where the three values under the reconstruction represents the PSNR, SSIM and SAM, respectively.

the CAVE and 10 image in the Harvard dataset as the testing samples and the rest for training. For simulating the 2D snapshot image, we synthesize the transmission function $T(h, w)$ of the coded aperture in the HS imaging system via randomly generating a binary matrix according to a Bernoulli distribution with p = 0.5, and then create the snapshot measurements by transforming the original HSI with the synthesized transformation function. To prepare training samples, we extract the corresponding snapshot/HSI patches with spatial size of 48×48 from the training images. We implement our overall network by stacking 9 DFHM modules following the same number of stages in the conventional deep models: HSCNN [36] and DeepSSPrior [31] for fair comparison in model parameter and computational cost. Moreover, we quantitatively evaluate the HS image reconstruction performance using three metrics including the peak signal-to-noise ratio (PSNR), structural similarity (SSIM) and spectral angle mapper (SAM).

Table 2. Ablation study on the CAVE dataset. The best performance is labeled in **bold**, and the second best is labeled in underline.

Metrics	w/o SCAB ($S = 2$)	w/o SCAB ($S = 3$)	w/o SCAB ($S = 4$)	SHB + SCAB ($S = 2$)
PSNR	26.98	26.63	26.13	27.44
SSIM	0.813	0.805	0.813	0.830
SAM	0.307	0.324	0.327	0.302
Parameter(MB)	40	35	32	42

Comparison with the SoTA Methods: We compare our proposed method with several state-of-the-art HSI reconstruction methods, including three traditional methods with hand-crafted prior modeling, *i.e*, TwIST with TV prior [40], 3DNSR and SSLR with NLS prior [11,30], and four deep learning-based methods, *i.e.*, HSCNN [36], HyperReconNet [35], λ-net [22] and Deep Spatial Spectral Prior (DeepSSPrior) [31]. Our lightweight model was implemented by stacking 9 MFHM modules with the compression ratio: $S = 2$ (in the SHB). The compared quantitative results are illustrated in Table 1, which verifies that our proposed lightweight models can not only achieve promising reconstruction performance but also greatly reduce the parameter as well as computational cost. Moreover, we also provide the compared visualization results of our lightweight model with the HSCNN [36], HyperReconNet [35] and DeepSSPrior [31] in Fig. 3, which also demonstrated the better reconstruction performance by our method.

Ablation Study: As introduced in Sect. 2, we compress the spectral channel from C to $\frac{C}{S}$, and then hallucinate more spectral features using cheap depth-wise convolution operation in the SHB, where the hyper-parameter S can be adjusted according to the compression ratio of the parameter in the SHB. What is more, the following SCAB is incorporated for exploiting multi-scale spatial contexts with cheap operation, which can be plugged in or ignored in the DFHM module. To verify the effect of the compression ratio S and the additional SCAB, we carried out experiments by varying S from 2 to 4, and w/o the incorporation of the SCAB on the CAVE dataset. The ablation results are shown in Table 2. From Table 2, we observe that the compressive ratio 2 achieves the best performance while increasing the compression ratio will yield a little performance drop but with smaller parameter number. Moreover, the incorporation of the SCAB can further boost the reconstruction performance, whilst causes few parameter raising.

5 Conclusions

This study proposed a novel lightweight model for efficiently and effectively reconstruct a full hyperspectral image from a compressive snapshot measurement. Although the existing deep learning based models have achieved remarkable performance improvement compared with the traditional model-based methods for hyperspectral image reconstruction, it still confronts the difficulties to embed the deep models in real HSI systems due to large-scale model size. To this end, we exploited an efficient deep feature hallucination module (DFHM) to construct our lightweight models. Specifically, we elaborated the DFHM by a vanilla convolution-based spectral reduced layer and a depth-wise convolution-based spectral hallucination layer to learn sufficient feature maps with cheap operation. Moreover, we further incorporated a spatial context aggregation block to exploit multi-scale context in various receptive fields for boosting reconstruction performance. Experiments on two datasets demonstrated that our proposed method achieved superior performance over the SoTA models as well as greatly reduced the parameters and computational cost.

Acknowledgements. This work was supported in part by MEXT under the Grant No. 20K11867, and JSPS KAKENHI Grant Number JP12345678.

References

1. Arce, G., Brady, D., Carin, L., Arguello, H., Kittle, D.: Compressive coded aperture spectral imaging: An introduction. IEEE Signal Process. Mag. **31**, 105–115 (2014)
2. Bioucas-Dias, J., Figueiredo, M.A.T.: A new twist: Two-step iterative shrinkage/thresholding algorithms for image restoration. IEEE Trans. Image Process. **16**, 2992–3004 (2007)
3. Bjorgan, A., Randeberg, L.L.: Towards real-time medical diagnostics using hyperspectral imaging technology (2015)
4. Borengasser, M., Hungate, W.S., Watkins, R.: Hyperspectral Remote Sensing: Principles and Applications. CRC Press, Boca Raton (2007)
5. Cao, X., Du, H., Tong, X., Dai, Q., Lin, S.: A prism-mask system for multispectral video acquisition. IEEE Trans. Pattern Anal. Mach. Intell. **33**, 2423–2435 (2011)
6. Cao, X., Yue, T., Lin, X., Lin, S., Yuan, X., Dai, Q., Carin, L., Brady, D.: Computational snapshot multispectral cameras: toward dynamic capture of the spectral world. IEEE Signal Process. Mag. **33**, 95–108 (2016)
7. Chakrabarti, A., Zickler, T.E.: Statistics of real-world hyperspectral images. In: CVPR, pp. 193–200 (2011)
8. Choi, I., Jeon, D.S., Nam, G., Gutierrez, D., Kim, M.H.: High-quality hyperspectral reconstruction using a spectral prior. ACM Trans. Graphics **36**, 1–13 (2017)
9. Cui, Q., Park, J., Smith, R.T., Gao, L.: Snapshot hyperspectral light field imaging using image mapping spectrometry. Opt. Lett. **45**(3), 772–775 (2020)
10. Devassy, B.M., George, S.: Forensic analysis of beverage stains using hyperspectral imaging. Sci. Rep. **11**, 6512 (2021)
11. Fu, Y., Zheng, Y., Sato, I., Sato, Y.: Exploiting spectral-spatial correlation for coded hyperspectral image restoration. In: 2016 IEEE Conference on Computer Vision and Pattern Recognition (CVPR), pp. 3727–3736 (2016)
12. Gao, L., Kester, R., Hagen, N., Tkaczyk, T.: Snapshot image mapping spectrometer (IMS) with high sampling density for hyperspectral microscopy. Opt. Express **18**, 14330–14344 (2010)
13. Gehm, M., John, R., Brady, D., Willett, R., Schulz, T.: Single-shot compressive spectral imaging with a dual-disperser architecture. Opt. Express **1521**, 14013–27 (2007)
14. Goetz, A., Vane, G., Solomon, J., Rock, B.: Imaging spectrometry for earth remote sensing. Science **228**, 1147–1153 (1985)
15. Gregor, K., LeCun, Y.: Learning fast approximations of sparse coding (2010)
16. Han, K., Wang, Y., Tian, Q., Guo, J., Xu, C., Xu, C.: Ghostnet: More features from cheap operations (2019)

17. Kittle, D.S., Choi, K., Wagadarikar, A.A., Brady, D.J.: Multiframe image estimation for coded aperture snapshot spectral imagers. Appl. Opt. **49**(36), 6824–33 (2010)
18. Lin, X., Liu, Y., Wu, J., Dai, Q.: Spatial-spectral encoded compressive hyperspectral imaging. ACM Trans. Graph. **33**, 233:1–233:11 (2014)
19. Liu, Y., Yuan, X., Suo, J.L., Brady, D., Dai, Q.: Rank minimization for snapshot compressive imaging. IEEE Trans. Pattern Anal. Mach. Intell. **41**, 2990–3006 (2019)
20. Lu, G., Fei, B.: Medical hyperspectral imaging: a review. J. Biomed. Opt. **19**, 10901 (2014)
21. Ma, J., Liu, X.Y., Shou, Z., Yuan, X.: Deep tensor admm-net for snapshot compressive imaging. In: 2019 IEEE/CVF International Conference on Computer Vision (ICCV), pp. 10222–10231 (2019)
22. Miao, X., Yuan, X., Pu, Y., Athitsos, V.: lambda-net: Reconstruct hyperspectral images from a snapshot measurement. In: 2019 IEEE/CVF International Conference on Computer Vision (ICCV), pp. 4058–4068 (2019)
23. Nguyen, H.V., Banerjee, A., Chellappa, R.: Tracking via object reflectance using a hyperspectral video camera. In: IEEE Computer Vision and Pattern Recognition Workshops, pp. 44–51 (2010)
24. Pan, Z., Healey, G., Prasad, M., Tromberg, B.: Face recognition in hyperspectral images. IEEE Trans. Pattern Anal. Mach. Intell. **25**, 1552–1560 (2003)
25. Rajwade, A., Kittle, D.S., Tsai, T.H., Brady, D.J., Carin, L.: Coded hyperspectral imaging and blind compressive sensing. SIAM J. Imaging Sci. **6**, 782–812 (2013)
26. Schechner, Y., Nayar, S.: Generalized mosaicing: wide field of view multispectral imaging. IEEE Trans. Pattern Anal. Mach. Intell. **24**, 1334–1348 (2002)
27. Tan, J., Ma, Y., Rueda, H., Baron, D., Arce, G.: Compressive hyperspectral imaging via approximate message passing. IEEE J. Select. Topics Signal Process. **10**, 389–401 (2016)
28. Tan, M., Le, Q.V.: MixcCnv: mixed depthwise convolutional kernels (2019)
29. Wagadarikar, A.A., John, R., Willett, R.M., Brady, D.J.: Single disperser design for coded aperture snapshot spectral imaging. Appl. Opt. **4710**, B44-51 (2008)
30. Wang, L., Xiong, Z., Shi, G., Wu, F., Zeng, W.: Adaptive nonlocal sparse representation for dual-camera compressive hyperspectral imaging. IEEE Trans. Pattern Anal. Mach. Intell. **39**, 2104–2111 (2017)
31. Wang, L., Sun, C., Fu, Y., Kim, M.H., Huang, H.: Hyperspectral image reconstruction using a deep spatial-spectral prior. In: 2019 IEEE/CVF Conference on Computer Vision and Pattern Recognition (CVPR), pp. 8024–8033 (2019)
32. Wang, L., Sun, C., Zhang, M., Fu, Y., Huang, H.: DNU: deep non-local unrolling for computational spectral imaging. In: 2020 IEEE/CVF Conference on Computer Vision and Pattern Recognition (CVPR), pp. 1658–1668 (2020)
33. Wang, L., Xiong, Z., Gao, D., Shi, G., Wu, F.J.: Dual-camera design for coded aperture snapshot spectral imaging. Appl. Opt. **54**(4), 848–58 (2015)
34. Wang, L., Xiong, Z., Gao, D., Shi, G., Zeng, W., Wu, F.: High-speed hyperspectral video acquisition with a dual-camera architecture. In: 2015 IEEE Conference on Computer Vision and Pattern Recognition (CVPR), pp. 4942–4950 (2015)
35. Wang, L., Zhang, T., Fu, Y., Huang, H.: Hyperreconnet: Joint coded aperture optimization and image reconstruction for compressive hyperspectral imaging. IEEE Trans. Image Process. **28**, 2257–2270 (2019)
36. Xiong, Z., Shi, Z., Li, H., Wang, L., Liu, D., Wu, F.: HSCNN: CNN-based hyperspectral image recovery from spectrally undersampled projections. In: 2017 IEEE International Conference on Computer Vision Workshops (ICCVW), pp. 518–525 (2017)
37. Yang, Y., Sun, J., Li, H., Xu, Z.: ADMM-CSNet: a deep learning approach for image compressive sensing. IEEE Trans. Pattern Anal. Mach. Intell. **42**, 521–538 (2020)

38. Yasuma, F., Mitsunaga, T., Iso, D., Nayar, S.: Generalized assorted pixel camera: Postcapture control of resolution, dynamic range, and spectrum. IEEE Trans. Image Process. **19**, 2241–2253 (2010)
39. Yuan, X., Tsai, T., Zhu, R., Llull, P., Brady, D., Carin, L.: Compressive hyperspectral imaging with side information. IEEE J. Select. Topics Signal Process. **9**, 964–976 (2015)
40. Yuan, X.: Generalized alternating projection based total variation minimization for compressive sensing. In: 2016 IEEE International Conference on Image Processing (ICIP), pp. 2539–2543 (2016)
41. Zhang, J., Ghanem, B.: ISTA-Net: interpretable optimization-inspired deep network for image compressive sensing. In: 2018 IEEE/CVF Conference on Computer Vision and Pattern Recognition, pp. 1828–1837 (2018)
42. Zhang, S., Wang, L., Fu, Y., Zhong, X., Huang, H.: Computational hyperspectral imaging based on dimension-discriminative low-rank tensor recovery. In: 2019 IEEE/CVF International Conference on Computer Vision (ICCV), pp. 10182–10191 (2019)
43. Zhang, X., Lian, Q., Yang, Y.C., Su, Y.: A deep unrolling network inspired by total variation for compressed sensing MRI. Digit. Signal Process. **107**, 102856 (2020)
44. Zhou, S., He, Y., Liu, Y., Li, C.: Multi-channel deep networks for block-based image compressive sensing. ArXiv abs/1908.11221 (2019)

A Transformer-Based Model for Preoperative Early Recurrence Prediction of Hepatocellular Carcinoma with Muti-modality MRI

Gan Zhan[1](✉), Fang Wang[2], Weibin Wang[1], Yinhao Li[1], Qingqing Chen[2], Hongjie Hu[2], and Yen-Wei Chen[1]

[1] College of Information Science and Engineering, Ritsumeikan University, Kyoto, Japan
{gr0502vs,gr0342he}@ed.ritsumei.ac.jp, yin-li@fc.ritsumei.ac.jp, chen@is.ritsumei.ac.jp

[2] Department of Radiology, Sir Run Run Shaw Hospital, Zhejiang University School of Medicine, Hangzhou, China
{wangfang11,qingqingchen,hongjiehu}@zju.edu.cn

Abstract. Hepatocellular carcinoma (HCC) is the most common primary liver cancer which accounts for a high mortality rate in clinical, and the most effective treatment for HCC is surgical resection. However, patients with HCC are still at a huge risk of recurrence after tumor resection. In this light, preoperative early recurrence prediction methods are necessary to guide physicians to develop an individualized preoperative treatment plan and postoperative follow-up, thus prolonging the survival time of patients. Nevertheless, existing methods based on clinical data neglect information on the image modality; existing methods based on radiomics are limited by the ability of its predefined features compared with deep learning methods; and existing methods based on CT scans are constrained by the inability to capture the details of images compared with MRI. With these observations, we propose a deep learning transformer-based model on multi-modality MRI to tackle the preoperative early recurrence prediction task of HCC. Enlightened by the vigorous capacity of context modeling of the transformer architecture, our proposed model exploits it to dig out the inter-modality correlations, and the performance significantly improves. Our experimental results reveal that our transformer-based model can achieve better performance than other state-of-the-art existing methods.

Keywords: HCC early recurrence prediction · Multi-modality MRI · Transformer

G. Zhan and F. Wang—First authors.

Y. Zheng et al. (Eds.): ACCV 2022, LNCS 13848, pp. 185–194, 2023.
https://doi.org/10.1007/978-3-031-27066-6_13

1 Introduction

Hepatocellular carcinoma, which is also known as HCC, is one of the primary liver cancer that accounts for a high mortality rate in clinical. It is the fifth most common malignancy and leads to the second most common death related to cancer around the world [1], especially in East Asia and sub-Saharan Africa, HCC occupies 82% of liver cancer cases [2]. There are many mature treatment choices for patients with HCC including liver transplantation, targeted therapy, immunotherapy, transarterial chemoembolization, surgical resection, and radiofrequency ablation. Among these treatment choices, surgical resection is the first-line treatment choice and is widely recommended by the clinical practice guidelines for patients with a well-preserved liver function [3–5]. However, patients with HCC are still facing a huge risk of recurrence after surgery, the recurrence rate can reach about >10% in 1 year, 70–80% in 5 years after tumor resection [6]. And the overall survival time varies in patients with early recurrence and late recurrence, patients with late recurrence tend to live longer than patients with early recurrence [7]. Therefore, it is vital to identify those HCC patients at high risk of early recurrence, which can guide physicians to develop a preoperative individualized treatment plan and postoperative follow-up, thus prolonging the survival time of patients.

So far, there are many existing methods [8,10,14,25] being proposed for preoperative HCC early recurrence prediction after tumor resection; existing methods based on clinical data mainly utilize machine learning algorithms to project quantitative and qualitative variables to the prediction space. For example, Chuanli Liu et al. [8] construct 5 machine learning algorithms based on 37 patients' characteristics, and they select K-Nearest Neighbor as their optimal algorithm for HCC early recurrence prediction task after comparison. The biggest flaw in clinical-based methods is it provides limited clinical features, and it lacks the medical image information which could well reflect the heterogeneity of the tumor. Besides those subjective assessments of the lesion provided by physicians are irreproducible due to human bias, thus the performance of prediction has bad generalization. Radiomics is brought up by Gillies et al. [9] as a quantitative tool for feature extraction from medical images, it can save time for physicians on repetitive tasks and facilitate a non-invasive tool for feature extraction. Existing methods based on radiomics provide the image feature that reveals the heterogeneity of the tumor across HCC patients, and it could achieve promising results. For example, Ying Zhao et al. [10] extract 1146 radiomics features from each image phase, and radiomics models of each phase and their combination are constructed by the multivariate logistic regression algorithm to well complete the HCC early recurrence prediction task. The drawback of radiomics-based methods is that radiomics features are predefined, it does not directly serve our prediction task, and thus the performance of image information to our prediction task could not be fully tapped.

Recently, deep learning has yielded brilliant performance in the medical image area compared with conventional approaches [11]. The reason behind these successes is that deep learning can automatically explore robust and generalized image features directly related to the task without human intervention. And this process is mainly conducted by the convolutional neural networks [12], through the hierarchical structure, the convolutional neural network continuously combines low-level pixel features to obtain the ultimate high-level semantic feature that classification tasks demand. The disadvantage of deep learning methods is it needs a huge amount of data, which is normally not feasible for medical images. But with the technology of finetune [13], we now can also exploit the performance of neural networks on small sample medical datasets. Deep learning has been applied to the HCC early recurrence prediction task, Weibin Wang et al. [14] proposed the Phase Attention prediction model based on multi-phase CT (computed tomography) scans. But certain study [15] has shown that MRI(magnetic resonance imaging) is superior to CT scans in the detection of HCC given the preponderance of capturing the image details, especially the soft tissue of the tumor. Thus utilizing MRI has the advantage of image information over CT scans in our prediction task.

With these observations, we aim to develop a deep learning method based on MRI to complete our early recurrence prediction task on HCC. Compared with single-modality MRI, multi-modality MRI can observe the tumor and analyze the internal composition of the tumor in a more comprehensive way, which is more conducive to judging the degree of malignancy [16], thus we conduct our study on multi-modality MRI. Considering the unique characteristics of each modality in our prediction task, how to well combine them is the biggest challenge in our task. Enlightened by the vigorous capacity of context modeling of transformer model [17], we formulate each modality image feature as a token of the sequence, then we utilize transformer architecture to dig out the inter-modality correlations. Our contributions are as follows: (1) We propose a transformer-based model for the preoperative early recurrence prediction of HCC, which can effectively use MRI images for computer-aided diagnosis. (2) Our model utilizes the transformer architecture to efficiently combine image features from multiple MRI modality, and detailed experimental results show that it achieves superior performance to other existing methods. To the best of our knowledge, we are the first to propose a deep learning model based on multi-modality MRI to tackle the early recurrence prediction task for HCC.

2 Proposed Method

We propose an end-to-end deep learning method based on multi-modality MRI to tackle our early recurrence prediction task. Since portal venous (PV) phase and arterial (ART) phase are helpful for physicians to judge benign and malignant tumors, diffusion-weighted with $b = 1000\,\mathrm{s/mm^2}$ (DWI1) is useful to judge the water content and hemorrhage and necrosis of tumors, and outline of tumors is generally clearer on axial T2-weighted imaging with fat suppression(T2) [18,19]. We select them as our research subjects, and we propose our model based on these 3 modalities (PV phase and ART phase belong to the same modality).

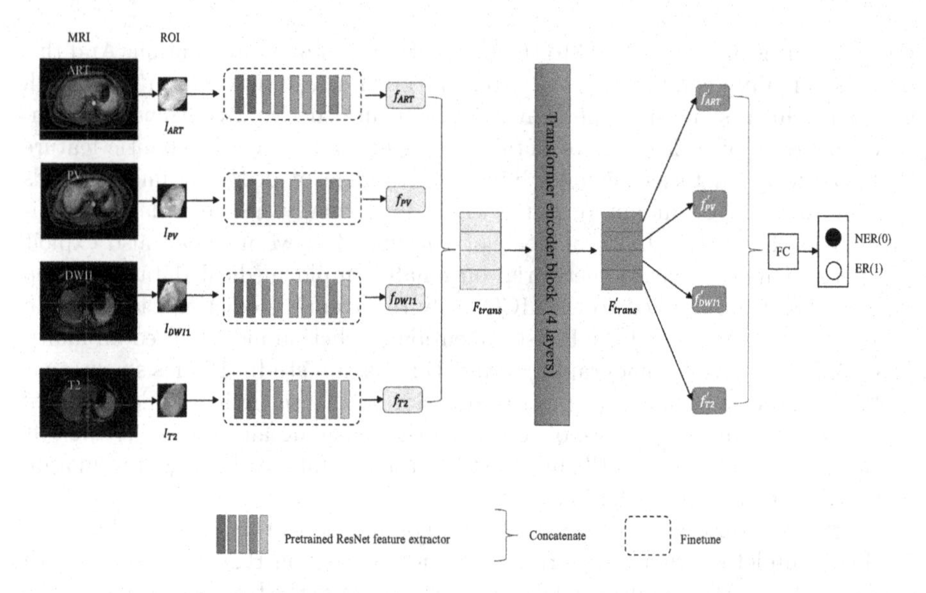

Fig. 1. The overall pipeline of our transformer-based model on multi-modality MRI. Image features are obtained through pre-trained ResNet feature extractor and transformer encoder block to complete our prediction task.

Figure 1 shows the overall pipeline of our proposed method. We feed the ROI of each modality into a pre-trained ResNet feature extractor and transformer encoder block to obtain the demand modality image feature, and then these image features will be combined and projected to the prediction space by an fully connected (FC) layer. In the following sections, we will introduce our prediction model in detail in terms of the pre-trained ResNet feature extractor and transformer encoder block.

2.1 Pre-trained ResNet Feature Extractor

ResNet [20] designed to solve the degradation problem of deep neural networks has achieved remarkable success in computer vision, so we select it to construct our modality image feature extractor. Since the last FC layer in ResNet serves as the classifier, we remove it to obtain our ResNet feature extractor for modality image feature extraction. Considering the small sample set in our study, we specifically select the 18-layer ResNet (ResNet18) pre-trained in ImageNet [21] as our feature extraction backbone, ResNet18 has few parameters, so it better fit our dataset size, and with the finetune technology, we can give our modality image feature extractor a very good initialization point.

Considering the pre-trained ResNet is originally designed and trained for natural images, which well accept input image with shape of 224 × 224 × 3 (both width and height are 224, number of channels is 3). For each modality, we selected the slice which has the largest tumor area and its two adjacent slices from each MRI modality, and we crop the ROI of the tumor region from these 3 slices, which can well represent the HCC charateristic of this input MRI modality volume. After we obtain this 3-channel ROI, we resize its width and height to 224×224, to obtain the input of 3 modalities $\{I_{PV}, I_{ART}, I_{DWI1}, I_{T2}\}$, then we feed them to our pre-trained ResNet feature extractor. After pre-trained ResNet feature extractor, the output features in the ART phase are denoted as f_{ART}, so are the f_{PV} in PV phase, f_{DWI1} in DWI1 modality and f_{T2} in T2 modality, and shape of each modality image feature, that we obtained before transformer encoder block, are all $1 \times 1 \times 512$.

2.2 Transformer Encoder Block

Transformer model [17] was first designed for natural language processing tasks, which accept tokens in the sequence and utilize the self-attention mechanism (Eq. 1) to dig out the inter-token correlations. Transformer has been widely used in computer vision, for example, images are divided into a collection of patches in vision transformer [22], and each patch could be served as the token in the image sequence, then the semantic of this image could be formulated as the composition of tokens and token grammar. Inspired by these intuitions, we formulate each modality image feature obtained in the pre-trained ResNet feature extractor as a token in the modality sequence, and inter-modality correlations serve as the modality grammar for our prediction task. We construct our transformer encoder block with 4 transformer encoder layers, and we first concatenate the 3 modality image features $\{f_{ART}, f_{PV}, f_{DWI1}, f_{T2}\}$ to obtain F_{trans}, then through linear projections, the concatenated modality image features are mapped to the key(K) vector, query(Q) vector and value(V) vector in self-attention described below:

$$Attention(Q, K, V) = Softmax(\frac{QK^T}{\sqrt{d_k}})V \tag{1}$$

Then the inter-modality correlations can be calculated by the dot product between Q vector and K^T vector, after the softmax operation scale the attention score value to 0–1, it will multiply the V, that is to encode the modality grammar into modality tokens, due to the large values of dk, the dot products grow large in magnitude, pushing the softmax function into regions where it has extremely small gradients, so we scale the dot products by $\frac{1}{\sqrt{d_k}}$, hereby we obtain the image feature F'_{trans}, and we split it by the token dimension to obtain $\{f'_{ART}, f'_{PV}, f'_{DWI1}, f'_{T2}\}$ as the ultimate image features. And finally, we concatenate these image features by the channel dimension and utilize one FC layer to complete our prediction task.

Table 1. Dataset arrangement of 5-fold cross-validation.

Fold-1	Fold-2	Fold-3	Fold-4	Fold-5	Total
57	58	58	58	58	289

3 Experiments

3.1 Patient Selection

From 2012 to 2019, 659 HCC patients from Run Run Shaw Hospital Affiliated to Medical College of Zhejiang University, who has undergone liver resection, pathologically confirmed as hepatocellular carcinoma and received enhanced MRI examination before surgery were recruited in this retrospective study. Under the following exclusion criteria: (1) patient received other anti-tumor treatments before surgery, such as TACE, RFA; (2) the interval between preoperative MR examination and surgery is more than 30 days; (3) image quality is not good, and (4) less than 2 years of follow-up after surgery; a total of 289 patients are included in our study. Patients who have a recurrence in 2 years are denoted as early recurrence (ER) [23], and patients who have a recurrence of more than 2 years or no recurrence are denoted as Non-early recurrence (NER) in our study.

3.2 Dataset Preparation and Metrics

For the MRI data, considering the affection of different types of artifacts to the MRI, we mainly use the preprocessing method described by Jose Vicente Manjon [24] on our multi-modality MRI, which includes denoise, bias field correction, resampling and normalization.

To fairly measure the performance of our proposed method, we use the 5-fold cross-validation on our 289 patients, we split it into 5 groups along the timeline, the arrangement of dataset is in Table 1.

We select AUC (area under the ROC curve), ACC (accuracy), SEN (sensitivity), SPE (specificity), PPV (positive predictive value), and NPV (negative predictive value) to comprehensively evaluate the model performance, among which, AUC, ACC, and SEN are the most important indicator in our task, especially SEN, we normally use SEN to validate the model's ability on identifying patients, and it is of great clinical value in our research if our model could identify more positive patients. Since we conduct 5-fold cross-validation, we calculate the average value of each metric as the indicator value; Cross-entropy loss is our model's loss function, and for every fold, we train 50 epochs, and we select the checkpoint that has the minimal training loss as our trained model.

3.3 Ablation Study

Selecting early fusion or late fusion as our baseline method for multiple modality image inputs is controversial, thus it is necessary to do this ablation on our task.

Table 2. Ablation on baseline method.

Model	AUC	ACC	SEN	SPE	PPV	NPV
Early fusion	0.6447	0.6469	0.4031	0.8101	0.5777	0.6926
Late fusion	**0.6829**	**0.6365**	**0.4307**	**0.7639**	**0.5237**	**0.6900**

Table 3. Ablation on transformer encoder block.

Model	Transformer	AUC	ACC	SEN	SPE	PPV	NPV
Baseline		0.6829	0.6365	0.4307	0.7639	0.5237	0.6900
Trans-based	✓	**0.6907**	**0.6782**	**0.4360**	**0.8296**	**0.5944**	**0.7117**

The comparison result between early fusion and late fusion is shown in Table 2. For the early fusion, we simply concatenate the 3 modality ROI images by the channel dimension, and feed it into a single pre-trained ResNet18 model, since the new input has 12 channels, we create a new convolution layer that would accept this input to replace the first convolution in Pre-trained ResNet18. And to fit our binary task, we replace the last FC layer which outputs 1000 neurons with a new FC layer outputting 1 neuron, and we will use the value 0.5 as the threshold to do the binary prediction. For the late fusion, we utilized 4 pre-trained ResNet feature extractors to extract image features from each modality, then we simply concatenate these 3 modality image features and utilize 1 FC layer to complete our binary prediction task.

We can see in Table 2 that the late fusion model is more appropriate for our task. In the indicators AUC and SEN, which we care about most, the late fusion model obtains a large gain compared with the early fusion model, so we select it as our baseline method. And based on the late fusion model, we add a transformer encoder block to construct our transformer-base model, to measure the effectiveness of our transformer encoder block, we also conduct a detailed ablation study, the results of which are shown in Table 3.

As we can see, when we dig out the inter-modality correlations using the transformer encoder block, our transformer-based model (denoted as Trans-based in Table 3) can achieve better performance on all metrics, which strongly proves the effectiveness of our transformer encoder block, and it is indeed necessary to dig out the inter-modality correlations in our study.

3.4 Comparison with Existing Methods

To further prove the effectiveness of our proposed method, we compare our proposed method with existing methods on the HCC early recurrence prediction task, including the radiomics-based method and deep learning-based methods. The comparison result is shown in Table 4.

Table 4. Comparison with existing methods

Model	AUC	ACC	SEN	SPE	PPV	NPV
Radiomics [10]	0.6861	0.6364	0.3337	0.8252	0.5502	0.6721
PhaseNet [25]	0.6450	0.6643	0.3148	0.8770	0.6068	0.6820
DPANet [14]	0.6818	0.6711	0.4308	0.8105	0.5817	0.7075
Trans-based	**0.6907**	**0.6782**	**0.4360**	**0.8296**	**0.5944**	**0.7117**

We can see from the Table 4, established on the multi-modality MRI only, our transformer-based model can achieve better performance than radiomics-based method (denoted as Radiomics in Table 4) and deep learning method (PhaseNet [25] and DPANet [14]).

4 Conclusion

In this paper, we construct our transformer-based model on the multi-modality MRI to tackle the preoperative early recurrence prediction of HCC patients. Our proposed model formulates each modality image feature as a token in the sequence and obtains the modality correlations related to our prediction task by the transformer encoder block. Detailed experiments reveal the effectiveness of our proposed method and it could achieve better performance compared with existing methods, including state-of-the-art radiomics-based method and deep learning-based method.

Acknowledgments. This work was supported in part by the Grant in Aid for Scientific Research from the Japanese Ministry for Education, Science, Culture and Sports (MEXT) under the Grant Nos. 20KK0234, 21H03470, and 20K21821, and in part by the Natural Science Foundation of Zhejiang Province (LZ22F020012), in part by Major Scientific Research Project of Zhejiang Lab (2020ND8AD01), and in part by the National Natural Science Foundation of China (82071988), the Key Research and Development Program of Zhejiang Province (2019C03064), the Program Co-sponsored by Province and Ministry (No. WKJ-ZJ-1926) and the Special Fund for Basic Scientific Research Business Expenses of Zhejiang University (No. 2021FZZX003-02-17).

References

1. Elsayes, K.M., Kielar, A.Z., Agrons, M.M.: Liver imaging reporting and data system: an expert consensus statement. J. Hepatocel. Carcinoma **4**, 29–39 (2017)
2. Zhu, R.X., Seto, W.K., Lai, C.L.: Epidemiology of hepatocellular carcinoma in the Asia-Pacific region. Gut Liver **10**, 332–339 (2016)
3. Thomas, M.B., Zhu, A.X.: Hepatocellular carcinoma: the need for progress. J. Clin. Oncol. **23**, 2892–2899 (2005)
4. association, E.: EASL clinical practice guidelines: management of hepatocellular carcinoma. J. Hepatol. **69**, 182–236 (2018)

5. Marrero, J.A., et al.: Diagnosis, staging and management of hepatocellular carcinoma: 2018 practice guidance by the American Association for the study of liver diseases. Hepatology **68**, 723–750 (2018)
6. Bray, F., Ferlay, J., Soerjomataram, I., Siegel, R.L., Torre, L.A., Jemal, A.: Global cancer statistics 2018: Globocan estimates of incidence and mortality worldwide for 36 cancers in 185 countries. CA Cancer J. Clin. **68**, 394–424 (2018)
7. Cheng, Z., Yang, P., Qu, S.: Risk factors and management for early and late intrahepatic recurrence of solitary hepatocellular carcinoma after curative resection. HPB **17**, 422–427 (2015)
8. Liul, C., Yang, H., Feng, Y.: A k-nearest neighbor model to predict early recurrence of hepatocellular carcinoma after resection. J. Clin. Translat. Hepatol. **10**, 600–607 (2022)
9. Gillies, R., Kinahan, P.E., Hricak, H.: Radiomics: images are more than pictures, they are data. radiology. Radiology **278**, 563–577 (2015)
10. Zhao, Y., Wu, J., Zhang, Q.: Radiomics analysis based on multiparametric MRI for predicting early recurrence in hepatocellular carcinoma after partial hepatectomy. J. Magn. Reson. Imaging **53**, 1066–1079 (2021)
11. Litjens, G., Kooi, T., Bejnordi, B.E.: A survey on deep learning in medical image analysis. Med. Image Anal. **42**, 60–88 (2017)
12. Krizhevsky, A., Sutskever, I., HintonImagenet, G.E.: ImageNet classification with deep convolutional neural networks. Communications ACM **60**, 84–90 (2017)
13. Kolesnikov, A., Beyer, L., Zhai, X., Puigcerver, J., Yung, J., Gelly, S., Houlsby, N.: Big Transfer (BiT): general visual representation learning. In: Vedaldi, A., Bischof, H., Brox, T., Frahm, J.-M. (eds.) ECCV 2020. LNCS, vol. 12350, pp. 491–507. Springer, Cham (2020). https://doi.org/10.1007/978-3-030-58558-7_29
14. Wang, W., et al.: Phase attention model for prediction of early recurrence of hepatocellular carcinoma with multi-phase CT images and clinical data. Front. Radiol. **8** (2022)
15. Burre, M., et al.: MRI angiography is superior to helical CT for detection of HCC prior to liver transplantation: an explant correlation. Hepatology **38**, 1034–1042 (2003)
16. Armbruster, M., et al.: Measuring HCC tumor size in MRI-the sequence matters! Diagnostics **11** (2002)
17. Vaswan, A., et al.: Attention is all you need. In: 31st Conference on Neural Information Processing Systems (NIPS 2017), Long Beach, CA, USA (2017)
18. Lee, Y., et al.: Benign versus malignant soft-tissue tumors: differentiation with 3t magnetic resonance image textural analysis including diffusion-weighted imaging. Investig. Magn. Resonance Imaging **25**, 118–128 (2021)
19. Chartampilas, E., Rafailidis, V., Georgopoulou, V., Kalarakis, G., Hatzidakis, A., Prassopoulos, P.: Current imaging diagnosis of hepatocellular carcinoma. Cancers **14**, 3997 (2022)
20. He, K., Zhang, X., Ren, S.: Deep residual learning for image recognition. In: Proceedings of the IEEE Conference on Computer Vision and Pattern Recognition, pp. 770–778 (2016)
21. Deng, J., Dong, W., Socher, R., Li, L.-J., Li, K., Li, F.-F.: ImageNet: a large-scale hierarchical image database. In: IEEE Conference on Computer Vision and Pattern Recognition, pp. 248–255 (2009)
22. Dosovitskiy, A., et al.: An image is worth 16x16 words: transformers for image recognition at scale. arXiv preprint arXiv:2010.11929 (2020)

23. Xing, H., Zhang, W.G., Cescon, M.: Defining and predicting early recurrence after liver resection of hepatocellular carcinoma: a multi-institutional study. HPB **22**, 677–689 (2020)
24. Manjon, J.V.: MRI preprocessing. In: Imaging Biomarkers, pp. 53–63 (2017)
25. Wang, W., et al.: Deep learning-based radiomics models for early recurrence prediction of hepatocellular carcinoma with multi-phase CT images and clinical data. In: 2019 41st Annual International Conference of the IEEE Engineering in Medicine and Biology Society (EMBC), pp. 4881–4884 (2019)

CaltechFN: Distorted and Partially Occluded Digits

Patrick Rim[1(✉)], Snigdha Saha[1], and Marcus Rim[2]

[1] California Institute of Technology, Pasadena, CA 91125, USA
{patrick,snigdha}@caltech.edu
[2] Vanderbilt University, Nashville, TN 37235, USA
marcus.g.rim@vanderbilt.edu

Abstract. Digit datasets are widely used as compact, generalizable benchmarks for novel computer vision models. However, modern deep learning architectures have surpassed the human performance benchmarks on existing digit datasets, given that these datasets contain digits that have limited variability. In this paper, we introduce Caltech Football Numbers (CaltechFN), an image dataset of highly variable American football digits that aims to serve as a more difficult state-of-the-art benchmark for classification and detection tasks. Currently, CaltechFN contains 61,728 images with 264,572 labeled digits. Given the many different ways that digits on American football jerseys can be distorted and partially occluded in a live-action capture, we find that in comparison to humans, current computer vision models struggle to classify and detect the digits in our dataset. By comparing the performance of the latest task-specific models on CaltechFN and on an existing digit dataset, we show that our dataset indeed presents a far more difficult set of digits and that models trained on it still demonstrate high cross-dataset generalization. We also provide human performance benchmarks for our dataset to demonstrate the current gap between the abilities of humans and computers in the tasks of classifying and detecting the digits in our dataset. Finally, we describe two real-world applications that can be advanced using our dataset. CaltechFN is publicly available at https://data.caltech.edu/records/33qmq-a2n15, and all benchmark code is available at https://github.com/patrickqrim/CaltechFN.

1 Introduction

The task of classifying digits was one of the first computer vision tasks successfully "solved" by deep learning architectures. Released in 1998, the MNIST dataset [1] serves as a benchmark for model performance in the task of classifying digits. However, deep learning models have been able to achieve human levels of performance in the task of classifying the digits in the MNIST dataset [2–5]. Due to the standardized nature of the handwritten digits in MNIST, there

Supplementary Information The online version contains supplementary material available at https://doi.org/10.1007/978-3-031-27066-6_14.

Y. Zheng et al. (Eds.): ACCV 2022, LNCS 13848, pp. 195–212, 2023.
https://doi.org/10.1007/978-3-031-27066-6_14

is low variability between the digits, which makes it easy for modern computer vision architectures to learn the characteristic features of each digit [6,7].

The Street View House Numbers (SVHN) dataset [8] consists of digits from house numbers obtained from Google Street View images, which pose a more difficult challenge than MNIST. Due to the natural settings and diversity in the designs of the house numbers, there is a far higher variability between the digits in SVHN than in MNIST. Thus, when SVHN was published in 2011, there was initially a large gap between human performance and model performance in the task of classifying its digits [9]. Because of this disparity, SVHN began to serve as a more difficult benchmark for novel image classification and object detection models [10,11]. However, newer models have since been able to achieve a classification accuracy on SVHN exceeding 98% [12–15], which is the published human performance benchmark. Some recent models have even achieved an accuracy exceeding 99% [16,17]. With minimal room left for improvement in performance on SVHN, there is a need for a more difficult digit dataset to benchmark the progress of future classification and detection models.

In this paper, we present **Caltech F**ootball **N**umbers (CaltechFN), a new benchmark dataset of digits from American football jerseys. Samples of the digits in our dataset are displayed in Fig. 1. We demonstrate that the latest image classification and object detection models are not able to achieve human performance on our dataset. This performance gap can be explained by the significantly increased variability of CaltechFN compared to current benchmark digit datasets. Due to the nature of American football jerseys, many of the digits in the dataset are wrinkled, stretched, twisted, blurred, unevenly illuminated, or otherwise distorted [18–21]. Sample images containing distorted digits are displayed in Fig. 2(a). These possibilities introduce a substantial number of ways that each digit can differ in appearance from the other digits in its class. We demonstrate that even the latest models struggle to learn the characteristic features of each digit when trained on such highly variable digits. This is likely due to the scarcity of digits that are distorted and occluded in the same way. As improved few-shot learning methods are developed, we expect an improvement in model performance on our dataset.

Furthermore, due to the nature of American football games, many images of digits on jerseys in live-action will be partially occluded [18–20]. For example, another player or the ball may be present between the camera and the subject player, or the player may be partially turned away from the camera such that parts of digits are not visible. Sample images containing partially occluded digits are displayed in Fig. 2(b). CaltechFN contains many such images of digits that are partially occluded, yet identifiable by human beings. This can be explained by recent neuroscience studies that have demonstrated the capability of the human brain to "fill in" visual gaps [22–25]. On the other hand, computer vision models struggle to fill in these visual gaps [26,27] since they are often unique and not represented in the training set. In other words, there are a large number of unique ways in which a certain digit may be partially occluded. Compounding this with the number of ways that a digit can be distorted, it is difficult for models trained on our dataset to learn the characteristic features of each digit [28].

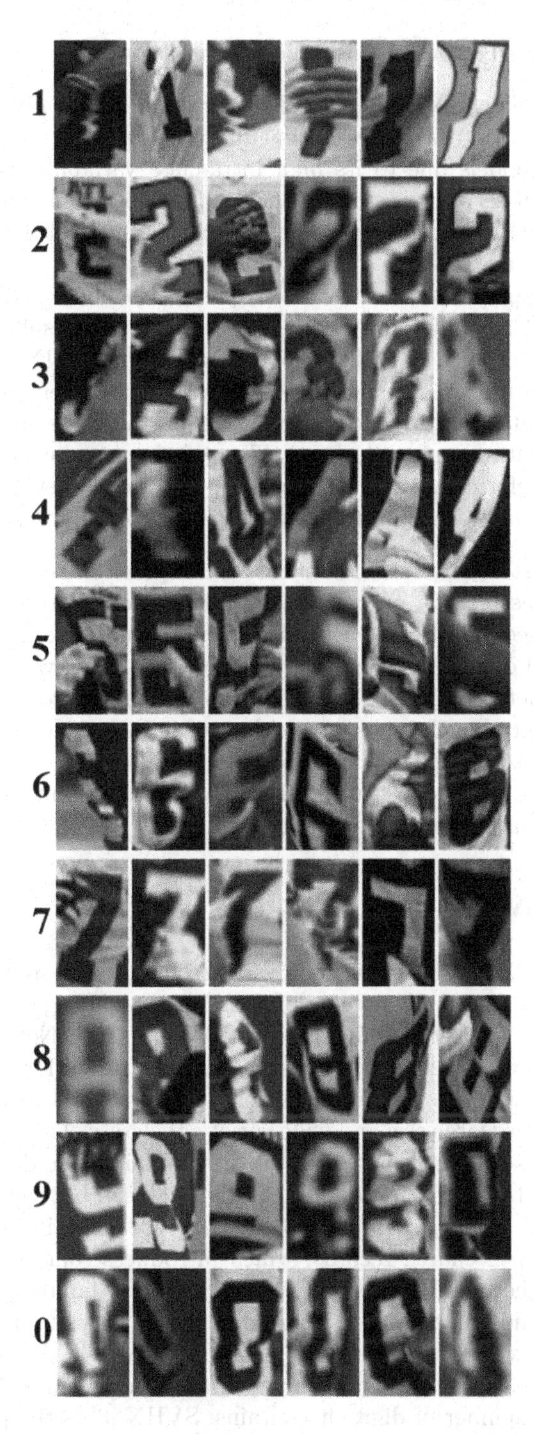

Fig. 1. Samples of cropped digits from the CaltechFN dataset that are distorted (e.g. wrinkled and stretched) and partially occluded.

The main contributions of this paper are as follows:

1. We present CaltechFN, a dataset of distorted and partially occluded digits. The CaltechFN dataset poses a difficult challenge for even the latest computer vision models due to its high intra-class variability. For this reason, CaltechFN can serve as a state-of-the-art benchmark for future image classification, object detection, and weakly supervised object detection (WSOD) models.
2. We perform experiments to measure cross-dataset model performance benchmarks on the CaltechFN and SVHN datasets. The results illustrate that CaltechFN is indeed a more difficult benchmark than SVHN and that models trained on CaltechFN demonstrate high cross-dataset generalization.
3. We record human performance benchmarks on the CaltechFN dataset using experiments with human volunteers. The existing gap between the best current model performance and our human performance benchmark will hopefully catalyze innovations in the construction of computer vision models.

This paper is structured as follows: Sect. 2 compares and contrasts CaltechFN with related datasets. The properties and goals of our dataset are introduced in Sect. 3. This section also describes the process undertaken to construct the dataset. Section 4 details and compares the performance of various image classification, object detection, and WSOD models on our dataset. We then provide human performance benchmarks on our dataset in Sect. 5. In Sect. 6, we present examples of real-world tasks that can be better solved by models trained on our dataset. In Sect. 7, we discuss future directions that can be taken to utilize the richness of information in the images in our dataset.

2 Related Work

Digit Datasets. Digit datasets are advantageous in their simplicity and their ease of use, containing a small number of classes and requiring little preprocessing and formatting to begin the training process. While ImageNet [29], MS-COCO [30], and PascalVOC [31] are the most popular datasets for image classification and object detection, they lack the compactness and the simplicity of digit datasets due to their large size and wide variety of classes. There are several popular digit datasets available, but many of the digits in these datasets are handwritten. MNIST [1] was the first prominent digit dataset, but the standardization of the digits limits variability. The ARDIS dataset [32] contains handwritten digits from old Swedish church records, which introduces some variability due to age-induced weathering. However, the variability is still limited by the standardized nature of handwritten digits. There do exist datasets that instead contain digits in natural settings. Roughly 10% of the Chars74k dataset [33] is from real-life, outdoor images. However, Chars74k also contains non-digit characters, which limits the number of digits it contains. SVHN [8] is the primary dataset consisting exclusively of digits in real-world settings. However, the images in SVHN have few distortions besides natural blur, and close to no occlusions since they are house numbers intended to be visible from the street. Meanwhile, CaltechFN consists of many examples of distorted and partially occluded digits, which constitutes a more difficult set of digits than any existing digit dataset.

Datasets with Distorted Objects. Distortions in many popular image datasets are limited in their complexity. For instance, the SmartDoc-QA dataset [34] contains images of documents distorted by blur, perspective, and illumination effects. All of these distortions fall under the same general domain and do not cover the wide variety of distortions in the real world. The dataset of soccer jersey numbers by Gerke, Müller, and Schäfer [35] consists of images taken from soccer videos with image-level annotations of jersey numbers. The distortions in these images are similar to the ones found in CaltechFN since they were also captured from sports settings. However, unlike CaltechFN, this dataset does not contain bounding box annotations, meaning that the dataset cannot be used to benchmark object detection tasks. Furthermore, there is more physical contact between players in American football than in soccer, meaning that there are more distorted digits in CaltechFN than in the soccer dataset, which can be empirically confirmed when observing the two datasets.

Datasets with Partially Occluded Objects. There are also many existing datasets with partially occluded objects. However, most of these datasets are not focused on digits, but on a larger range of objects. For example, the Occluded REID [36] and the Caltech Occluded Faces in the Wild [37] datasets present the challenge of identifying humans and faces, respectively, when partially occluded by other objects. Similarly, the Pascal3D+ dataset [38] augments images from the PascalVOC and ImageNet datasets with 3D annotations, partially occluding the target objects. All of these datasets lack the simplicity and convenience of digit datasets. Chars74k does contain some digits and characters that are partially occluded. However, unlike CaltechFN, the Chars74k dataset does not provide bounding boxes and thus cannot be used to benchmark object detection tasks.

3 Caltech Football Numbers (CaltechFN) Dataset

3.1 Dataset Construction

Image Collection. The first step of the data construction process was to collect candidate images. In order to construct a representative and unbiased dataset, we chose to sample an equal number of images of each jersey number in American football, which ranges from 1 to 99. Since each digit from 0 to 9 is roughly equally represented in this range, we sampled uniformly across each jersey number. This is to ensure that models trained on our dataset do not overfit to any over-represented digits [39].

We collected our candidate images by querying the Google Image Search database. Using the query "Team Name" + "Number", we collected 50 images for each combination of the 32 teams in the National Football League and the

Fig. 2. Samples of full images from the CaltechFN dataset that contain labeled bounding boxes around digits that are **(a)** distorted (e.g. wrinkled and stretched), and **(b)** partially occluded.

99 jersey numbers, for a total of 1600 images for each number and a grand total of 158,400 images. We chose this query because it seemed to be neither too general nor too specific: queries that were too general returned too many irrelevant images, while queries that were too specific did not return a sufficient number of relevant images.

Image Filtering. We then completed a filtering process to remove unwanted and duplicate images. First, we made two full passes through the set of candidate images to remove images that did not contain any digits. Almost half of the candidate images were removed in this step. Then, we removed any images where the digits contained were distorted or occluded to the extent that we were not able to identify them. We first made another two full passes through the set of candidate images to identify and mark images that contained digits that were not immediately identifiable. We then carefully sorted through each of these marked images, only keeping those that contained at least one identifiable digit. Finally, we used a deduplication tool [40] to identify and remove duplicate images. The result of this cleaning process is our current set of 61,728 images and 264,572 labeled digits.

Image Annotation. As done by many previous studies [41], we utilized the Amazon Mechanical Turk (AMT) platform [42] to label each individual digit in each of the images. For each digit, including those that were partially occluded, partially cut off, or rotated, AMT workers were asked to draw and label a maximally tight bounding box that contained every visible pixel of the digit. We then worked through the results and fixed any errors; specifically, we labeled identifiable digits that were not already labeled, removed erroneous boxes, and corrected any incorrect labels.

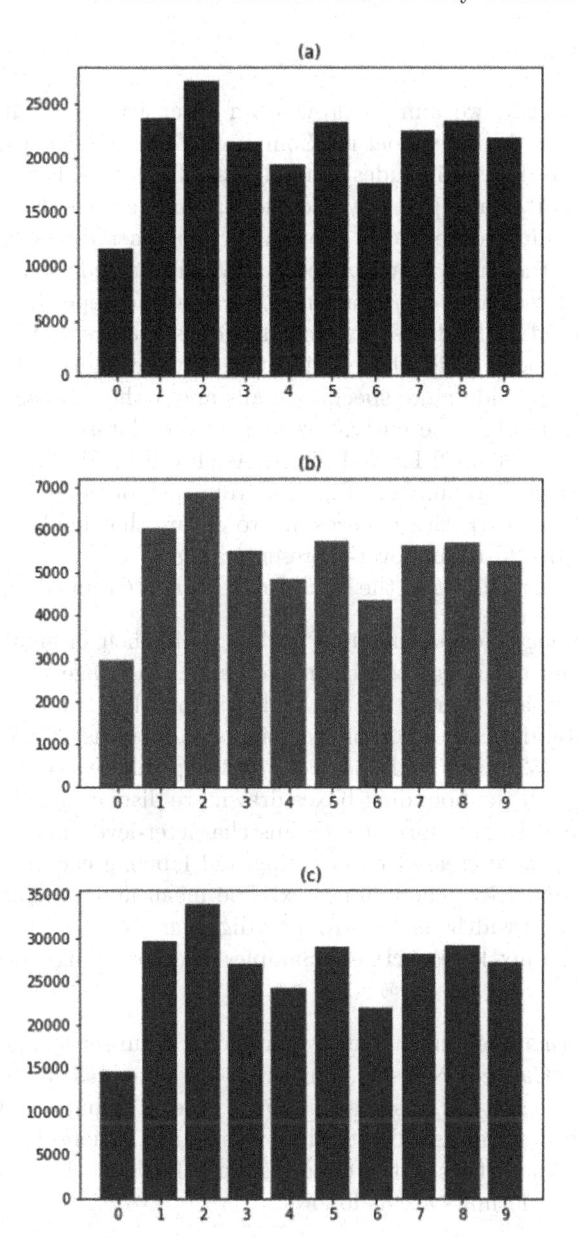

Fig. 3. Distribution of digits in **(a)** the train set, **(b)** the test set, and **(c)** the entire dataset.

3.2 Properties

Through CaltechFN, we aim to provide an extensive set of digits with high variability to provide a new goal for Computer Vision models to work towards. To that end, the dataset includes a highly variable set of digits, with many of them being distorted and partially occluded in unique ways. For instance, some digits are partially blocked by a football, while other digits are twisted and wrinkled due to the jersey being pulled on. We also include some easier images, such as stationary images of unobstructed jerseys. We hope that the variability of digits presented in the dataset will challenge researchers to design innovations that will allow models to identify similarities between digits of the same class.

We will now provide more specific details about the CaltechFN dataset. As described in Sect. 3.1.2, the current version of our dataset contains a total of 61,728 images and 264,572 labeled digits. As shown in Fig. 3, our dataset contains a roughly uniform number of images from each of the ten digit classes. As mentioned in Sect. 3.1.1, this is necessary to ensure that models trained on our dataset do not overfit to any over-represented digits.

We publish our dataset in the "Full Image" and "Cropped Digits" formats:

- The "Full Image" format includes all images in their original resolutions as obtained from the image collection process. Each image is accompanied by bounding box annotations for each identifiable digit that it contains. The mean and standard deviation of the heights and widths of the full images are 181.571 ± 15.452 pixels and 234.692 ± 54.365 pixels respectively. Samples of full images with the bounding boxes drawn are displayed in Fig. 2.
- The "Cropped Digits" format contains character-level images of each digit. These images were created by cropping and labeling each region of the full images contained by a bounding box. The mean and standard deviation of the heights and widths of the cropped digits are 32.360 ± 18.042 pixels and 21.334 ± 9.375 pixels respectively. Samples of cropped digits are displayed in Fig. 1.

For both formats of our dataset, we provide a train set ("CaltechFN-train") and a test set ("CaltechFN-test"). This train-test split was created using a random, uniform 80–20 split. As seen in Fig. 3, the distribution of digits in the train set and test set are similar to the overall distribution of digits across the entire dataset. The details of the train-test split for both the "Full Image" and "Cropped Digits" formats are as follows:

- "Full Image": train set contains 49,383 images (80.0% of total), test set contains 12,345 images (20.0% of total).
- "Cropped Digits": train set contains 211,611 digits (80.0% of total), test set contains 52,911 digits (20.0% of total).

Table 1. Model performance when **(A)** trained on CaltechFN-train, tested on CaltechFN-test; (B) trained on CaltechFN-train, tested on SVHN-test; (C) trained on SVHN-train, tested on CaltechFN-test; (D) trained on SVHN-train, tested on SVHN-test. Image classification models are evaluated using classification accuracy. Object detection and WSOD models are evaluated using mAP.

Image classification				
Model	(A)	(B)	(C)	(D)
MobileNet (CVPR '18) [43]	86.0 ± 0.6	93.1 ± 0.5	76.0 ± 0.6	98.2 ± 0.5
DenseNet121 (CVPR '17) [15]	87.9 ± 0.4	95.0 ± 0.3	77.9 ± 0.3	98.6 ± 0.4
ResNet50 (CVPR '16) [44]	86.9 ± 0.4	94.2 ± 0.6	77.1 ± 0.5	98.3 ± 0.4
Object detection				
Model	(A)	(B)	(C)	(D)
YOLOv5 ('21) [45]	54.4 ± 0.5	61.2 ± 0.5	37.5 ± 0.9	67.9 ± 0.4
RetinaNet (ICCV '17) [46]	52.7 ± 0.8	57.8 ± 0.7	30.0 ± 1.4	65.2 ± 0.7
SSD (ECCV '16) [47]	54.6 ± 0.4	61.1 ± 0.2	38.6 ± 1.0	67.2 ± 0.4
Faster-RCNN (NIPS '15) [48]	57.4 ± 0.3	60.9 ± 0.5	38.8 ± 0.6	68.5 ± 0.3
Weakly supervised object detection (WSOD)				
Model	(A)	(B)	(C)	(D)
Wetectron (CVPR '20) [49]	29.5 ± 0.6	37.5 ± 0.5	20.7 ± 1.8	42.6 ± 0.3
C-MIL (CVPR '19) [50]	26.3 ± 1.4	36.0 ± 1.2	17.2 ± 1.0	39.4 ± 0.8
WSOD2 (ICCV '19) [51]	21.1 ± 0.4	27.0 ± 0.8	14.8 ± 1.8	30.9 ± 0.3
PCL (CVPR '17) [52]	27.1 ± 0.9	34.5 ± 0.6	16.9 ± 1.0	37.3 ± 1.1

4 Model Performance

In the following experiments, we will compare the performance on CaltechFN and SVHN of some of the latest models built for the tasks of image classification, object detection, and weakly supervised object detection. We compare performance on our dataset to performance on SVHN because it is the most similar existing digit dataset, as we explained in Sect. 2. We will show that CaltechFN is a significantly more difficult dataset than SVHN, while also showing that models trained on CaltechFN perform at least as well as models trained on SVHN. For each model, we will present results for the following four experiments, labeled as follows:

(A) Training on CaltechFN-train, testing on CaltechFN-test,
(B) Training on CaltechFN-train, testing on SVHN-test,
(C) Training on SVHN-train, testing on CaltechFN-test,
(D) Training on SVHN-train, testing on SVHN-test.

The experimental results are presented in Table 1. We note that the results labeled (A) serve as benchmark performance scores for CaltechFN in the three tasks we detailed. The evaluation metric for the image classification results is classification accuracy, while the evaluation metric for the object detection and WSOD results is mAP. All experimental details including the hyperparameter search effort, the compute resources used, and a description of the evaluation metrics are discussed at length in the Supplementary Material.

We now explain the relevance of the experimental results:

1. **We demonstrate the comparative difficulty of CaltechFN.** When trained on CaltechFN, models perform worse on CaltechFN (A) than on SVHN (B). The same models trained on SVHN also perform worse on CaltechFN (C) than on SVHN (D). Regardless of which dataset is used as the training set, models perform worse on CaltechFN than on SVHN.
2. **We demonstrate that models trained on CaltechFN demonstrate high cross-dataset generalization.** Models perform worse on CaltechFN when training on SVHN (C) than when training on CaltechFN (A) itself, but the performance of the same models on SVHN does not significantly drop when trained on CaltechFN (B) instead of on SVHN (D). This shows that computer vision models are able to learn robust, generalizable features by training on CaltechFN.

5 Human Performance Benchmark

To demonstrate the potential for improvement in current computer vision architectures, we provide human performance benchmarks on our dataset in the same classification and detection tasks performed by computer vision models in Sect. 3.

To measure human performance in the task of classifying the cropped digits in CaltechFN, we asked five human volunteers to label a subset of 15,000 cropped digits ("All Samples"). We calculate mean human performance by computing the accuracies of the volunteer-generated labels.

To measure human performance in the task of detecting digits in the full images in CaltechFN, we asked the same five human volunteers to draw bounding boxes on a subset of 5,000 images ("All Samples"). We calculate mean human performance using the same mAP metric used for the object detection models in Sect. 3.

Furthermore, we calculate the human performance in both tasks for only the subset of "Difficult Samples" that even the best models in Sect. 4 were unable to classify/detect. The results, which we provide as benchmarks for human performance in the tasks of image classification and object detection on our dataset, are presented in Table 2.

We see that humans are able to achieve high levels of performance, even on the samples that the best models were unable to classify/detect. This clear disparity between human performance and the best model performances in both tasks demonstrates that there potentially exist certain techniques not yet learned by models that humans use to identify difficult digits. Evidently, there is still a need for innovations in the construction of computer vision models for computers to be able to achieve human levels of performance in the aforementioned tasks.

Table 2. Human performance on the CaltechFN dataset. Image classification performance is evaluated using classification accuracy, while object detection performance is evaluated using mAP.

Human performance benchmarks		
Task	All samples	Difficult samples
Image classification	99.1 ± 0.8	97.8 ± 2.3
Object detection	87.2 ± 4.5	83.0 ± 5.9

However, it is clear that even humans find it difficult to identify every digit in our dataset. Even though the digits that are included in our dataset are the ones that we approved as identifiable, it is not necessarily true that other humans will also be able to identify each of these digits. This may be due to a bias stemming from the fact that we collected the images, or simply due to variability in performance across different humans. Ultimately, this leaves open the possibility that significant advancements in computer vision techniques may result in models being able to achieve even higher performance on classification and detection tasks on our dataset than humans.

Figure 4 provides a visual representation of Tables 1 and 2. The comparative difficulty of CaltechFN and the high cross-dataset generalization of models trained on CaltechFN, as well as the gap between human performance and model performance on CaltechFN, are clearly illustrated.

6 Applications of CaltechFN

We will now present two potential real-world applications that involve detecting and classifying digits. Our dataset provides a rich source of difficult digits that are distorted and partially occluded in the same manner that they are in these two real-world settings. We will explain the benefits of using the images in our dataset to train models that can perform these real-world tasks.

6.1 Player Detection and Tracking in Sports

Currently, coaches in sports must watch footage of a past game to chart the personnel (players on the field) over the duration of the game, which is an important task in sports analytics [19]. Computer vision models have thus far struggled to outperform humans at this task due to the distortions and occlusions of jersey numbers, which are the primary identifying features of players [18]. Our dataset can be used to train a model that can identify players in game footage using their jersey numbers even under such conditions, given that it includes many training examples of distorted and partially occluded jersey numbers. While a potentially negative societal impact of such a model would be that this would relieve coaches of this responsibility, coaches could instead focus on analyzing the personnel information.

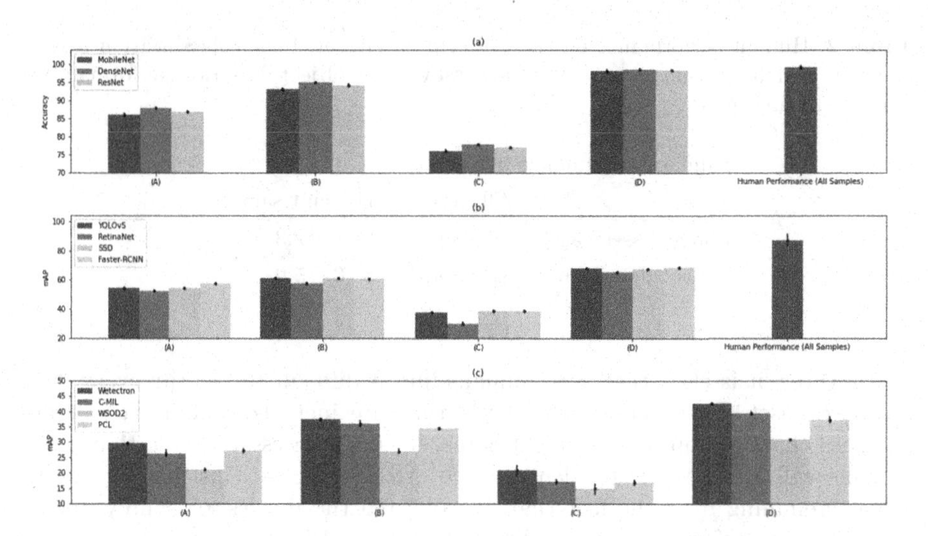

Fig. 4. Visual representation of Tables 1 and 2 for **(a)** image classification; **(b)** object detection; and **(c)** weakly supervised object detection. (A), (B), (C), and (D) are as defined in Sect. 4. The models for each subplot are, in the color order displayed, as follows: **(a)** MobileNet, DenseNet, ResNet; **(b)** YOLOv5, RetinaNet, SSD, Faster-RCNN; **(c)** Wetectron, C-MIL, WSOD2, PCL.

Another useful application of a model trained on our dataset is that high school and college sports coaches may be able to track the movements of their players using game footage. While many professional sports teams use microchips to track the movements of their players, high school and college teams often do not have access to this level of technology. Our model would be able to locate players using their jersey numbers at each frame and be able to chart their movements, which is useful information that enables complex sports analysis.

6.2 Self-driving Cars

In order to stay within speed limits, self-driving cars rely on internal vision systems to detect and read speed limits on the road [53,54], which are most often printed on signs or painted directly onto the road. The consequences of a self-driving car being unable to read the speed limit may be dire, especially in an area where the speed limit warnings are few and far between.

Some speed limit signs may be old and worn-out, causing the digits to be faded or otherwise distorted, while speed limits painted on the road may be chipped and eroded. A self-driving car may not detect such speed limit postings, especially since it may be traveling past them at high speeds. By training its vision systems on our dataset, which contains many examples of distorted digits, a self-driving car may be better equipped to read speed limits even under imperfect conditions. In other cases, speed limit signs and speed limits painted on the road may be partially occluded by other cars, pedestrians, or other obstacles. A self-driving car trained

Fig. 5. Sample detections on worn-out and partially occluded speed limit postings using Faster-RCNN trained on CaltechFN.

on our dataset, which contains many examples of partially occluded digits, would be better able to read such speed limit postings. To demonstrate the viability of this application, we applied the Faster-RCNN model mentioned in Sect. 4 to two sample images containing distorted (worn out) and partially occluded speed limit postings. The bounding box predictions are shown in Fig. 5.

7 Discussion and Future Work

In this paper, we have introduced CaltechFN, a new dataset containing digits found on American football jerseys. This dataset is novel in its variability: each digit is distorted or partially occluded in a unique way such that current models have difficulty learning the representative features of each digit class. We queried the Google Image Search database to collect our images, deleted images with no identifiable digits, then utilized Amazon Mechanical Turk to create annotations for the digits in each image. Through our experiments with various image classification, object detection, and WSOD models, we have demonstrated that CaltechFN is indeed a more difficult benchmark than SVHN and that models trained on CaltechFN demonstrate high cross-dataset generalization. Furthermore, we recorded human performance benchmarks on the CaltechFN dataset using experiments with human volunteers. In doing so, we illustrated the existing gap between model performance and human performance on our dataset. With this dataset, we aim to introduce a new state-of-the-art benchmark that will be used to foster the development of novel computer vision models. We hope that innovations in computer vision research will allow future models to achieve human levels of performance on our dataset.

We believe that models that perform well on our dataset will be better equipped to perform real-world tasks. Two such real-world applications of our dataset were described in Sect. 5. Models trained on our dataset could be used to automate the process of charting the personnel on the field at a given moment in a game. Also, self-driving cars can train its vision systems on our dataset to be better equipped to read speed limit postings under imperfect conditions.

7.1 Further Labeling

The images in the CaltechFN dataset contain rich information yet to be annotated, beyond the existing digit annotations that are the focus of this paper. Thus, we believe that the primary future direction of the CaltechFN dataset is to expand upon the existing annotations for the following applications:

Scene Recognition. Scene recognition refers to the computer vision task of identifying the context of a scene within an image. There is extensive research being conducted on the development of models to improve performance in this task [55–57]. While there do exist several large datasets for scene recognition, the task remains challenging and largely unsolved. These challenges have been attributed to class overlaps and high variance within classes [58]. Class overlaps occur when there are several classes that are not sufficiently distinct from each other, while high variance within classes means that classes have a wide variety of scenes attached to them. We believe that the unexploited qualities of our dataset may be able to address these challenges and improve model performance in the task of scene recognition. Within American football, each scene has a distinctive action being performed: tackling, throwing, catching, running, kicking, and so on. There is very little ambiguity between which of these actions is being performed in which scene. This lack of variance thus addresses the challenge of having class overlap-no two scene classes are the same and models should be better able to distinguish between the scene classes in this dataset.

Instance Segmentation. Another commonly studied computer vision task is that of instance segmentation [59–61]. This task is similar to object detection, with the difference being that the output consists of the set of pixels contained within the object rather than a rectangular box that bounds the object. Each object in an image segmentation dataset is annotated with its exact pixel-level boundary, rather than a simple rectangular bounding box. The images in our dataset contain many distinct objects that can be delineated in this way, including players, jerseys, helmets, and balls. We see that many of these objects are shaped in the form of some part of a human being. Thus, by adding pixel-level boundary annotations for these objects, our dataset could be used as a benchmark for the advancement of human detection and tracking techniques.

Acknowledgments. We would like to thank Elijah Cole, Yisong Yue, and A. E. Hosoi for their valuable comments and suggestions. We would also like to thank the members of our community who participated in our human performance benchmark experiments. Funding for all crowdsource worker tasks was provided by the George W. Housner Fund.

References

1. LeCun, Y., Cortes, C., Burges, C.: The MNIST database of handwritten digits (1999). http://yann.lecun.com/exdb/mnist/. Accessed 15 May 2022
2. Ciresan, D., Meier, U., Schmidhuber, J.: Multi-column deep neural networks for image classification. In: 2012 IEEE Conference on Computer Vision and Pattern Recognition (2012)
3. Kussul, E., Baidyk, T.: Improved method of handwritten digit recognition tested on MNIST database. Image Vis. Comput. **22**(12), 971–981 (2004)
4. Hasanpour, S.H., Rouhani, M., Fayyaz, M., Sabokrou, M.: Lets keep it simple, using simple architectures to outperform deeper and more complex architectures. arXiv (2016)
5. Ciresan, D.C., Meier, U., Gambardella, L.M., Schmidhuber, J.: Convolutional neural network committees for handwritten character classification. In: 2011 International Conference on Document Analysis and Recognition (2011)
6. Shorten, C., Khoshgoftaar, T.M.: A survey on image data augmentation for deep learning. J. Big Data **6**(1), 1–48 (2019)
7. Zhang, Y., Ling, C.: A strategy to apply machine learning to small datasets in materials science. npj Comput. Mater. **4**(1) (2018). Article number: 25
8. Netzer, Y., Wang, T., Coates, A., Bissacco, A., Wu, B., Ng, A.Y.: Reading digits in natural images with unsupervised feature learning. In: NIPS Workshop on Deep Learning and Unsupervised Feature Learning, December 2011
9. Goodfellow, I.J., Bulatov, Y., Ibarz, J., Arnoud, S., Shet, V.: Multi-digit number recognition from street view imagery using deep convolutional neural networks. arXiv, December 2013
10. Srivastava, N., Hinton, G., Krizhevsky, A., Sutskever, I., Salakhutdinov, R.: Dropout: a simple way to prevent neural networks from overfitting. J. Mach. Learn. Res. **15**, 1929–1958 (2014)
11. Radford, A., Metz, L., Chintala, S.: Unsupervised representation learning with deep convolutional generative adversarial networks. In: ICLR (Poster) (2016)
12. Lim, S., Kim, I., Kim, T., Kim, C., Kim, S.: Fast AutoAugment. In: Advances in Neural Information Processing Systems (2019)
13. Gowda, S.N., Yuan, C.: ColorNet: investigating the importance of color spaces for image classification. In: Jawahar, C.V., Li, H., Mori, G., Schindler, K. (eds.) ACCV 2018. LNCS, vol. 11364, pp. 581–596. Springer, Cham (2019). https://doi.org/10.1007/978-3-030-20870-7_36
14. DeVries, T., Taylor, G.W.: Improved regularization of convolutional neural networks with cutout. arXiv (2017)
15. Huang, G., Liu, Z., Van Der Maaten, L., Weinberger, K.Q.: Densely connected convolutional networks. In: 2017 IEEE Conference on Computer Vision and Pattern Recognition (CVPR) (2017)
16. Foret, P., Kleiner, A., Mobahi, H., Neyshabur, B.: Sharpness-aware minimization for efficiently improving generalization. In: International Conference on Learning Representations (2021)
17. Phong, N.H., Ribeiro, B.: Rethinking recurrent neural networks and other improvements for image classification. arXiv (2020)
18. Rahimian, P., Toka, L.: Optical tracking in team sports. J. Quant. Anal. Sports **18**(1), 35–57 (2022)
19. Moeslund, T.B., Thomas, G., Hilton, A.: Computer Vision in Sports. Springer, Cham (2015). https://doi.org/10.1007/978-3-319-09396-3

20. Bhargavi, D., Coyotl, E.P., Gholami, S.: Knock, knock. Who's there? - Identifying football player jersey numbers with synthetic data. arXiv (2022)
21. Atmosukarto, I., Ghanem, B., Ahuja, S., Muthuswamy, K., Ahuja, N.: Automatic recognition of offensive team formation in American football plays. In: 2013 IEEE Conference on Computer Vision and Pattern Recognition Workshops (2013)
22. Chong, E., Familiar, A.M., Shim, W.M.: Reconstructing representations of dynamic visual objects in early visual cortex. Proc. Natl. Acad. Sci. **113**(5), 1453–1458 (2015)
23. Kok, P., de Lange, F.P.: Shape perception simultaneously up- and downregulates neural activity in the primary visual cortex. Curr. Biol. **24**(13), 1531–1535 (2014)
24. Bosco, G., et al.: Filling gaps in visual motion for target capture. Front. Integr. Neurosci. **9**, 13 (2015)
25. Revina, Y., Maus, G.W.: Stronger perceptual filling-in of spatiotemporal information in the blind spot compared with artificial gaps. J. Vis. **20**(4), 20 (2020)
26. Chandler, B., Mingolla, E.: Mitigation of effects of occlusion on object recognition with deep neural networks through low-level image completion. Comput. Intell. Neurosci. **2016**, 1–15 (2016)
27. Ning, C., Menglu, L., Hao, Y., Xueping, S., Yunhong, L.: Survey of pedestrian detection with occlusion. Complex Intell. Syst. **7**(1), 577–587 (2020). https://doi.org/10.1007/s40747-020-00206-8
28. Hoiem, D., Chodpathumwan, Y., Dai, Q.: Diagnosing error in object detectors. In: Fitzgibbon, A., Lazebnik, S., Perona, P., Sato, Y., Schmid, C. (eds.) ECCV 2012. LNCS, vol. 7574, pp. 340–353. Springer, Heidelberg (2012). https://doi.org/10.1007/978-3-642-33712-3_25
29. Deng, J., Dong, W., Socher, R., Li, L.-J., Li, K., Fei-Fei, L.: ImageNet: a large-scale hierarchical image database. In: 2009 IEEE Conference on Computer Vision and Pattern Recognition (2009)
30. Lin, T.-Y., et al.: Microsoft COCO: common objects in context. In: Fleet, D., Pajdla, T., Schiele, B., Tuytelaars, T. (eds.) ECCV 2014. LNCS, vol. 8693, pp. 740–755. Springer, Cham (2014). https://doi.org/10.1007/978-3-319-10602-1_48
31. Everingham, M., Van Gool, L., Williams, C.K.I., Winn, J., Zisserman, A.: The PASCAL visual object classes (VOC) challenge. Int. J. Comput. Vis. **88**, 303–308 (2010)
32. Kusetogullari, H., Yavariabdi, A., Cheddad, A., Grahn, H., Hall, J.: ARDIS: a Swedish historical handwritten digit dataset. Neural Comput. Appl. **32**(21), 16505–16518 (2019). https://doi.org/10.1007/s00521-019-04163-3
33. de Campos, T., Babu, B.R., Varma, M.: Character recognition in natural images. In: VISAPP 2009 - Proceedings of the Fourth International Conference on Computer Vision Theory and Applications, vol. 2 (2009)
34. Nayef, N., Luqman, M.M., Prum, S., Eskenazi, S., Chazalon, J., Ogier, J.-M.: SmartDoc-QA: a dataset for quality assessment of smartphone captured document images - single and multiple distortions. In: 2015 13th International Conference on Document Analysis and Recognition (ICDAR) (2015)
35. Gerke, S., Müller, K., Schäfer, R.: Soccer jersey number recognition using convolutional neural networks. In: 2015 IEEE International Conference on Computer Vision Workshop (ICCVW) (2015)
36. He, L., Wang, Y., Liu, W., Zhao, H., Sun, Z., Feng, J.: Foreground-aware pyramid reconstruction for alignment-free occluded person re-identification. In: 2019 IEEE/CVF International Conference on Computer Vision (ICCV) (2019)

37. Burgos-Artizzu, X.P., Perona, P., Dollar, P.: Robust face landmark estimation under occlusion. In: 2013 IEEE International Conference on Computer Vision (2013)
38. Xiang, Y., Mottaghi, R., Savarese, S.: Beyond PASCAL: a benchmark for 3D object detection in the wild. In: IEEE Winter Conference on Applications of Computer Vision (2014)
39. Oksuz, K., Cam, B.C., Kalkan, S., Akbas, E.: Imbalance problems in object detection: a review. IEEE Trans. Pattern Anal. Mach. Intell. **43**(10), 3388–3415 (2021)
40. Voxel51: Voxel51: developer tools for ML. https://voxel51.com/. Accessed 08 June 2022
41. Sorokin, A., Forsyth, D.: Utility data annotation with amazon mechanical turk. In: 2008 IEEE Computer Society Conference on Computer Vision and Pattern Recognition Workshops (2008)
42. Amazon Mechanical Turk. https://www.mturk.com/mturk/welcome. Accessed 16 May 2022
43. Howard, A.G., et al.: MobileNets: efficient convolutional neural networks for mobile vision applications. arXiv (2017)
44. He, K., Zhang, X., Ren, S., Sun, J.: Deep residual learning for image recognition. In: 2016 IEEE Conference on Computer Vision and Pattern Recognition (CVPR) (2016)
45. Bochkovskiy, A., Wang, C.-Y., Liao, H.-Y.M.: YOLOv4: optimal speed and accuracy of object detection. arXiv (2020)
46. Lin, T.-Y., Goyal, P., Girshick, R., He, K., Dollar, P.: Focal loss for dense object detection. In: 2017 IEEE International Conference on Computer Vision (ICCV) (2017)
47. Liu, W., et al.: SSD: single shot multibox detector. In: Leibe, B., Matas, J., Sebe, N., Welling, M. (eds.) ECCV 2016. LNCS, vol. 9905, pp. 21–37. Springer, Cham (2016). https://doi.org/10.1007/978-3-319-46448-0_2
48. Ren, S., He, K., Girshick, R., Sun, J.: Faster R-CNN: towards real-time object detection with region proposal networks. In: Advances in Neural Information Processing Systems (NIPS 2015), vol. 28 (2015)
49. Ren, Z., et al.: Instance-aware, context-focused, and memory-efficient weakly supervised object detection. In: 2020 IEEE/CVF Conference on Computer Vision and Pattern Recognition (CVPR) (2020)
50. Wan, F., Liu, C., Ke, W., Ji, X., Jiao, J., Ye, Q.: C-MIL: continuation multiple instance learning for weakly supervised object detection. In: 2019 IEEE/CVF Conference on Computer Vision and Pattern Recognition (CVPR) (2019)
51. Zeng, Z., Liu, B., Fu, J., Chao, H., Zhang, L.: WSOD2: learning bottom-up and top-down objectness distillation for weakly-supervised object detection. In: 2019 IEEE/CVF International Conference on Computer Vision (ICCV) (2019)
52. Tang, P., et al.: PCL: proposal cluster learning for weakly supervised object detection. In: 2017 IEEE Conference on Computer Vision and Pattern Recognition (2017)
53. Kanagaraj, N., Hicks, D., Goyal, A., Tiwari, S., Singh, G.: Deep learning using computer vision in self driving cars for lane and traffic sign detection. Int. J. Syst. Assur. Eng. Manag. **12**(6), 1011–1025 (2021). https://doi.org/10.1007/s13198-021-01127-6
54. Farag, W.: Recognition of traffic signs by convolutional neural nets for self-driving vehicles. Int. J. Knowl. Based Intell. Eng. Syst. **22**(3), 205–214 (2018)

55. Herranz, L., Jiang, S., Li, X.: Scene recognition with CNNs: objects, scales and dataset bias. In: 2016 IEEE Conference on Computer Vision and Pattern Recognition (CVPR) (2016)
56. Chen, G., Song, X., Wang, B., Jiang, S.: See more for scene: pairwise consistency learning for scene classification. In: 35th Conference on Neural Information Processing Systems (NeurIPS 2021) (2021)
57. Zhou, B., Lapedriza, A., Xiao, J., Torralba, A., Oliva, A.: Learning deep features for scene recognition using places database. In: Advances in Neural Information Processing Systems (NIPS 2014), vol. 27 (2014)
58. Matei, A., Glavan, A., Talavera, E.: Deep learning for scene recognition from visual data: a survey. arXiv (2020)
59. Ganea, D.A., Boom, B., Poppe, R.: Incremental few-shot instance segmentation. In: 2021 IEEE/CVF Conference on Computer Vision and Pattern Recognition (CVPR) (2021)
60. Wang, Y., Xu, Z., Shen, H., Cheng, B., Yang, L.: CenterMask: single shot instance segmentation with point representation. In: 2020 IEEE/CVF Conference on Computer Vision and Pattern Recognition (CVPR) (2020)
61. Xie, E., et al.: PolarMask: single shot instance segmentation with polar representation. In: 2020 IEEE/CVF Conference on Computer Vision and Pattern Recognition (CVPR) (2020)

Temporal Extension Topology Learning for Video-Based Person Re-identification

Jiaqi Ning[1(✉)], Fei Li[1], Rujie Liu[1], Shun Takeuchi[2], and Genta Suzuki[2]

[1] Fujitsu Research and Development Center Co., Ltd, Beijing, China
ningjiaqi@fujitsu.com
[2] Fujitsu Research, Fujitsu Limited, Kawasaki, Japan

Abstract. Video-based person re-identification aims to match the same identification from video clips captured by multiple non-overlapping cameras. By effectively exploiting both temporal and spatial clues of a video clip, a more comprehensive representation of the identity in the video clip can be obtained. In this manuscript, we propose a novel graph-based framework, referred as Temporal Extension Adaptive Graph Convolution (TE-AGC) which could effectively mine features in spatial and temporal dimensions in one graph convolution operation. Specifically, TE-AGC adopts a CNN backbone and a key-point detector to extract global and local features as graph nodes. Moreover, a delicate adaptive graph convolution module is designed, which encourages meaningful information transfer by dynamically learning the reliability of local features from multiple frames. Comprehensive experiments on two video person re-identification benchmark datasets have demonstrated the effectiveness and state-of-the-art performance of the proposed method.

Keywords: Person ReID · Graph Convolution Network

1 Introduction

Person re-identification (ReID) [1,2] is an efficient computer vision technique to retrieve a specific person from multiple non-overlapping cameras. Person ReID has a wide range of applications such as security, video surveillance, etc., and has received extensive attention from researchers. Although many research results have been achieved, this task is still challenging due to background disturbances, occlusions, perspective changes, pose changes and other problems.

There are generally two kinds of person ReID processing methods [1]. One is image-based methods [3–9], which exploit temporally incoherent static images to retrieve pedestrians. The other is video-based methods [10–14], where both training and test data consist of temporally continuous image sequences. In recent years, impressive progress has been made in image-based person ReID. Some practical solutions are proposed especially for complex problems, such as occlusion [9,15,16]. However, the information contained in a single image is limited. If the information contained in a short video clip of a pedestrian could be effectively mined, it would significantly benefit the robustness of retrieval

Y. Zheng et al. (Eds.): ACCV 2022, LNCS 13848, pp. 213–225, 2023.
https://doi.org/10.1007/978-3-031-27066-6_15

results. Therefore, video-based person ReID methods concurrently utilize spatial and temporal information of the video clip and have the potential to better solve the difficult problems in person ReID.

There have been several typical video-based person ReID methods that aggregate temporal and spatial cues of video clips to obtain discriminative representations. Several elementary methods extract global features from each frame independently. Then the features of each frame are aggregated into the representation of a video clip by a temporal pooling layer or recurrent neural network (RNN) [17–19]. Due to the problems such as occlusion and background noise, these methods usually do not achieve excellent results. Recent works began to focus on the role of local features. Some works divide video frames into rigid horizontal stripes or utilize an attention mechanism to extract local appearance features [20–24]. However, it is hard to align local features learned from videos precisely. Some methods adopt pose estimation model to detect key points of identity in order to obtain well aligned local features [15,25,26]. Nevertheless, the noise will be introduced into the extracted local features due to occlusion and inaccurate key point detection. Some works use the Graph Convolution Network (GCN) technique to enhance the description of local features by setting local features as the nodes of GCN [27–29]. In the GCN, information could be transferred between nodes through edges, and the information of nodes can be enhanced or supplemented. However, in the occluded regions, the features are often unintelligible [9]. If all the local parts are considered to have the same reliability for information transmission, it brings in more noise and is terrible for extracting discriminative representations.

In most of graph-based video-based person ReID methods, the transfer metric among nodes is determined by the affinity between feature pairs. This may result in ignoring global contextual information from all other nodes and only considering undirected dependency [14,27,28]. Some of them utilize more than one graph to realize temporal and spatial dimension information extraction, which increases the complexity of the method [14,28,29].

A novel Temporal Extension Adaptive Graph Convolution (TE-AGC) framework is proposed for video-based person ReID in this manuscript. TE-AGC extracts global and local semantic features from multiple images as graph nodes. Then, the TE-AGC learns the reliability of each local feature extracted from multiple images, encouraging high-reliability nodes to transfer more information to low-reliability nodes, and inhibiting information passing of low-reliability nodes. Further, TE-AGC considers the dependencies of body parts within a frame or across different frames in both temporal dimension and spatial dimension using only one single graph. This way, it could mine comprehensive and discriminative features from the video clip by performing the designed graph convolution. The validity of the TE-AGC is revealed by the experiments on two benchmark datasets, MARS and DukeMTMC-VideoReID.

The main contributions of this paper are as follows: (1) A novel Temporal Extension Adaptive Graph Convolution (TE-AGC) framework for video-based person ReID is proposed. (2) We learned the reliability of local features and adaptively pass information from more meaningful nodes to less meaningful nodes. (3) We mine the temporal and spatial dimension information of video clips with one convolution graph.

2 Related Work

2.1 Image-Based Person ReID

There have been many works for image-based person ReID [3–9]. Benefit from the continuous advance of deep learning technology, the rank-1 accuracy of most image-based person ReID methods on the benchmark dataset is higher than that of human beings [1]. With utilizing the local semantic features and attention mechanisms [16,30], the performance of person ReID is further improved. In recent years, more researchers have paid attention to the occlusion problem of ReID and achieved fruitful results [9,15,16].

2.2 Video-Based Person ReID

Video-based person ReID can extract richer spatial-temporal clues than image-based person ReID and is expected for more accurate retrieval [10]. Some works extract features for each image of the video clip then aggregate them using temporal pooling or RNN [17–19]. To learn robust representation against pose changes and occlusions, the local semantic features and attention mechanisms are also be used in the video-based person ReID to improve the performance [15,20–26]. Different from the image-based person ReID, the time dimension is added, and both spatial attention and temporal attention are used to mine the information of a video clip.

2.3 Graph Convolution

Graphs are often used to model the relationship between different nodes. Graph Convolution Network (GCN) simply utilizes the convolution operation of image processing in graph structure data processing for the first time [31]. Great success has been achieved in many computer vision tasks, like skeleton-based action recognition [32], object detection [33] and person ReID [9,27–29]. Some methods have been proposed to use the GCN in person ReID. Some treat the image as nodes of graph, ignoring the relationship between different body parts within or across frames. In addition, some recent works model the temporal and spatial relationships of nodes in two or more graphs. For example, the Spatial-Temporal Graph Convolutional Network (STGCN) [28] constructs two graph convolution branches. The spatial relation of human body and the temporal relation from the adjacent frames are learned in two different graphs.

3 Method

We aim to develop an efficient spatial-temporal representation for video-based person ReID. To this end, the Temporal Extension Adaptive Graph Convolution (TE-AGC) framework is proposed in this manuscript, as shown in Fig. 1. In general, the whole frame contains two parts. One is to extract preliminary semantic features, and the other is to obtain an improved discriminative representation of a video clip.

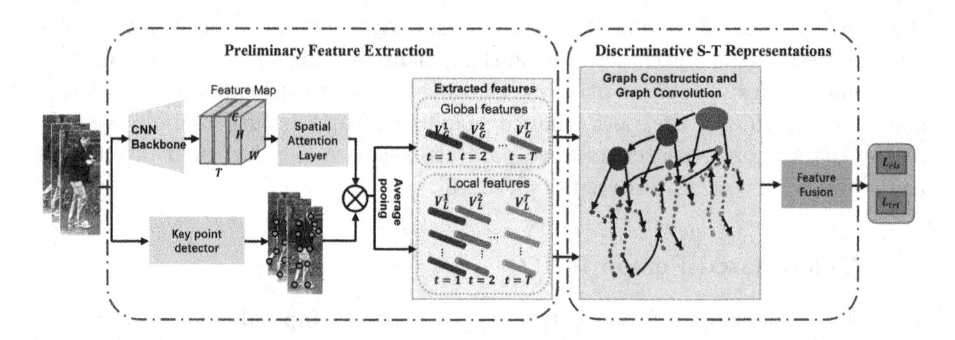

Fig. 1. The overall architecture of the proposed TE-AGC. It includes a backbone network, a spatial attention layer and a key-point detector to extract the global and local semantic features. Moreover, it also includes a graph construction and graph convolution layer and a feature fusion block to obtain discriminative spatial-temporal representation for each video clip.

3.1 Semantic Feature Extraction

First, we perform preliminary semantic feature extraction. It has been demonstrated that part features are effective for person ReID [1,2]. Inspired by this idea, we aim to extract both global and local semantic features in this module. To better resist the viewpoint variation and misalignment, a key-point detector is utilized to locate key points. Then we extract local features from different key points. It should be noted that, although human key-point detection is a relative mature technique, there still exist key point position errors and key point confidence errors in some cases [9]. Thus, the module introduced in **3.2** is necessary and will enhance the features.

Given a video clip, we randomly sample T frames. These randomly sampled frames are denoted as $\{I^t\}_{t=1}^T$ and t is the index of the frame. The backbone network is used to generate the initial feature maps for each frame. Then we use a spatial attention layer to enhance the spatial feature and suppress the interference information. The spatial attention layer is implemented by a convolutional layer followed by a sigmoid activation function. Then the initial feature maps are weighted by the attention layer. The set of feature maps after spatial attention operation are denoted as $\mathcal{F} = \{F^t\}_{t=1}^T$, where $F^t \in \mathbb{R}^{C \times H \times W}$ and H, W, C denote the height, width and channel number respectively.

We use a key point detector to help extracting aligned local features. For each frame, the number of extracted key point is K. The local and global semantic features of each frame are computed as follows. For distinction, we denote the features at this stage as V and denote the features output by the module introduced in 3.2 as $V^{'}$.

$$V_L{}^t = \{v_k^t\}_{k=1}^K = \{g_{GAP}\left(F^t \bigotimes m_k^t\right)\}_{k=1}^K \tag{1}$$

$$V_G{}^t = g_{GAP}\left(F^t\right) \tag{2}$$

where $V_L{}^t \in \mathbb{R}^{K \times C}$denotes the local features of the frame t, which include semantic local features of K key points. $v_k^t \in \mathbb{R}^{1 \times C}$is the local feature around k^{th} key point of frame t. m_k^t is derived from the heatmap of the kth key point of frame t by normalizing original heatmap with a SoftMax function. $g_{GAP}(\cdot)$ refers to global average pooling operation. $\bigotimes$ is element by element multiplication operation. $V_G{}^t \in \mathbb{R}^{1 \times C}$ denotes the global feature of the frame t. Therefore, the preliminary semantic of this video clip contains local feature $\{V_L{}^t\}_{t=1}^T$ and global feature $\{V_G{}^t\}_{t=1}^T$.

3.2 Temporal Extension Adaptive Graph Convolution Layer

After extracting each frame's preliminary global and local features, we employ advanced GCN to mine spatial-temporal representation from video frames.

A graph convolution can be operated as [31]:

$$O = \hat{A}XW \tag{3}$$

where $\hat{A}$ is normalized version of the adjacent matrix A, and X is the feature matrix which contains features of all nodes. W refers to parameters to be learned. O is the output after graph convolution operation.

In our method, the preliminary global feature and local feature of each frame within one video clip are treated as the graph nodes. For a video clip, the number of nodes is $N = T \times (K+1)$ including features of T frames, K local features and 1 global feature per frame. The feature matrix X with the size of $N \times C$ is constructed by the concatenation of $\{V_L{}^t\}_{t=1}^T$ and $\{V_G{}^t\}_{t=1}^T$ in vertical direction. The adjacent matrix $A \in \mathbb{R}^{N \times N}$ illustrates the topology of the graph. $A(i,j)$ represents the information propagation metric from node j to node i. When $A(i,j)$ equals to zero, no information transfers from node j to node i.

Most of the GCN-based person ReID methods obtain the $A(i,j)$ by calculating the feature affinity between node j and node i. However, this calculation method might introduce noise when some nodes are unreliable. Considering that, we propose a Temporal Extension Adaptive Graph Convolution (TE-AGC) layer to calculate the adjacent matrix A.

Our method efficiently utilizes one single graph to transmit information in both temporal and spatial dimensions. In addition to local features, we introduce global features of each frame as graph nodes, considering complex spatial-temporal dependencies of different body parts and the whole body within a frame or across frames. Linkage modes between nodes are shown in Fig. 2. They include local features within single frame (① in Fig. 2), local features and global features within single frame (④ in Fig. 2), corresponding local features across frames (② in Fig. 2), non-corresponding local features across frames (③ in Fig. 2) and global features across frames (⑤ in Fig. 2). It should be noted that we just illustrate the modes of connections and do not draw all the connections in Fig. 2. The connections between different local features including both within and across frames are defined by human skeleton. Information transfer will be performed among key points in adjacent positions.

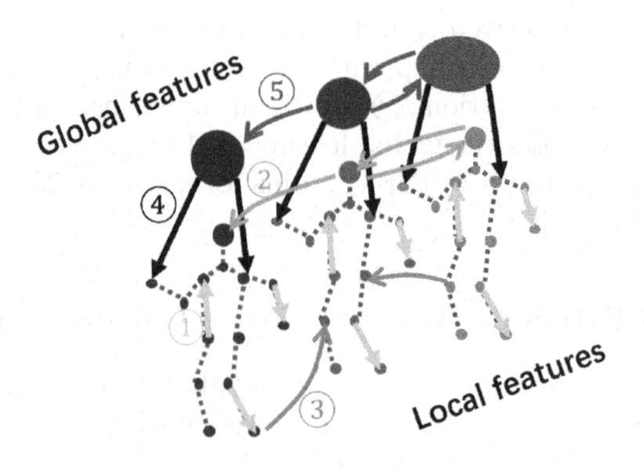

Fig. 2. Linkage modes among nodes.

After determining whether there is a connection relationship between each node, our method delicately designs the information propagation metric. Inspired by the assumption proposed in [9] that the meaningful local feature is more similar to the global feature than the meaningless local feature, a method to learn the reliability of local features and compute the value of $A(i, j)$ is proposed. Suppose node i is local feature of kth key point of frame t, the reliability of node i, referred as D_i is learned as follows:

$$D_i = FC\left(BN\left(abs\left(v_k^t - V_G{}^t\right)\right)\right) \tag{4}$$

where $abs(\cdot)$ and $BN(\cdot)$ are absolution and batch normalization. $FC(\cdot)$ is fully connected layer mapping a vector with size $1 \times C$ to a real number. The reliabilities of local features are normalized by a SoftMax operation.

If both node j and node i are local features and exist information transfer (types ① ② ③ in Fig. 2), $A(i, j)$ is calculated as follow:

$$A(i, j) = ReLU(1 + \alpha(D_j - D_i)) \times \alpha(D_i + D_j) \tag{5}$$

where α is a hyperparameter larger than zero to balance the transfer metric between global features and local features. D_i and D_j are the reliability of node i and node j calculated by (4). Obviously, if node j is more reliable than node i, node j will transmit more information to node i. If node j is global feature and node i is local feature and exist information transfer (type ④ in Fig. 2), the transfer metric is set as $ReLU(1 - \beta \times D_i)$, where β is another hyperparameter. And the pass-through metrics between global feature and global feature (type ⑤ in Fig. 2) is set as $1/(T-1)$. The matrix A is normalized as $\hat{A}$ by applying $L1$ normalization operation to each row of A.

By (3), output feature O after the graph convolution can be obtained. For stabilize training, we fuse O and the input features X as in the ResNet [34]. The output of the TE-AGC layer is the improved feature $V^{'}$.

$$V^{'} = FC(X) + ReLU(O) \tag{6}$$

where $ReLU(\cdot)$ is the activation function that if the input is greater than zero, it remains unchanged; otherwise the output is zero. The $V^{'}$ contains improved local features $\{V_L^{'t}\}_{t=1}^T$ and improved global features $\{V_G^{'t}\}_{t=1}^T$. The $V_L^{'t} = \{v_{\ k}^{'t}\}_{k=1}^K$, $v_{\ k}^{'t}$ is the improved local feature around k^{th} key point of frame t.

With this adaptive method we proposed, the linkage among nodes is decided by the input features. Nodes with high reliability will transfer more information to nodes with low reliability. Therefore, the information can be transmitted more effectively through graph convolution.

3.3 Model Optimizing

After obtaining the improved features and preliminary semantic features, we will further incorporate the global and local representations. We employ a temporal average pooling layer $g_{TAP}(\{\cdot\}_{t=1}^T)$ to generate the time average (TA) feature vector

$$V_G{}^{TA} = g_{TAP}(\{V_G{}^t\}_{t=1}^T) \tag{7}$$

$$V_G^{'TA} = g_{TAP}(\{V_G^{'t}\}_{t=1}^T) \tag{8}$$

And we obtain the local and global combined TA feature vector by

$$V_C{}^{TA} = g_{TAP}(\{\sum_{k=1}^{K} v_k^t\}_{t=1}^T) + V_G{}^{TA} \tag{9}$$

$$V_C^{'TA} = g_{TAP}(\{\sum_{k=1}^{K} v_{\ k}^{'t}\}_{t=1}^T) + V_G^{'TA} \tag{10}$$

The model is optimized by lose function. We utilized identification loss and triplet loss for $V_G^{'TA}$, $V_C^{'TA}$, $V_G{}^{TA}$ and $V_C{}^{TA}$. We combine the identification loss and triplet loss as the total loss with a weighted parameter λ to balance weights of different kind of loss. We adopt triplet loss with hard mining strategy [35] and identification loss with label smoothing regularization [36] to optimize the loss function.

4 Experiments

4.1 Dataset and Implementation

Datasets. Two benchmarks of video-based person ReID datasets, MARS and DukeMTMC-VideoReID, are utilized to evaluate the TE-AGC. MARS, the largest video-based person ReID dataset, contains 17503 tracklets from 1261 identities and 3248 distractor sequences. 625 identities are contained in training set and 636 identities are contained in test set. DukeMTMC-Video is derived from the DukeMTMC dataset, with 4832 tracklets from 1812 identities. There are 408, 702, 702 identities for distraction, training and testing respectively.

Evaluation Protocols. We adopt the mean average precision (mAP) and the Cumulative Matching Characteristic (CMC) to evaluate the performance of our method.

Implementation Details. We set $T = 3$, which means we randomly select three frames as an input sample from a variable-length video clip. Each image is resized to 256×128 pixels. Random horizontal flips and random erasing are performed in the image augmentation process. We employ the ResNet-50 [34] pre-trained on ImageNet [37] as the backbone network after removing the global average pooling and full connected layers. We use HR-Net [38] pretrained on the COCO dataset [39] as the human key points detector. In our method, 13 body key-points are used. During the training period, the learning rate is initialized as 3.5×10^{-4} and decayed by 5 after every 70 epochs. The optimizer is Adam with weight decay 5×10^{-4}. The model is totally trained for 500 epochs. During inference, the representation of a video clip to calculate the similar scores is the $V_C^{\prime TA}$. It should be noted that, when we set $T = 4$ or more, the ReID performance is very similar to $T = 3$. Considering the calculation cost, $T = 3$ is suitable for our method. Therefore, the following experimental analysis is completed under the setting of $T = 3$.

4.2 Comparison with State-of-the-Arts

Table 1 makes a comparison between our method and state-of-the-art algorithms on MARS and DukeMTMC-VideoReID datasets. Our method has achieved state-of-the-art performance.

Results on MARS. Our method is compared with 12 state-of-the-art methods on MARS dataset. Among these methods, AGRL, STGCN and MGH are three other graph-based methods. Compared with these graph-based methods, our method achieves higher Rank-1, Rank-5 accuracy and mAP. There are two main reasons for this improvement. One is our method considers the complex spatial-temporal relation among different body parts and whole body within a frame or across frames. On the other hand, instead of using pair-wise feature affinity, the information pass metrics we designed encourage reliable nodes to pass more information to other modes.

Results on DukeMTMC-Video. Our method is compared with 11 state-of-the-art methods on DukeMTMC-Video dataset. Our method has gotten 97.2% rank-1 results and 96.3% mAP, which exceeds the vast majority of state-of-the-art methods. The comparison verifies the effectiveness of our method.

Table 1. Performance comparison to the state-of-the-art methods on MARS and DukeMTMC-VideoReID dataset.

Methods	MARS			DukeMTMC-VideoReID		
	Rank-1	Rank-5	mAp	Rank-1	Rank-5	mAP
GLTR [40]	87.0	95.8	78.5	96.3	99.3	93.7
COSAM [23]	84.9	95.5	79.9	95.4	99.3	94.1
VRSTC [41]	88.5	96.5	82.3	95.0	99.1	93.5
RGSAT [42]	89.4	96.9	84.0	97.2	**99.4**	95.8
AGRL [27]	89.8	96.1	81.1	96.7	99.2	94.2
TCLNet [24]	89.8	-	85.1	96.9	-	96.2
STGCN [28]	90.0	96.4	83.7	**97.3**	99.3	95.7
MGH [29]	90.0	96.7	85.8	-	-	-
AP3D [43]	90.1	-	85.1	96.3	-	95.6
AFA [44]	90.2	96.6	82.9	97.2	**99.4**	95.4
BiCnet-TKS [13]	90.2	-	**86.0**	96.3	-	96.1
TE-AGC (ours)	**90.7**	**97.5**	85.8	97.2	**99.4**	**96.3**

4.3 Model Component Analysis

The architecture of our method has only one branch, which is both concise and effective. Here the contribution of each part of TE-AGC is evaluated and results on MARS dataset are reported.

Table 2 reports the experimental results of the ablation studies for TE-AGC. The 1st line can be regarded as baseline result of our method. In the 1st line, for each frame, we use the backbone network and key point detector to get global and local features. Then we combine all features like (7) and (9). Compared with 1st line, the 2nd line shows the result of adding the Spatial Attention (SA) layer. In addition, the 3rd line adds GCN layer we designed but does not use the spatial attention layer, which means only removing the spatial attention layer compared with the entire architecture (the 4th line).

Comparing the 1st line with the 2nd line or comparing the 3rd line with the 4th line, the effectiveness of the spatial attention can be directly proved. Though it

Table 2. Component analysis of the effectiveness of each component of TE-AGC on MARS dataset.

Number	Methods	MARS		
		Rank-1	Rank-5	mAP
1	TE-AGC -GCN -SA	88.7	96.1	83.8
2	TE-AGC -GCN	89.0	96.5	84.1
3	TE-AGC -SA	90.3	96.9	85.3
4	TE-AGC	**90.7**	**97.5**	**85.8**

is relatively simple, it still has a certain inhibitory effect on background noise. The comparison of the 2nd & 4th lines and the comparison of the 1st & 3rd lines show the effect of the graph convolution layer we designed. The information of each node is effectively transferred and enhanced during the graph convolution. Finally, the powerful spatio-temporal representation for the video is obtained.

5 Conclusion

In this paper, a novel Temporal Extension Adaptive Graph Convolution (TE-AGC) framework is proposed for video-based person ReID. The TE-AGC could mine features in spatial and temporal dimensions in one graph convolution operation effectively. The TE-AGC utilizes a CNN backbone, a simple spatial attention layer and a key-point detector to extract global and local features. A delicate adaptive graph convolution module is designed to encourage meaningful information transfer by dynamically learning the reliability of local features from multiple frames. We combine the global feature and local features of each frame with a future fusion module to obtain discriminative representations of each video clip. The effectiveness of the TE-AGC method is verified by a large number of experiments on two video datasets.

References

1. Ye, M., Shen, J., Lin, G., Xiang, T., Hoi, S.: Deep learning for person re-identification: a survey and outlook. IEEE Trans. Pattern Anal. Mach. Intell. 1–1 (2021)
2. Zheng, L., Yang, Y., Hauptmann, A.G.: Person re-identification: past, present and future. Sensors (Basel) **22**(24), 9852 (2016)
3. Farenzena, M., Bazzani, L., Perina, A., Murino, V., Cristani, M.: Person re-identification by symmetry-driven accumulation of local features. In: Proeedings of IEEECONFERENCE on Computer Vision & Patternrecognition. pp. 2360–2367 (2010)
4. Liu, C., Gong, S., Loy, C.C., Lin, X.: Person re-identification: what features are important? In: Fusiello, A., Murino, V., Cucchiara, R. (eds.) ECCV 2012. LNCS, vol. 7583, pp. 391–401. Springer, Heidelberg (2012). https://doi.org/10.1007/978-3-642-33863-2_39

5. Liao, S., Yang, H., Zhu, X., Li, S.Z.: Person re-identification by local maximal occurrence representation and metric learning. In: 2015 IEEE Conference on Computer Vision and Pattern Recognition (CVPR). (2015)
6. Matsukawa, T., Okabe, T., Suzuki, E., Sato, Y.: Hierarchical gaussian descriptor for person re-identification. In: 2016 IEEE Conference on Computer Vision and Pattern Recognition (CVPR) (2016)
7. Xiong, F., Gou, M., Camps, O., Sznaier, M.: Person re-identification using kernel-based metric learning methods. In: Fleet, D., Pajdla, T., Schiele, B., Tuytelaars, T. (eds.) ECCV 2014. LNCS, vol. 8695, pp. 1–16. Springer, Cham (2014). https://doi.org/10.1007/978-3-319-10584-0_1
8. Zheng, W.S., Xiang, L., Tao, X., Liao, S., Lai, J., Gong, S.: Partial person re-identification. In: IEEE International Conference on Computer Vision. (2016)
9. Wang, G., et al.: High-order information matters: Learning relation and topology for occluded person re-identification. In: 2020 IEEE/CVF Conference on Computer Vision and Pattern Recognition (CVPR)(2020)
10. Chung, D., Tahboub, K., Delp, E.J.: A two stream Siamese convolutional neural network for person re-identification. In: 2017 IEEE International Conference on Computer Vision (ICCV) (2017)
11. Chen, D., Li, H., Tong, X., Shuai, Y., Wang, X.: Video person re-identification with competitive snippet-similarity aggregation and co-attentive snippet embedding. In: 2018 IEEE/CVF Conference on Computer Vision and Pattern Recognition (CVPR) (2018)
12. Xu, S., Yu, C., Kang, G., Yang, Y., Pan, Z.: Jointly attentive spatial-temporal pooling networks for video-based person re-identification. In: 2017 IEEE International Conference on Computer Vision (ICCV). (2017)
13. Hou, R., Chang, H., Ma, B., Huang, R., Shan, S.: BiCnet-TKS: learning efficient spatial-temporal representation for video person re-identification (2021)
14. Liu, J., Zha, Z.J., Wu, W., Zheng, K., Sun, Q.: Spatial-temporal correlation and topology learning for person re-identification in videos. In: Proceedings of the IEEE/CVF Conference on Computer Vision and Pattern Recognition (CVPR), pp. 4370–4379 (2021)
15. Miao, J., Wu, Y., Liu, P., Ding, Y., Yang, Y.: Pose-guided feature alignment for occluded person re-identification. In: 2019 IEEE/CVF International Conference on Computer Vision (ICCV) (2019)
16. Sun, Y., et al.: Perceive where to focus: Learning visibility-aware part-level features for partial person re-identification. In: 2019 IEEE/CVF Conference on Computer Vision and Pattern Recognition (CVPR) (2020)
17. Mclaughlin, N., Rincon, J., Miller, P.: Recurrent convolutional network for video-based person re-identification. In: 2016 IEEE Conference on Computer Vision and Pattern Recognition (CVPR) (2016)
18. Gao, J., Nevatia, R.: Revisiting temporal modeling for video-based person reID (2018)
19. Zheng, L., Bie, Z., Sun, Y., Wang, J., Su, C., Wang, S., Tian, Q.: MARS: a video benchmark for large-scale person re-identification. In: Leibe, B., Matas, J., Sebe, N., Welling, M. (eds.) ECCV 2016. LNCS, vol. 9910, pp. 868–884. Springer, Cham (2016). https://doi.org/10.1007/978-3-319-46466-4_52
20. Fu, Y., Wang, X., Wei, Y., Huang, T.S.: STA: spatial-temporal attention for large-scale video-based person re-identification. In: National Conference on Artificial Intelligence (2019)

21. Li, S., Bak, S., Carr, P., Wang, X.: Diversity regularized spatiotemporal attention for video-based person re-identification. In: Proceedings of the IEEE Conference on Computer Vision and Pattern recognition, pp. 369–378 (2018)
22. Ouyang, D., Zhang, Y., Shao, J.: Video-based person re-identification via spatio-temporal attentional and two-stream fusion convolutional networks. Pattern Recog. Lett. **117**, 153–160 (2018)
23. Subramaniam, A., Nambiar, A., Mittal, A.: Co-segmentation inspired attention networks for video-based person re-identification. In: Proceedings of the IEEE/CVF International Conference on Computer Vision, pp. 562–572 (2019)
24. Hou, R., Chang, H., Ma, B., Shan, S., Chen, X.: Temporal complementary learning for video person re-identification. In: Vedaldi, A., Bischof, H., Brox, T., Frahm, J.-M. (eds.) ECCV 2020. LNCS, vol. 12370, pp. 388–405. Springer, Cham (2020). https://doi.org/10.1007/978-3-030-58595-2_24
25. Jones, M.J., Rambhatla, S.: Body part alignment and temporal attention for video-based person re-identification. In: BMVC (2019)
26. Zhao, H., et al.: Spindle net: person re-identification with human body region guided feature decomposition and fusion. In: Proceedings of the IEEE Conference on Computer Vision and Pattern Recognition, pp. 1077–1085 (2017)
27. Wu, Y., Bourahla, O.E.F., Li, X., Wu, F., Tian, Q., Zhou, X.: Adaptive graph representation learning for video person re-identification. IEEE Trans. Image Process. **29**, 8821–8830 (2020)
28. Yang, J., Zheng, W.S., Yang, Q., Chen, Y.C., Tian, Q.: Spatial-temporal graph convolutional network for video-based person re-identification. In: Proceedings of the IEEE/CVF Conference on Computer Vision and Pattern Recognition, pp. 3289–3299 (2020)
29. Yan, Y., Qin, J., Chen, J., Liu, L., Zhu, F., Tai, Y., Shao, L.: Learning multi-granular hypergraphs for video-based person re-identification. In: Proceedings of the IEEE/CVF Conference on Computer Vision and Pattern Recognition, pp. 2899–2908 (2020)
30. Zhang, Z., Lan, C., Zeng, W., Jin, X., Chen, Z.: Relation-aware global attention for person re-identification. In: Proceedings of the IEEE/CVF Conference on Computer Vision and Pattern Recognition, pp. 3186–3195 (2020)
31. Kipf, T.N., Welling, M.: Semi-supervised classification with graph convolutional networks. arXiv preprint arXiv:1609.02907 (2016)
32. Obinata, Y., Yamamoto, T.: Temporal extension module for skeleton-based action recognition. In: 2020 25th International Conference on Pattern Recognition (ICPR), IEEE, pp. 534–540 (2021)
33. Shi, W., Rajkumar, R.: Point-GNN: graph neural network for 3D object detection in a point cloud. In: Proceedings of the IEEE/CVF Conference on Computer Vision and Pattern Recognition, pp. 1711–1719 (2020)
34. He, K., Zhang, X., Ren, S., Sun, J.: Deep residual learning for image recognition. In: Proceedings of the IEEE Conference on Computer Vision and Pattern Recognition, pp. 770–778 (2020)
35. Hermans, A., Beyer, L., Leibe, B.: In defense of the triplet loss for person re-identification. arXiv preprint arXiv:1703.07737 (2017)
36. Szegedy, C., Vanhoucke, V., Ioffe, S., Shlens, J., Wojna, Z.: Rethinking the inception architecture for computer vision. In: Proceedings of the IEEE Conference on Computer Vision and Pattern Recognition, pp. 2818–2826 (2016)
37. Deng, J., Dong, W., Socher, R., Li, L.J., Li, K., Fei-Fei, L.: ImageNet: a large-scale hierarchical image database. In: 2009 IEEE Conference on Computer Vision and Pattern Recognition, IEEE, pp. 248–255 (2009)

38. Sun, K., Xiao, B., Liu, D., Wang, J.: Deep high-resolution representation learning for human pose estimation. In: Proceedings of the IEEE/CVF Conference on Computer Vision and Pattern Recognition, pp. 5693–5703 (2019)
39. Lin, T.Y., et al.: Microsoft COCO: common objects in context. In: Fleet, D., Pajdla, T., Schiele, B., Tuytelaars, T. (eds.) ECCV 2014. LNCS, vol. 8693, pp. 740–755. Springer, Cham (2014). https://doi.org/10.1007/978-3-319-10602-1_48
40. Li, J., Wang, J., Tian, Q., Gao, W., Zhang, S.: Global-local temporal representations for video person re-identification. In: Proceedings of the IEEE/CVF International Conference on Computer Vision, pp. 3958–3967 (2019)
41. Hou, R., Ma, B., Chang, H., Gu, X., Shan, S., Chen, X.: VRSTC: occlusion-free video person re-identification. In: Proceedings of the IEEE/CVF Conference on Computer Vision and Pattern Recognition, pp. 7183–7192 (2019)
42. Li, X., Zhou, W., Zhou, Y., Li, H.: Relation-guided spatial attention and temporal refinement for video-based person re-identification. In: Proceedings of the AAAI Conference on Artificial Intelligence. **34**, 11434–11441 (2020)
43. Gu, X., Chang, H., Ma, B., Zhang, H., Chen, X.: Appearance-preserving 3D convolution for video-based person re-identification. In: Vedaldi, A., Bischof, H., Brox, T., Frahm, J.-M. (eds.) ECCV 2020. LNCS, vol. 12347, pp. 228–243. Springer, Cham (2020). https://doi.org/10.1007/978-3-030-58536-5_14
44. Chen, G., Rao, Y., Lu, J., Zhou, J.: Temporal coherence or temporal motion: which is more critical for video-based person re-identification? In: Vedaldi, A., Bischof, H., Brox, T., Frahm, J.-M. (eds.) ECCV 2020. LNCS, vol. 12353, pp. 660–676. Springer, Cham (2020). https://doi.org/10.1007/978-3-030-58598-3_39

Deep RGB-Driven Learning Network for Unsupervised Hyperspectral Image Super-Resolution

Zhe Liu and Xian-Hua Han(✉)

Graduate School of Sciences and Technology for Innovation, Yamaguchi University, 1677-1 Yoshida, Yamaguchi 753-8511, Japan
{a501wbu,hanxhua}@yamaguchi-u.ac.jp

Abstract. Hyperspectral (HS) images are used in many fields to improve the analysis and understanding performance of captured scenes, as they contain a wide range of spectral information. However, the spatial resolution of hyperspectral images is usually very low, which limits their wide applicability in real tasks. To address the problem of low spatial resolution, super-resolution (SR) methods for hyperspectral images (HSI) have attracted widespread interest, which aims to mathematically generate high spatial resolution hyperspectral (HR-HS) images by combining degraded observational data: low spatial resolution hyperspectral (LR-HS) images and high resolution multispectral or RGB (HR-MS/RGB) images. Recently, paradigms based on deep learning have been widely explored as an alternative to automatically learn the inherent priors for the latent HR-HS images. These learning-based approaches are usually implemented in a fully supervised manner and require large external datasets including degraded observational data: LR-HS/HR-RGB images and corresponding HR-HS data, which are difficult to collect, especially for HSI SR scenarios. Therefore, in this study, a new unsupervised HSI SR method is proposed that uses only the observed LR-HS and HR-RGB images without any other external samples. Specifically, we use an RGB-driven deep generative network to learn the desired HR-HS images using a encoding-decoding-based network architecture. Since the observed HR-RGB images have a more detailed spatial structure and may be more suitable for two-dimensional convolution operations, we employ the observed HR-RGB images as input to the network as a conditional guide and adopt the observed LR-HS/HR-RGB images to formulate the loss function that guides the network learning. Experimental results on two HS image datasets show that our proposed unsupervised approach provides superior results compared to the SoTA deep learning paradigms.

Keywords: Hyperspectral image · Super-resolution · Unsupervised learning

Y. Zheng et al. (Eds.): ACCV 2022, LNCS 13848, pp. 226–239, 2023.
https://doi.org/10.1007/978-3-031-27066-6_16

1 Introduction

With hyperspectral (HS) imaging, detailed spectral direction traces and rich spectral features in tens or even hundreds of bands can be obtained at every spatial location in a scene, which can significantly improve the performance of different HS processing systems. Existing HS image sensors typically collect HS data at a low spatial resolution, which severely limits their wide applicability in the real world. Therefore, generating high-resolution hyperspectral (HR-HS) images by combining the degraded observational data: low-resolution hyperspectral (LR-HS) and high-resolution multispectral/RGB (HR-MS/RGB) images, known as HS super-resolution (HSI SR) images, has attracted great attention in the field of computer vision [29,32], medical diagnosis [19,21,33], mineral exploration [24,30] and remote sensing [3,20,23]. According to the reconstruction principles, HSI SR is mainly divided into two categories: traditional mathematical model-based methods and deep learning-based methods. In the following, we will describe these two types of methods in detail.

1.1 Traditional Mathematical Model-Based Methods

In the past decades, most HSI SR methods have focused on studying various manually computed a prior parameters to develop a mathematical model and use optimization techniques to solve the problem. Specifically, such methods have focused on developing a mathematical formulation to model the process of degrading HR-HS images into LR-HS images and HR-RGB images. Since the known variables of the observed LR-HS/HR-RGB images are much smaller than the underestimated variables of the HR-HS images, this task is extremely challenging and direct optimization of the formulated mathematical model would lead to a very unstable solution. Therefore, existing effort usually exploits various priors to regularize the mathematical model, i.e., to impose constraints on the solution space. Depending on the priors to be investigated, existing studies are generally classified into three different approaches: spectral unmixing-based [16], sparse representation-based [6] and tensor factorization-based methods [4]. In the spectral un-mixing-based method, Yokoya et al. [31] proposed a coupled non-negative matrix decomposition (CNMF) method, which alternately blends LR-HS images and HR-RGB images to estimate HR-HS images. Recently, Lanaras et al. [16] proposed a similar framework to jointly extract two observed images by decomposing the original optimization problem into two constrained least squares problems. A similar framework was proposed by Dong et al. [6], which employed the alternating direction method of multipliers (ADMM) to solve the spectral hash model for the robust reconstruction of the base image of HR-HS images.

In addition, sparse representation is widely used as an alternative mathematical model for HSI SR, where a spectral dictionary is first learned from the observed HR-LS image, and then the sparse coefficients of the HR-RGB image are calculated to reconstruct the HR-HS image. For example, Zhao et al. [34] used the K-SVD method to learn the dictionary, and then adopted the sparse matrix decomposition to combine the LR-HS and HR-RGB images to reconstruct the HR-HS image. Inspired by the spectral similarity of neighboring pixels in latent HS images, Akhtar et al. [1] proposed to perform group sparsity and non-negativity representations, while Kawakami et al. [15] used a sparse regularizer to decompose the spectral dictionary. In addition, Han et al. [9] proposed a non-negative sparse coding algorithm that effectively exploits pixel sparsity and non-local spatial similarity in HR-HS images. Furthermore, the tensor factorization-based approach was shown to be feasible for the HSI SR problem. Motivated by the inherent low dimensionality of spectral signatures and the 3D structure of HR-HS images, He et al. [13] employed matrix factorization to decomposed the HR-HS image into two low-rank constraint matrices and showed impressive super-resolution results. Despite some improvements achieved using manually designed priors, super-resolution performance tends to be unstable and sensitive to the content varying in the under-studying investigated images as well as may lead to significant spectral distortions due to the insufficient representation of empirically designed priors.

1.2 Deep Learning-Based Methods

Deep Supervised Learning-Based Methods. Due to the high success of DCNN in different vision tasks, DCNN-based approaches have been proposed for HSI SR tasks to automatically learn specific priors for the latent HR-HS images. Han et al. [10] firstly conducted a pioneering work for merging the HS/RGB image to estimating the latent HR-HS image using deep learning network, which contained three simple convolutional layers but demonstrated very impressive performance, and then employed more complex CNN architectures such as ResNet and DenseNet [8] for performance improvement. Dian et al. [5] proposed an optimization and learning integration strategy for the fusion task by first solving the Sylvester equation and then exploring a DCNN-based approach to improve the initialization results. Han et al. [12] further investigated a multi-level and multi-scale spatial and spectral fusion network to effectively fuse the observed LR-HS and HR-RGB images with large spatial structure differences. Xie et al. [28] studied the MS/HS fusion network using a low-resolution imaging model and spectral low-rank property of the HR-HS images, and solved the proposed MS/HS fusion network using an approximate gradient strategy. In addition, Zhu et al. [35] investigated a lightweight deep neural network, dubbed as the progressive zero centric residual network (PZRes-Net), to achieve efficiency and performance in solving the HS image reconstruction problem. Despite the significant improvement in reconstruction performance, all the above DCNN-based approaches require training with large external datasets, including the degraded LR-HS/HR-RGB images and corresponding HR-HS images, which are difficult to collect, especially for HSI SR tasks.

Deep Unsupervised Learning-Based Methods. As mentioned above, in practice, it is very difficult to collect enough training triples, especially the latent

HR-HS images for a good-generalization CNN model. Therefore, the quality and quantity of the collected training datasets (external datasets) usually become the bottleneck of DCNN-based approaches. To alleviate the heavy reliance on external datasets, several deep unsupervised learning methods have been investigated to take advantage of the powerful modeling capabilities of deep networks [17]. Qu et al. [22] attempted to solve the HSI super-resolution problem with unsupervised learning strategy and developed an encoder-decoder architecture that exploits the approximate structure of the low-rank spectral prior in the latent HR-HS images. Although this approach does not require external training samples to construct a CNN-based end-to-end model for the recovery of the HR-HS images, it requires careful designing alternative optimization procedure for two sub-problems, and tends to produce unstable super-resolution results. Liu et al. [18] proposed a deep unsupervised fusion learning (DUFL) method to generate the HR-HS image from a random noisy input using the observed HR-RGB and LR-HS images only. However, DUFL aims to use generative networks to learn HR-HS images from a random noise and therefore does not take full use of the available information in the observations such as the HR-RGB image with high spatial resolution structures. Subsequently, Uezato et al. [25] used a deep decoder network to generate the latent HR-HS images from both noisy input data as well as the observed HR-RGB observations as guided information, called a guided deep decoder (GDD) network. In addition, Fu et al. [7] propose to conduct joint optimization of the optimal CSF and the potential HR-HS images from the observed observations only. Although these current unsupervised methods illustrate the potential for plausible HR-HS image generation, most of them do not fully exploit the high-resolution spatial structure of the observed HR-RGB images. Therefore, there is still room for improvement in terms of performance.

To handle the above mentioned issues, this study proposes a new deep RGB-driven generative network for unsupervised HSI SR that uses the observed HR-RGB images instead of a random noise as the network input. Specifically, by leveraging the observed HR-RGB images with high-resolution spatial structure as the input, we design an encoder-decoder based two-dimensional convolutional network to learn the latent HR-HS image, and then follow the specially designed convolutional layers to implement the spatial and spectral degradation procedure for obtaining the approximated LR-HS and HR-RGB images. Thus the loss functions for training the generative network can be formulated using the observed LR-HS and HR-RGB image only where no external samples are required in the end-to-end unsupervised learning model. Experimental results on two HS datasets show that our method outperforms the state-of-the-art methods. The main advantages of this study are summarized as follows.

I. We propose a RGB-driven generative network by making full use of the high-resolution spatial structure in the observed RGB image as the conditional input for robust HR-HS image estimation.

II. We learn the specific CNN model directly from the observations without the need for a labeled training set.

III. We construct a simple convolution-based degradation modules using the specifically designed depth-wise and point-wise convolution layers for imitating the observation procedure, which are easy to be optimized with generative networks.

2 Proposed Method

In this section, we first introduce the problem of formulating the HSI SR task and then describe our proposed deep RGB-driven generative network.

2.1 Problem Formulation

The goal of the HSI SR task is to generate HR-HS images by combining the LR-HS and HR-RGB images: $\mathbf{I_y} \in \Re^{\mathbf{w \times h \times L}}$ and $\mathbf{I_x} \in \Re^{\mathbf{W \times H \times 3}}$, where $W(w)$ and $H(h)$ denote the width and height of the HR-HS (LR-HS) image and $L(3)$ denotes the spectral number. In general, the degradation process of the observations: $\mathbf{I_x}$ and $\mathbf{I_y}$ from the HR-HS image $\mathbf{I_z}$ can be mathematically expressed as follows.

$$\mathbf{I_x} = \mathbf{k^{(Spa)}} \otimes \mathbf{I_z^{(Spa)}} \downarrow + \mathbf{n_x}, \mathbf{I_y} = \mathbf{I_z} * \mathbf{C^{(Spec)}} + \mathbf{n_y}, \tag{1}$$

where $\otimes$ denotes the convolution operator, $\mathbf{k^{(Spa)}}$ is the two-dimensional blur kernel in the spatial domain, and $(Spa) \downarrow$ denotes the down-sampling operator in the spatial domain. $\mathbf{C^{(Spec)}}$ is the spectral sensitivity function of the RGB camera (three filters in the one-dimensional spectral direction), which converts L spectral bands into RGB bands, and $\mathbf{n_x}$, $\mathbf{n_y}$ are additive white Gaussian noise (AWGN) with a noise level of σ. For simplicity, we rewrite the mathematical degradation model in Eq. 1 as the following matrix form.

$$\mathbf{I_x} = \mathbf{D I_z} + \mathbf{n_x}, \mathbf{I_y} = \mathbf{I_z C} + \mathbf{n_y}, \tag{2}$$

where $\mathbf{D}$ is the spatial degradation matrix containing the spatial blurring matrix and the down-sampling matrix, and $\mathbf{C}$ is the spectral transformation matrix representing the camera spectral sensitivity function (CSF). By assuming the known spatial and spectral degradations, the HSI SR problem can be solved intuitively through minimizing the following reconstruction errors.

$$\mathbf{I_z}^* = \arg\min_{\mathbf{I_z}} \|\mathbf{I_x} - \mathbf{D I_z}\|_F^2 + \|\mathbf{I_y} - \mathbf{I_z C}\|_F^2, \tag{3}$$

where $|\cdot|_F$ denotes the Frobenius norm. However, Eq. 3 may have several optimal solutions that yield minimal reconstruction errors, and thus direct optimization would lead to a very unstable solution. Most existing methods typically use different hand-crafted priors to model the potential HR-HS to impose the constraints on Eq. 3 for narrowing the solution space, and demonstrate great performance improvement with elaborated priors. The prior-regularized mathematical model approach is expressed as follows.

$$\mathbf{I_z}^* = \arg\min_{\mathbf{I_z}} \|\mathbf{I_x} - \mathbf{D I_z}\|_F^2 + \|\mathbf{I_y} - \mathbf{I_z C}\|_F^2 + \alpha\phi(\mathbf{I_z}), \tag{4}$$

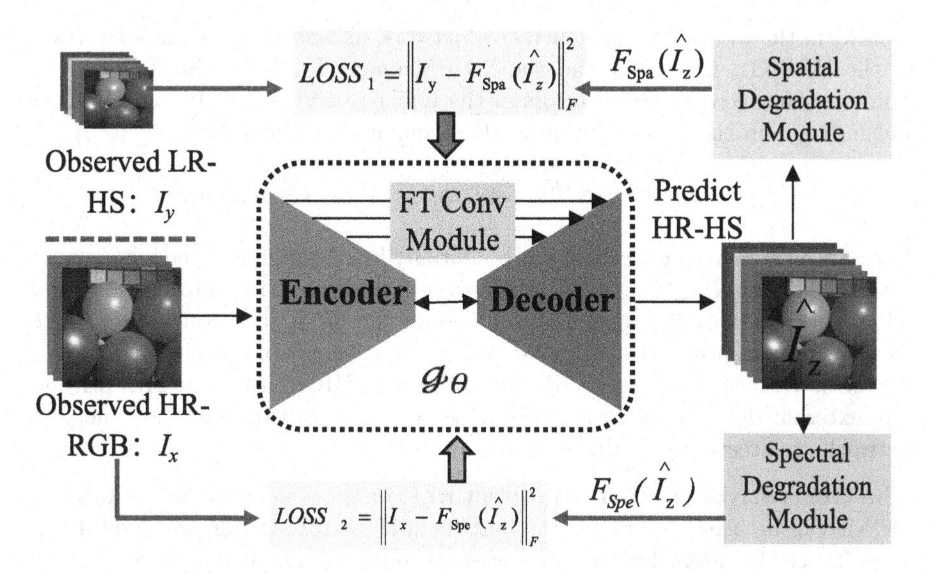

Fig. 1. Proposed framework of deep RGB-driven generative network. FT Conv Module denotes feature transfer convolution module. Spatial Degradation module is implemented by a specifically designed depth-wise convolution layer while Spectral Degradation module is realized by a point-wise convolution layer.

where $\phi(\mathbf{I_z})$ is used as the regularization term for modeling the prior in the latent HR-HS image, while α is the hyper-parameter for adjusting the contribution of the regularization term and the reconstruction error. However, the investigated priors in the existing methods are designed empirically and usually face difficulty to sufficiently modeling the complicated spatial and spectral structures.

2.2 Proposed Deep RGB-Driven Generative Network

As shown in many vision tasks, deep convolutional networks have powerful modeling capabilities to capture the inherent prior knowledge of different visual data (images), and in this study, deep learning networks are used to automatically learn the prior knowledge embedded in HR-HS images. Specifically, we use an encoder-decoder-based generative network to automatically reconstruct HR-HS images. In the absence of the ground-truth HR-HS images for training the generative network as in the conventional fully supervised learning networks, we employ the observed HR-RGB and LR-HS images to formulate the loss functions expressed in Eq. 3.

Specifically, given the predicted HR-HS image $\hat{\mathbf{I}}_\mathbf{z} = g_\theta(\cdot)$, where g_θ denotes the generative network and θ is its parameter, we specifically designed a depth-wise convolutional layer to implement a spatially degradation model as $F_{Spa}(\hat{\mathbf{I}}_\mathbf{z})$ and a point-wise convolutional layer to implement the spectral transform $F_{Spe}(\hat{\mathbf{I}}_\mathbf{z})$. By simply fixing the weights of the specially designed convolutional layers in the spatial degradation matrix $\mathbf{D}$ and the CSF matrix $\mathbf{C}$, we can transform the output of the generative network into the approximated versions of the HR-RGB image: $\mathbf{I_x}$ and the LR-HS image: $\mathbf{I_y}$. According to Eq. 3, we minimize the reconstruction errors of the under-studying LR-HS and HR-RGB images to train the generative network, formulated as the follows:

$$\theta^* = \arg\min_{\theta} \|\mathbf{I}_x - F_{Spe}(g_\theta(\cdot))\|_F^2 + \|\mathbf{I}_y - F_{Spa}(g_\theta(\cdot))\|_F^2, \tag{5}$$

Since the generative network g_θ can potentially learn and model the inherent priors in the latent HR-HS image, it is not necessary to explicitly impose prior modeling constraints as the regularization termin the Eq. 5. The conceptual framework of the proposed deep RGB-driven generativenetwork is shown in Fig. 1. It can be trained using the observed LR-HS and HR-RGB image only without any external data. In the following subsections, we will present the generative network architecture and the network inputs.

The Generative Network Architecture: For the generative network g_θ, any DCNN can be used to serve as the beseline architecture in our proposed framework. Since the latent HR-HS images often contain various structures, rich textures, and complex spectra, the employed generative network g_θ has to possess enough modeling ability to ensure reliable HR-HS image representation. Several generative architectures [2] have been investigated and significant progress has been made in generating high-quality natural images [14], for example in the context of the adversarial learning research. Since our unsupervised framework requires training a specific CNN model for each under-studying observation, shallower networks are preferred to reduce the training time. Moreover, it is known that a deeper network architecture, which can capture feature representation in a large receptive field can improve the super-resolution performance. Therefore, a shallow network with sufficient representation modeling capability in a larger receptive field would be suitable for our network structure.

It is well known that encoder-decoder networks have a shallow structure being possible to learn feature representation in large-scale spatial context due to down-sampling operations between adjacent scales, and thus we employ the encoder-decoder structure as our generative network. In detail, the generative network consists of an encoder subnet and a decoder subnet, and both encoder and decoder include multiple blocks with different scales that can capture feature at different receptive fields. The outputs of all blocks in the encoder subnet are transferred to the corresponding blocks in the decoder using a convolution-based feature transfer module (FT Conv module) to reuse the learned detailed features. Each block consists of three convolutional/RELU pairs, where a max pooling layer with 2×2 kernels are used to reduce the feature map size between the blocks of the encoder, and an up-sampling layer is used to double the feature map size between the blocks of the decoder. Finally, a vanilla convolution-based output layer is used to estimate the underlying HR-HS image.

The RGB-Guided Input: Most generative neural networks are trained to synthesize the target images with the specific defined concept from noisy vectors, which are randomly generated based on a distribution function (e.g., Gaussian

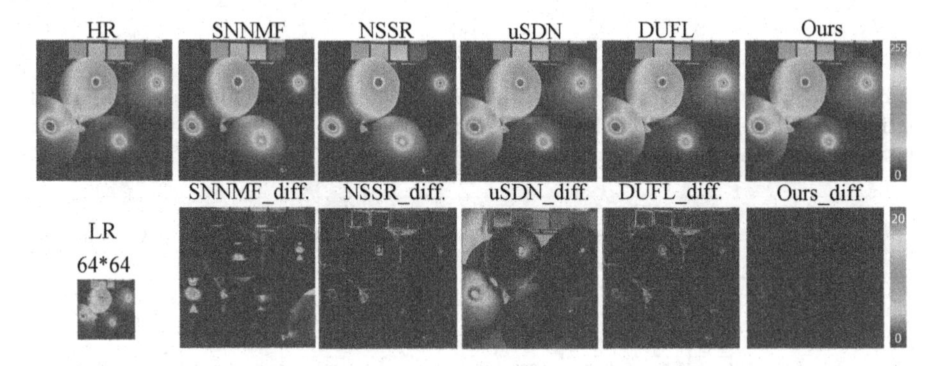

Fig. 2. Visual results of mathematical optimization-based methods: SNNMF [27], NSSR [6] and deep learning-based methods: uSDN [22], DUFL [18] on the CAVE dataset and our method with the up-scale factor 8.

or uniform distribution). As recent studies have confirmed, randomly generated noisy inputs usually produce sufficiently diverse and unique images. Our HSI SR task aims to use the observed LR-HS and HR-RGB images to learn the corresponding HR-HS images. Simply using noise as input does not take full advantage of the existed information in the observations. Therefore, we attempt to employ the available observation as a conditional guide for our generative network.

The observed HR-RGB images are known to have a high spatial resolution structure, and it is expected to assist the two-dimension convolution-based generative network learning more effective representation for reliable HR-HS image recovery. The observed LR-HS images can also be used as conditional inputs to the network. However, the low-resolution spatial structure may lead to local minimization of the network training process, which may have a negative impact on the prediction results. More importantly, the magnification factor, e.g., 10 for 31 spectral bands estimation from RGB in the spectral domain,, is usually much smaller than in the spatial domain (64 in total (8×8) with an up-sampling factor of 8), so we use the observed HR-RGB image as conditional network input, denoted as $\mathbf{Z}^* = g_\theta(\mathbf{I}_y)$.

3 Experiment Result

3.1 Experimental Setting

We evaluated our method on two commonly used datasets, Cave and Harvard datasets. The Cave dataset contains 32 HS images taken in real material and object space, all with the same spatial resolution, e.g. 512×512 with 31 adjacent spectral bands ranging from 400 nm to 700 nm. The Harvard dataset contains 50 HS images taken during daylight hours, both outdoors and indoors, all with the same spatial resolution of 1392×1040 and 31 spectral bands ranging from 420 nm

Table 1. Compared results with the SoTA methods including mathematical optimization-based and deep learning-based methods on both CAVE and Harvard datasets with the up-scale factors 8 and 16.

		CAVE					Harvard				
		RMSE↓	PSNR↑	SSIM↑	SAM↓	ERGAS↓	RMSE↓	PSNR↑	SSIM↑	SAM↓	ERGAS↓
		Up-scale factor = 8									
Mathematical optimization	GOMP	5.69	33.64	–	11.86	2.99	3.79	38.89	–	4.00	1.65
	MF	2.34	41.83	–	3.88	1.26	1.83	43.74	–	2.66	0.87
	SNNMF	1.89	43.53	–	3.42	1.03	1.79	43.86	–	2.63	0.85
	CSU	2.56	40.74	0.985	5.44	1.45	1.40	46.86	0.993	1.77	0.77
	NSSR	1.45	45.72	0.992	2.98	0.80	1.56	45.03	0.993	2.48	0.84
Deep learning	SSFNet	1.89	44.41	0.991	3.31	0.89	2.18	41.93	0.991	4.38	0.98
	DHSIS	1.46	45.59	0.990	3.91	0.73	1.37	46.02	0.981	3.54	1.17
	ResNet	1.47	45.90	0.993	2.82	0.79	1.65	44.71	0.984	2.21	1.09
	uSDN	4.37	35.99	0.914	5.39	0.66	2.42	42.11	0.987	3.88	1.08
	DUFL	2.08	42.50	0.975	5.36	1.16	2.38	42.16	0.965	2.35	1.09
	Ours	1.35	46.20	0.992	3.05	0.77	1.07	49.17	0.994	1.59	0.72
	Up-scale factor = 16										
Mathematical optimization	GOMP	6.08	32.96	–	12.60	1.43	3.83	38.56	–	4.16	0.77
	MF	2.71	40.43	–	4.82	0.73	1.94	43.30	–	2.85	0.47
	SNNMF	2.45	42.21	–	4.61	0.66	1.93	43.31	–	2.85	0.45
	CSU	2.87	39.83	0.983	5.65	0.79	1.60	45.50	0.992	1.95	0.44
	NSSR	1.78	44.01	0.990	3.59	0.49	1.65	44.51	0.993	2.48	0.41
Deep learning	SSFNet	2.18	41.93	0.991	4.38	0.98	1.94	43.56	0.980	3.14	0.98
	DHSIS	2.36	41.63	0.987	4.30	0.49	1.87	43.49	0.983	2.88	0.54
	ResNet	1.93	43.57	0.991	3.58	0.51	1.83	44.05	0.984	2.37	0.59
	uSDN	3.60	37.08	0.969	6.19	0.41	9.31	39.39	0.931	4.65	1.72
	DUFL	2.61	40.71	0.967	6.62	0.70	2.81	40.77	0.953	3.01	0.75
	Ours	1.71	44.15	0.990	3.63	0.48	1.28	47.37	0.992	1.92	0.49

Table 2. Ablation studies of different numbers of employed blocks in the generative network and loss terms on CAVE dataset with the up-scale factor 8.

Number of employed blocks	Loss	RMSE↓	PSNR↑	SSIM↑	SAM↓	ERGAS↓
2	Both	1.45	45.49	0.992	3.47	0.81
3		1.42	45.69	0.992	3.28	0.81
4		1.38	46.05	0.993	3.13	0.77
5	Loss1	26.27	19.85	0.601	43.53	16.19
	Loss2	3.30	38.57	0.972	3.68	1.88
	Both	1.35	46.20	0.992	3.05	0.77

Table 3. Ablation studies of different network inputs on CAVE dataset with the up-scale factor 8.

Block number: 5 Loss: both					
Input	RMSE↓	PSNR↑	SSIM↑	SAM↓	ERGAS↓
Noise	2.10	42.53	0.978	5.30	1.12
Combined	1.46	45.47	0.992	3.27	0.81
Combined + noise	1.44	45.61	0.992	3.72	0.80
Ours (RGB)	1.35	46.20	0.992	3.05	0.77

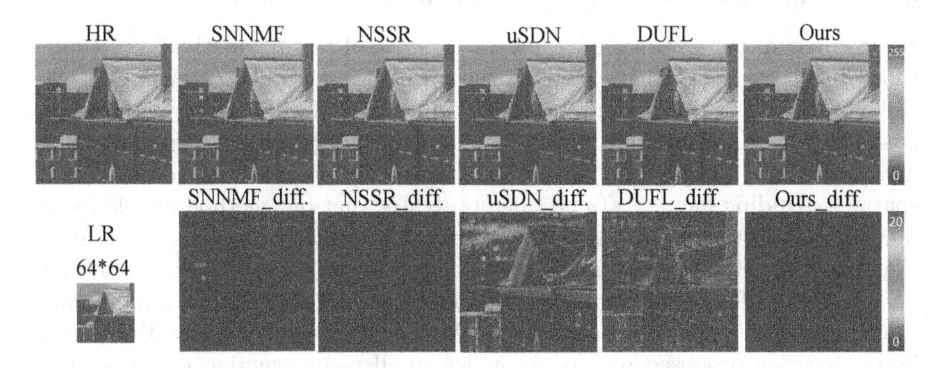

Fig. 3. Visual results of mathematical optimization-based methods: SNNMF [27], NSSR [6] and deep learning-based methods: uSDN [22], DUFL [18] on the Harvard dataset and our method with the up-scale factor 8.

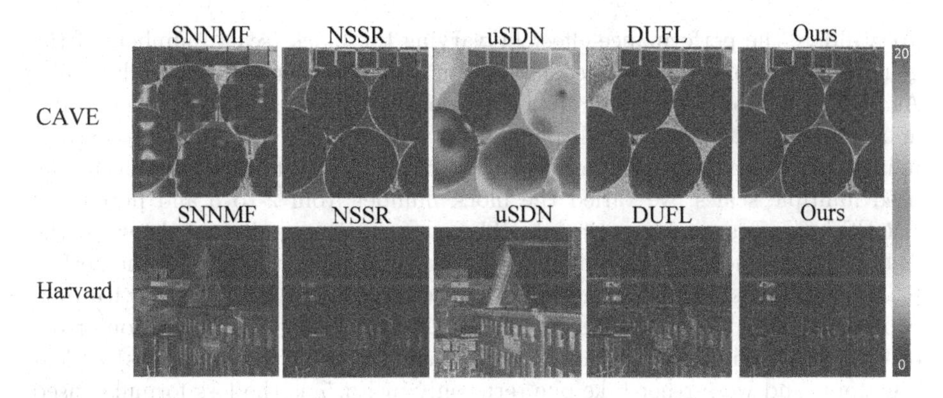

Fig. 4. SAM visual results of mathematical optimization-based methods: SNNMF [27], NSSR [6] and deep learning-based methods: uSDN [22], DUFL [18] on the CAVE and Harvard datasets and our method with the up-scale factor 8.

to 720 nm. For both datasets, we transformed the corresponding HS images using the spectral response function of the Nikon D700 camera to obtain HR-RGB HS images, while LR-HS images were obtained by bicubic downsampling of the HS images. To objectively evaluate the performance of different HSI SR methods, we adopted five widely used metrics, including root mean square error (RMSE), peak signal to noise ratio (PSNR), structural similarity index (SSIM), spectral angle mapper (SAM), and relative dimensional global errors (ERGAS).

First, we conducted experiments using the generative network with five blocks, all two terms of loss in Eq. 5 and the RGB input for comparing with the state-of-the-art methods, and then we performed an ablation study by varying the number of blocks, loss terms, and inputs to the generative network.

3.2 Comparisons with the State-of-the-Art Methods

We compared our approach with various state-of-the-art methods, including those based on mathematical optimization-based methods: GOMP [26], MF [15], SNNMF [27], CSU [31], NSSR [6], supervised deep learning-based methods: SSFNet [9], DHSIS [5], ResNet [11], and unsupervised deep learning-based methods: uSDN [22], DUFL [18]. Table 1 shows the comparative results for the spatial expanding factor 8. Table 1 demonstrates that our method is able to significantly improve the performance in term of all evaluation metrics. In addition, Fig. 2 and Fig. 3 show the visualization difference results of two representative images with different deep unsupervised learning methods. Figure 4 illuminates the SAM visualization results on both CAVE and Harvard datsets. It also manifests that our proposed method provides small reconstruction errors in most spatial locations.

3.3 Ablation Study

We validate the performance effect by varying the block (scale) numbers of the generative network, the used reconstruction error term, and the network inputs. As mentioned above, we used an encoder-decoder structure where both encoder and decoder paths contain multiple blocks as our specific CNN model to extract multi-scale contexts in different receptive fields. To test the efficiency of the used multiple scales, we varied the block number from 2 to 5 and performed HR-HS image learning experiments. The comparative results are shown in the Table 2, where more blocks demonstrates the improvement in term of the performance, while the generative network achieves impressive result even with only two blocks. In addition, as described in Eq. 5, we use the reconstruction errors of both observed HR-RGB and LR-HS images (denoted as 'both' loss) as loss functions, and we further take one term only in Eq. 5 as the loss formulas used to train our generative the network, denoted as loss 1 and loss 2 for comparison. Table 2 illustrates the comparison results using different loss functions, which indicates that the proposed two loss terms perform much better.

Finally, we verified the effect of different inputs to the generative network. As mentioned above, it is popular in most generative networks to use randomly generated noisy inputs to synthesize different natural images. To make full use of the available data, we employed the observed HR-RGB image as the conditional input to guide the training of the proposed generative network. Without lack of generality, we also combined the HR-RGB image with the up-sampled LR-HS image together as the network input (marked as 'combined') and additionally disturb the combined input with a small level of noise in each training step to increase the robustness of model training. The comparison results of different network inputs are shown in Table 3, and the conditional input using the HR-RGB image manifests the best recovery performance.

4 Conclusion

In this study, we proposed a new deep RGB-driven generative network that learns the latent HR-HS image from its degraded observations without the need of any external data. To build an efficient and effective specific CNN model, we adopted an encoder-decoder-based generative network with a shallow structure but being able to perform multi-scale spatial context exploration in large receptive fields to learn high-representative feature of the latent HR-HS image with the conditioned HR-RGB image as a guide. Moreover, since the under-studying scene do not have the ground-truth HR-HS image, we specifically designed the convolution-based degradation modules to transform the predicted HR-HS image in the generative network, and then obtained the approximated observations to formulate the loss function for network training. Experimental results showed that our method significantly improves the performance over the SoTA methods.

Acknowledgement. This research was supported in part by the Grant-in Aid for Scientific Research from the Japanese Ministry for Education, Science, Culture and Sports (MEXT) under the Grant No. 20K11867, and JSPS KAKENHI Grant Number JP12345678.

References

1. Akhtar, N., Shafait, F., Mian, A.: Sparse spatio-spectral representation for hyperspectral image super-resolution. In: Fleet, D., Pajdla, T., Schiele, B., Tuytelaars, T. (eds.) ECCV 2014. LNCS, vol. 8695, pp. 63–78. Springer, Cham (2014). https://doi.org/10.1007/978-3-319-10584-0_5
2. Bach, S.H., He, B., Ratner, A., Ré, C.: Learning the structure of generative models without labeled data. In: International Conference on Machine Learning, pp. 273–282. PMLR (2017)
3. Bioucas-Dias, J.M., Plaza, A., Camps-Valls, G., Scheunders, P., Nasrabadi, N., Chanussot, J.: Hyperspectral remote sensing data analysis and future challenges. IEEE Geosci. Remote Sens. Mag. **1**(2), 6–36 (2013)
4. Dian, R., Fang, L., Li, S.: Hyperspectral image super-resolution via non-local sparse tensor factorization. In: Proceedings of the IEEE Conference on Computer Vision and Pattern Recognition, pp. 5344–5353 (2017)
5. Dian, R., Li, S., Guo, A., Fang, L.: Deep hyperspectral image sharpening. IEEE Trans. Neural Netw. Learn. Syst. **99**, 1–11 (2018)
6. Dong, W., et al.: Hyperspectral image super-resolution via non-negative structured sparse representation. IEEE Trans. Image Process. **25**(5), 2337–2352 (2016)

7. Fu, Y., Zhang, T., Zheng, Y., Zhang, D., Huang, H.: Hyperspectral image super-resolution with optimized RGB guidance. In: Proceedings of the IEEE/CVF Conference on Computer Vision and Pattern Recognition, pp. 11661–11670 (2019)
8. Han, X.H., Chen, Y.W.: Deep residual network of spectral and spatial fusion for hyperspectral image super-resolution. In: 2019 IEEE Fifth International Conference on Multimedia Big Data (BigMM), pp. 266–270. IEEE (2019)
9. Han, X.H., Shi, B., Zheng, Y.: Self-similarity constrained sparse representation for hyperspectral image super-resolution. IEEE Trans. Image Process. **27**(11), 5625–5637 (2018)
10. Han, X.H., Shi, B., Zheng, Y.: SSF-CNN: spatial and spectral fusion with CNN for hyperspectral image super-resolution. In: 2018 25th IEEE International Conference on Image Processing (ICIP), pp. 2506–2510. IEEE (2018)
11. Han, X.H., Sun, Y., Chen, Y.W.: Residual component estimating CNN for image super-resolution. In: 2019 IEEE Fifth International Conference on Multimedia Big Data (BigMM), pp. 443–447. IEEE (2019)
12. Han, X.H., Zheng, Y., Chen, Y.W.: Multi-level and multi-scale spatial and spectral fusion CNN for hyperspectral image super-resolution. In: Proceedings of the IEEE International Conference on Computer Vision Workshops (2019)
13. He, W., Zhang, H., Zhang, L., Shen, H.: Total-variation-regularized low-rank matrix factorization for hyperspectral image restoration. IEEE Trans. Geosci. Remote Sens. **54**(1), 178–188 (2015)
14. Huang, Q., Li, W., Hu, T., Tao, R.: Hyperspectral image super-resolution using generative adversarial network and residual learning. In: ICASSP 2019-2019 IEEE International Conference on Acoustics, Speech and Signal Processing (ICASSP), pp. 3012–3016. IEEE (2019)
15. Kawakami, R., Matsushita, Y., Wright, J., Ben-Ezra, M., Tai, Y.W., Ikeuchi, K.: High-resolution hyperspectral imaging via matrix factorization. In: CVPR 2011, pp. 2329–2336. IEEE (2011)
16. Lanaras, C., Baltsavias, E., Schindler, K.: Hyperspectral super-resolution by coupled spectral unmixing. In: Proceedings of the IEEE International Conference on Computer Vision, pp. 3586–3594 (2015)
17. Liu, Z., Zheng, Y., Han, X.H.: Unsupervised multispectral and hyperspectral image fusion with deep spatial and spectral priors. In: Proceedings of the Asian Conference on Computer Vision (2020)
18. Liu, Z., Zheng, Y., Han, X.H.: Deep unsupervised fusion learning for hyperspectral image super resolution. Sensors **21**(7), 2348 (2021)
19. Lu, G., Fei, B.: Medical hyperspectral imaging: a review. J. Biomed. Opt. **19**(1), 010901 (2014)
20. Mertens, S., et al.: Proximal hyperspectral imaging detects diurnal and drought-induced changes in maize physiology. Front. Plant Sci. **12**, 240 (2021)
21. Park, S.M., Kim, Y.L.: Spectral super-resolution spectroscopy for biomedical applications. In: Advanced Chemical Microscopy for Life Science and Translational Medicine 2021, vol. 11656, p. 116560N. International Society for Optics and Photonics (2021)
22. Qu, Y., Qi, H., Kwan, C.: Unsupervised sparse dirichlet-net for hyperspectral image super-resolution. In: Proceedings of the IEEE Conference on Computer Vision and Pattern Recognition, pp. 2511–2520 (2018)
23. Sakthivel, S.P., Sivalingam, J.V., Shanmugam, S., et al.: Super-resolution mapping of hyperspectral images for estimating the water-spread area of peechi reservoir, southern india. J. Appl. Remote Sens. **8**(1), 083510 (2014)

24. Saralıoğlu, E., Görmüş, E.T., Güngör, O.: Mineral exploration with hyperspectral image fusion. In: 2016 24th Signal Processing and Communication Application Conference (SIU), pp. 1281–1284. IEEE (2016)
25. Uezato, T., Hong, D., Yokoya, N., He, W.: Guided deep decoder: unsupervised image pair fusion. In: Vedaldi, A., Bischof, H., Brox, T., Frahm, J.-M. (eds.) ECCV 2020. LNCS, vol. 12351, pp. 87–102. Springer, Cham (2020). https://doi.org/10.1007/978-3-030-58539-6_6
26. Wang, J., Kwon, S., Shim, B.: Generalized orthogonal matching pursuit. IEEE Trans. Signal Process. **60**(12), 6202–6216 (2012)
27. Wycoff, E., Chan, T.H., Jia, K., Ma, W.K., Ma, Y.: A non-negative sparse promoting algorithm for high resolution hyperspectral imaging. In: 2013 IEEE International Conference on Acoustics, Speech and Signal Processing, pp. 1409–1413. IEEE (2013)
28. Xie, Q., Zhou, M., Zhao, Q., Meng, D., Zuo, W., Xu, Z.: Multispectral and hyperspectral image fusion by MS/HS fusion net. In: Proceedings of the IEEE/CVF Conference on Computer Vision and Pattern Recognition, pp. 1585–1594 (2019)
29. Xu, J.L., Riccioli, C., Sun, D.W.: Comparison of hyperspectral imaging and computer vision for automatic differentiation of organically and conventionally farmed salmon. J. Food Eng. **196**, 170–182 (2017)
30. Yokoya, N., Chan, J.C.W., Segl, K.: Potential of resolution-enhanced hyperspectral data for mineral mapping using simulated EnMAP and sentinel-2 images. Remote Sens. **8**(3), 172 (2016)
31. Yokoya, N., Zhu, X.X., Plaza, A.: Multisensor coupled spectral unmixing for time-series analysis. IEEE Trans. Geosci. Remote Sens. **55**(5), 2842–2857 (2017)
32. Yue, L., Shen, H., Li, J., Yuan, Q., Zhang, H., Zhang, L.: Image super-resolution: the techniques, applications, and future. Signal Process. **128**, 389–408 (2016)
33. Zhang, S., Liang, G., Pan, S., Zheng, L.: A fast medical image super resolution method based on deep learning network. IEEE Access **7**, 12319–12327 (2018)
34. Zhao, Y., Yang, J., Zhang, Q., Song, L., Cheng, Y., Pan, Q.: Hyperspectral imagery super-resolution by sparse representation and spectral regularization. EURASIP J. Adv. Signal Process. **2011**(1), 1–10 (2011)
35. Zhu, Z., Hou, J., Chen, J., Zeng, H., Zhou, J.: Residual component estimating CNN for image super-resolution, vol. 30, pp. 1423–1428 (2020)

Gift from Nature: Potential Energy Minimization for Explainable Dataset Distillation

Zijia Wang(✉), Wenbin Yang, Zhisong Liu, Qiang Chen, Jiacheng Ni, and Zhen Jia

Dell Technologies OCTO Research Office, Shanghai, China
Zijia_Wang@dell.com

Abstract. Dataset distillation aims to reduce the dataset size by capturing important information from original dataset. It can significantly improve the feature extraction effectiveness, storage efficiency and training robustness. Furthermore, we study the features from the data distillation and found unique discriminative properties that can be exploited. Therefore, based on Potential Energy Minimization, we propose a generalized and explainable dataset distillation algorithm, called Potential Energy Minimization Dataset Distillation (PEMDD). The motivation is that when the distribution for each class is regular (that is, almost a compact high-dimensional ball in the feature space) and has minimal potential energy in its location, the mixed-distributions of all classes should be stable. In this stable state, Unscented Transform (UT) can be implemented to distill the data and reconstruct the stable distribution using these distilled data. Moreover, a simple but efficient framework of using the distilled data to fuse different datasets is proposed, where only a lightweight finetune is required. To demonstrate the superior performance over other works, we first visualize the classification results in terms of storage cost and performance. We then report quantitative improvement by comparing our proposed method with other state-of-the-art methods on several datasets. Finally, we conduct experiments on few-shot learning, and show the efficiency of our proposed methods with significant improvement in terms of the storage size requirement.

Keywords: Dataset distillation · Potential energy

1 Introduction

Energy consumption in deep learning model training is a big concern in real-world applications [29], one of the solutions for this problem is knowledge distillation [10]. It transfers the knowledge from a deep complex model to a simple one, hence people can save the running time. Another direction is dataset distillation [9,30,38]. It tries to synthesize some data samples or features to summarize information in the original huge dataset and use these synthesized data to train models more efficiently.

However, most dataset distillation algorithms mainly focus on the information in the models and try to reproduce the ability by utilizing small models or

Y. Zheng et al. (Eds.): ACCV 2022, LNCS 13848, pp. 240–255, 2023.
https://doi.org/10.1007/978-3-031-27066-6_17

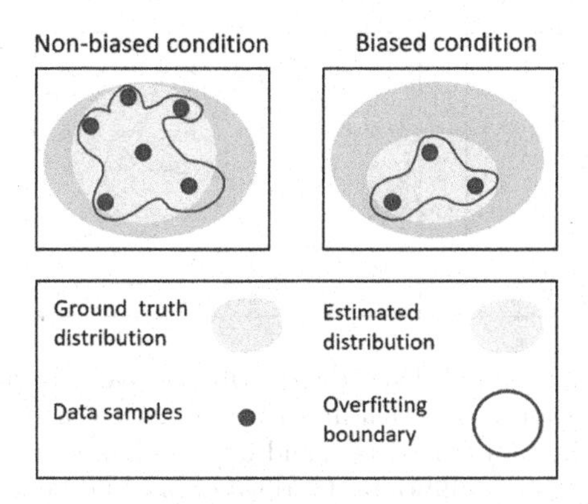

Fig. 1. If the distilled data is not biased (top left), even the overfitting boundary generated by deep learning models, which is common in few-shot learning, can better represent the ground truth distribution than the biased one(top right). For properly estimated distribution (yellow circle), the condition is the same. (Color figure online)

small datasets. The properties of dataset are ignored. In the naive dataset distillation [39], the dataset properties are explored by "black box" meta-training process, which is not robust and explainable. It should be noticed that, in the setting of dataset distillation, which is essentially a few-shot learning scenario, models tend to overfit such small dataset [42]. The obtained distilled dataset is only adequate for networks used in the training iterations of the distillation process.

Therefore, to avoid overfitting and the rigid training constraints, we hope to distill the data with the consideration of inherent dataset statistical properties. Noticed that the dimension in feature space is much lower, so it's easier to be calibrated [41]. Also, mean and variance of Gaussian distribution could be transferred across similar classes [26], all the transformation are performed in the feature space (Fig. 2 left), so we use Gaussian distribution to calibrate the feature distribution. In the final calibrated space, the features of different classes are expected to be far enough while features of similar classes should be in a distribution with low variance. The second one could be achieved in the distribution calibration stage, the ideal distance in the first requirement, however, is hard to determine. A large distance could make classification more accurate, but it makes the feature space sparse. On the other side, a small distance would cause the decrease in accuracy. Therefore, we choose to adapt the idea of potential energy stable equilibrium, this equilibrium exists if the net force is zero, any changes in the system would increase the potential energy [8]. This stable equilibrium describes the perfect distance between the center of each class (Fig. 2 right). In the stable system, the features from different classes could be easily classified using a simple classifier. What's more, the centers and edge points can be the distilled data which is the best subset of the original dataset considering the data distribution, then if new models are trained on these distilled data, the accuracy of models trained on original dataset could be recovered.

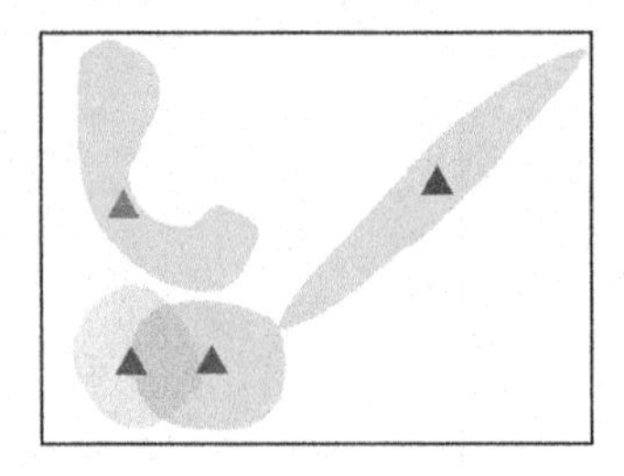

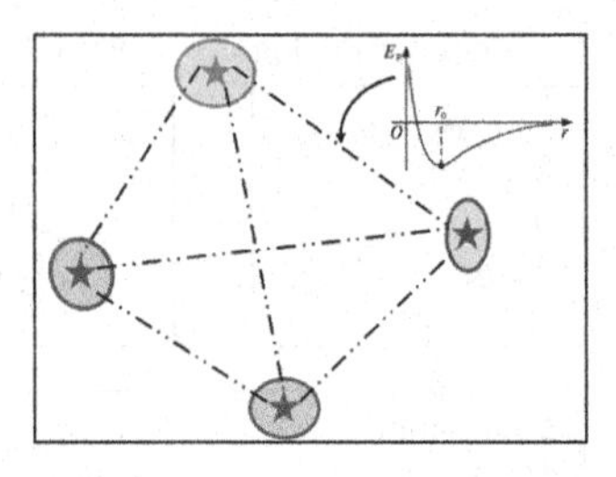

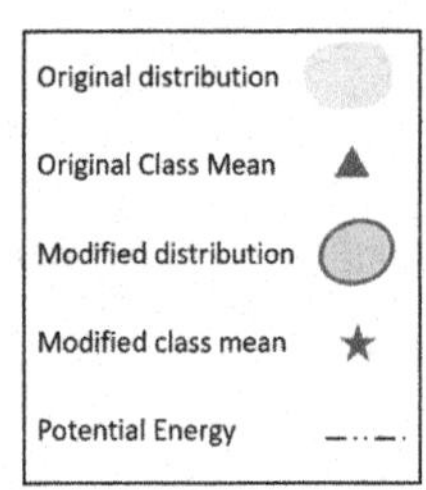

Fig. 2. In original feature space (left), the distributions are random and hard to classify. After the distribution calibration and transformation, a stable state is achieved (right) in which the distribution is tight and all the classes are 'stable' according to the potential energy stable equilibrium. Distilled data could be easily chosen based on this stable system, these distilled data, along with the transformation matrix, could be used to fuse dataset and evaluate data quality.

In experiments, besides the original ability of dataset distillation (accuracy recovery), visualization results show that our the chosen images are explainable with good diversity, this explainability can be further exploited to make more use of these distilled samples. We also use experiment results to show that simple classifier with our dataset distillation strategy can perfectly handle data fusion and data quality evaluation tasks. As an extension to current dataset distillation algorithm, our algorithm can also outperform SOTA results in few-shot learning. Overall, our contributions are:

- Potential Energy Minimization based Dataset Distillation (PEMDD). Applying the concept of PEM and distribution calibration to the feature vectors to find the stable state in feature space. Therefore, dataset could be distilled while, to the maximum extent, avoiding harming the distribution reproduction. In our model, only few parameters are added. What's more, this strategy is also shown to be useful in few-shot learning.
- Unscented Transformation (UT) for dataset distillation. UT is used to distill the data which could be used to reconstruct the stable distribution from distilled samples. Then the reconstructed distribution could be used to perform the up-sampling and other downstream tasks.
- Framework for the applications using dataset distillation and our stable system. Based on the distilled data and the transformation matrix of stable system, we propose frameworks to fuse dataset in three scenarios: (i) 2 datasets share same classes; (ii) new dataset contains new classes; (iii) 2 datasets share some same classes while new dataset contains new classes. What's more, PEM solution for few-shot learning is also tested in this paper and the results are good.

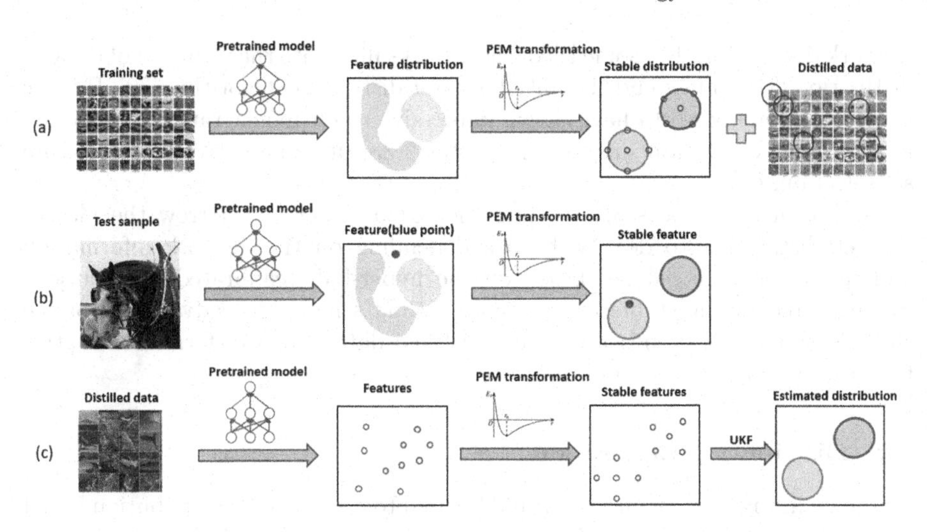

Fig. 3. Workflow for PEM-based dataset distillation. **(a) Training process.** This part shows the training process in which the PEM transformation model is trained, then the stable distribution is derived and the distilled data is selected from the original dataset based on the sampling strategy in stable distribution system. **(b) Testing process.** This part illustrates the testing process, the test sample is also first transformed into the feature space and then converted into the stable system using the trained PEM transformation model. **(c) Distilled data for applications.** This part demonstrates the basic strategy of using the distilled data, they could be used to reconstruct the distributions for each class while ideally, they are same to the distributions in the stable system. This strategy could be flexibly adapted into different scenarios.

In the following part of this paper, some preliminaries and related works are listed in Sect. 2, then the algorithm details of PEM-based dataset distillation is demonstrated in the beginning of Sect. 3, then the rest of that section shows the solution for few-shot learning and dataset fusion. Finally, experiment details are described in Sect. 4. Concretely, in the experiment, we use few-shot learning, where our algorithm could achieve competitive results compared with SOTA results, to show the power of stable state derived by PEMDD, then the dataset fusion experiment shows the advantages of our distilled data by almost recovering the classification accuracy with only few data samples kept.

2 Related Work

2.1 Dataset Distillation

Computational cost in deep learning becomes more and more expensive, model compression starts to attract much attention of researchers [1,11,23]. Dataset distillation [38] is one of them and was first introduced in the inspiration of network distillation [10], some theoretical works illustrate the intuition of dataset distillation [31] and extend the initial dataset distillation algorithm [30]. Besides these works, many works have shown impressive result in generating or selecting a small number of data samples [2,3,7,27,32,33] utilizing active learning, core set selection, etc.

Although our idea is also to find the core dataset, we borrow the idea of dataset distillation to deal with data depending on the network information and select the core dataset based on the hyper-ball (calibrated distribution) we generated in the stable feature space. What's more, the advantages of the stable system make it possible to use distilled data and transformation matrix to perform more downstream tasks.

2.2 Distribution Calibration

As shown in Fig. 1, some metrics could be used to estimate the distribution based on the distilled data [13,37], but most of them assume the kind of distribution is known. To better reconstruct the distribution based on distilled data, the data distribution should follow a specific distribution (Gaussian in this paper). Many papers tried to calibrate the distribution of the data for different purposes [24,28,42], the main idea is to calibrate the data distribution into a regular and tight distribution.

However, the calibrated distribution in these algorithms cannot be directly used in the dataset distillation setting. In dataset distillation, we want the distilled data to maximally contain the information in original dataset without being affected by other classes. Furthermore, in the application phase, the addition of new data samples would make the system unstable if the distance of each class if unstable, so we use the concept of potential energy to avoid such risk to the greatest extent possible.

2.3 Potential Energy

Potential energy is a simple concept in physics. In this theory, there exists an distance between two particles (r_0). If the distance becomes closer, the resultant force is attractive while the resultant force changes to be repulsive force. Only when the resultant force is zero, the potential energy is lowest, which means that the system is stable [4,18].

In this paper we choose to adapt this concept to find the perfect distance between 2 feature vectors. We use molecular potential energy to optimize the position of centroids to make them easy to classify while not being too far, then atomic potential energy is used to optimize the position of features of same class to make them close enough.

3 Method

In this section, the basic problem for dataset distillation is first defined in Sect. 3.1 and the solution is revealed in Sect. 3.2, then a concrete scenario is shown to demonstrate the application framework in Sect. 3.3. Finally, our algorithm is applied into the few-shot learning problems and a thorough analysis is demonstrated.

3.1 Problem Definition

Given a labelled dataset $\mathcal{D} = \{x_i, y_i\}$ where $x_i \in \mathbb{R}$ is the raw data sample and $y_i \in \mathbb{C}$ is the corresponding labels with $\mathbb{C}$ denoting the set of classes. Then assume that a pretrained model $\mathcal{M}$ which could extract features $\mathcal{F} = \{f_i\}$ from $\mathcal{D}$ where f_i is the feature vector of x_i. The goal of dataset distillation is to select a few data samples which capture the most important distribution properties of $\mathcal{D}$.

To realize the dataset distillation, we suggest to learn a transformation $\mathcal{T}$ to transfer $\mathcal{F}$ into an new embedding space where features from same class become more compact while features from different classes are in a 'moderate' distance. Then, based on the transferred distribution, a subset of original dataset $\mathcal{D}$ are collected as distilled dataset $\mathcal{S}$.

3.2 PEM-Based Transformation

Tukey's Ladder of Powers Transformation. To make the feature distributions more regular, i.e. be more like Gaussian, the first step in PEM-based transformation is adopting the Tukey's Ladder of Powers Transformation [34] to reduce the skewness of distributions.

The function of Tukey's Ladder of Powers Transformation can be varied based on the configuration of the power. The formulation of this transformation can be expressed as:

$$\hat{x} = \begin{cases} x^P & if\ P \neq 0 \\ log(x) & if\ P = 0 \end{cases} \tag{1}$$

where P is the hyper-parameter which could control the way of distribution regularization. To recover the feature distribution, P should be set to 1. If the P decreases, the distribution becomes less positively skewed and vice versa.

Potential Energy Minimization. Considering a linear transformation with weight $\mathbf{W}_T$ as

$$\mathcal{F}_s = \mathbf{W}_T \mathcal{F} \tag{2}$$

where $\mathcal{F}_s$ is the desired feature, $\mathcal{F}$ is the input feature. We hope to find a suitable distance among classes to ensure the diversity of the latent feature. To achieve this goal, recall the potential energy expression [4], which may have different forms. In this paper we use the following formula:

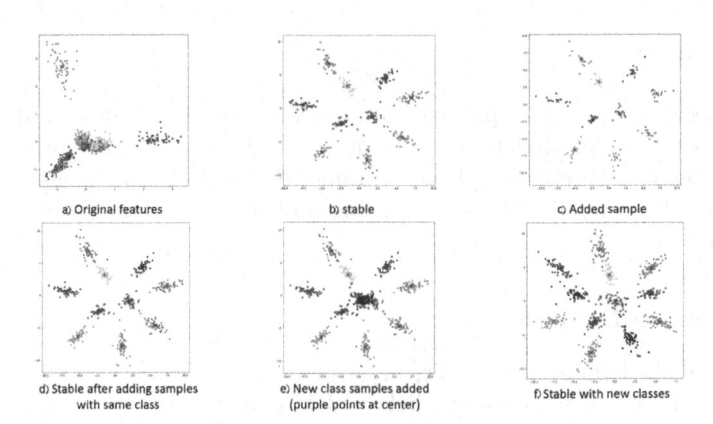

Fig. 4. Visualization for dataset fusion with PEM in CUB-200. a) shows the original feature distribution, after PEM transformation, they are stable in b). In c), some new samples whose class are same with the classes in b) are added, and in d) they are perfectly handled using PEM. In e), some samples of new classes (purple points) are added, then the PEM transformation results are shown in f) after fine-tune, they are in the stable equilibrium again. (Color figure online)

$$E(r) = \frac{1}{r^3} - \frac{1}{r^2} \tag{3}$$

where r is the distance between two particles. Here, r_0 is the optimal distance for minimal potential energy. Then we adapt this equation and derive our loss function to learn linear transformation $\mathbf{W}_T$ that minimizes the'potential energy' between the extracted features of every pair of data points:

$$L = \sum_{i=1}^{N-1} \sum_{j=i+1}^{N} \left[\frac{1}{(\gamma_{ij} d_{ij} + b_0)^3} - \frac{1}{(\gamma_{ij} d_{ij} + b_0)^2} \right] \tag{4}$$

where $d_{ij} = dis(\mathbf{W}_T f_i, \mathbf{W}_T f_j)$ with $dis(\cdot, \cdot)$ representing the Euclidean distance, f_i, f_j are the input features, N is the number of data samples. The hyperparameter b_0 $(0 < b_0 < r_0)$ is introduced to improve the numerical stability of PE. γ_{ij} is the function to control the properties of loss, which can be defined as:

$$\gamma_{ij} = \begin{cases} \tau_0 & if\ y_i = y_j \\ \tau_1 & if\ y_i \neq y_j \end{cases} \tag{5}$$

where τ_0, τ_1 are the inter class and inner class weights, respectively. In this paper, we set $0 < \tau_1 < \tau_0$.

Process of Dataset Distillation. In Fig. 3 (a), a co-stable system with desired distribution is derived after the optimized transformation. Then some data points can be sampled based on the stable equilibrium. Considering the inner-class

Gaussian-like properties, we use the data sampling strategy in UT [36,37] to get the distilled data points $\mathcal{S}$ (sigma points in UT) and their corresponding weights ω for distribution reconstruction. Here there are two sets of weights ω_m and ω_c, ω_m is used to recover the means of the original distribution while the ω_c is for the reconstruction of the covariance matrix.

For each class, the first sample in sigma points set is $\mathcal{S}[0] = \mu$ with μ representing the mean, then the other samples are sampled as following:

$$\mathcal{S}[i] = \begin{cases} \mu + V_i & for\ i = 1, ..., d-1 \\ \mu - V_{i-d} & for\ i = d, ..., 2d \end{cases} \tag{6}$$

where variance $V_i = \sqrt{(d+\lambda)\Sigma(:,i)}$, d stands for the dimension of the features, λ is the scaling parameter and $\Sigma(:,i)$ is the i_{th} column of the covariance matrix while the covariance matrix could be easily derived with data samples.

For this sequence $\mathcal{S}$, the corresponding weights ω_m for mean estimation with sigma set can be calculated as:

$$\omega_m^{[i]} = \begin{cases} \frac{\lambda}{d+\lambda} & if\ i = 0 \\ \frac{1}{2(d+\lambda)} & if\ i = 1, ..., 2d \end{cases} \tag{7}$$

while $\omega_m^{[i]}$ is the weight for the i_{th} element in $\mathcal{S}$, then the equation of the weight ω_c for calculating the covariance is:

$$\omega_c^{[i]} = \begin{cases} \omega_m^{[0]} + H & if\ i = 0 \\ \frac{1}{2(d+\lambda)} & if\ i = 1, ..., 2d \end{cases} \tag{8}$$

In Eqs. 7 and 8, $\lambda = \alpha^2(d+k) - d$ and $H = 1 - \alpha^2 + \beta$. To control the distance between the sigma points and the mean, we could adjust $\alpha \in (0, 1]$ and $k \geq 0$. In some literature [36,37], $\beta = 2$ is an optimal choice for Gaussian.

It should be noticed that, the sigma points $\mathcal{S}$ are just sampled in the latent feature space. Finally, for data sample selection, we suggest modeling the dataset distillation as an assignment problem and select the data samples according to the distance (such as the Euclidean distance) between real data features and the sigma points. When considering only one-to-one correspondences modeled as bipartite graph matching, Hungarian algorithm [12] can be used to solve the assignment problem in polynomial time. After the bipartite graph matching, sigma points will be assigned to real data samples.

Classification for New Samples. As shown in Fig. 3 (b), when a new test sample comes in, it is first transformed by the pretrained model (feature extractor) and our PEM model, then a simple classifier like Logistic classifier [21] could be used to classify this sample effectively and robustly.

3.3 How to Fuse Datasets

In this part, we extend the basic problem defined in Sect. 3.1 to demonstrate the solution in some more concrete application scenarios in dataset fusion.

Dataset Fusion Problem Definition. Recall the problem defined in Sect. 3.1, a distilled dataset $\mathcal{S}$ is derived from the labelled dataset $\mathcal{D} = \{x_i, y_i\}$. Now assuming a new dataset $\mathcal{D}^{new} = \{x_i^{new}, y_i^{new}\}$ appears, where $x_i^{new} \in \mathbb{R}$ is the raw data and $y_i^{new} \in \mathbf{C}^{new}$ is the data label with $\mathbf{C}^{new}$ representing the new class categories.

The goal in this section is to find a fine-tuned PEM transformation $\mathcal{T}^{new}$ to realize a new stable state for fused dataset $\mathcal{D}^{fuse} = \mathcal{S} \cup \mathcal{D}^{new}$. There are mainly 2 settings in this problem. The first setting is that two datasets share exactly same classes, i.e. $\mathbf{C}^{new} = \mathbf{C}$. The other one setting considers the $\mathbf{C}^{new} \neq \mathbf{C}$.

Distribution Fusion. In both setting, the fusion process is similar to the process described in Sect. 3.2. At first, to estimate the fused statistics more accurately, we will up-sampling some feature-points. Considering the distribution for each class after PEM based transformation are Gaussian-like, features can be easily generated with re-parameter trick. Then a PEM training process as shown in Fig. 3 (a) is performed on $\mathcal{D}^{new}$ to get the stable distribution.

3.4 Few-Shot Learning

In PEM, the inner-class compactness and inter-class diversity of the latent features is the foundation of our success in dataset distillation. Therefore, we extend the PEM strategy in few-shot learning setting to show our advantages in data property exploitation.

Few-Shot Learning Problem Definition. A few-shot learning problems can be a simple extension of the problem defined in previous section, where the samples in $\mathcal{D}^{new}$ are quite few. Tasks in few-shot learning could be called N-way-K-shot [35], where there are N classes in $\mathbf{C}^{new}$ and K labelled samples for each class.

Few-Shot Learning Solution. The solution for few-shot learning can be modeled as an simplified version of PEMDD. At first, the PEM transformation is trained using very few samples to get the stable state, then this transformation is used to transform the test features into the latent feature space. Then, a simple classifier, such as logistic regression, can be utilized for label prediction.

We will show more details about our implementation of few-shots learning in Sect. 4.

3.5 Analysis

Intuitively, the proposed PEM framework share some common ideas with the well-known Fisher Discriminant Analysis (FDA) [19]. Both methods try to increase the diversity between classes and reduce the diversity within classes. However, PEM is constructed based on the steady state through the PE function, which is discovered in physics and makes the separation between classes more balanced. Meanwhile, our PEM also guarantees the existence of an stable state for all systems. All these reasons above allow the PEM-based method to maintain good performance on few-sample data set compared with works like FDA.

Our method is also different from the dataset distillation in [39]. For the dataset distillation in [39], its aim is to replicate the performance of the entire dataset from the synthetic points, but our method is trying to select some informative samples and their corresponding linear transformation weights. Our method allows more flexible selection of data sample, while the dataset distillation must predefine the synthesized number before training. Also a Hungarian algorithm was used Sect. 3.2, one can also increase the sample to be selected by an unbalanced Hungarian algorithm. For a classical Hungarian problem for data sample selection, the complexity is just $O(d^3)$.

Last but not least, our method can also be adopted as a first principle to learn **diverse and discriminative** features for down-stream tasks.

4 Experiment and Discussion

4.1 Experiment Setup

Datasets. In this paper, miniImageNet [22] and CUB-200 [40] are used to evaluate our algorithm.

For miniImageNet, in few-shot learning validation experiment, all classes are split into 64 base classes, 16 validation classes and 20 new classes as [22] did in their work. However, in data fusion experiment and visualization part, only 10 classes are selected from the miniImageNet because of the reality and visualization simplicity, the train-test-split conditions are illustrated in the experiment part (Sect. 4.3).

CUB-200 is a fine-grained benchmark for few-shot learning. There are 11,788 images with size $84 \times 84 \times 3$ for 200 different classes of birds. These classes are spilt into 100 base classes, 50 validation classes and 50 novel classes [5].

Evaluation Metric. For different tasks, we adopted different metrics. In few shot learning setting, top-1 accuracy is used to evaluate our strategy in 5way1shot and 5way5shot settings for both 2 datasets [42]. In Data fusion setting, all experiments are variants of the image classification task, so top 1 accuracy is used to evaluate the performance of different strategy.

Fig. 5. Visualization of dataset distillation for dog.

Implementation Details. In all experiments, we use pretrained WideResNet [17] to extract the features, features are extracted from the layer before fully connected layer with a ReLU activation function which makes the output non-negative [42]. The parameters of τ_0,τ_1,b_0 are set to 1, 0.1, 0.3, respectively. The embedding dimension size is set to 12, therefore, the sigma points number for each class will be 25. We also follow the Tukey transformation settings in [42].

In few-shot learning part, we follow the solution described in previous section to deal with the extracted features, nearest center is used to be the classifier.

In data fusion part, we choose logistic classifier to represent the current SOTA results and compare our results with them. We use the logistic classifier implementation of scikit-learn [20] with the default settings.

4.2 Empirical Understanding of PEMDD

Table 1a shows the experiment results in a regular setting. In this setting, for each class, 450 samples are used to train the PEM model and 150 samples are used as test set. As shown in Table 1a, when the training samples are reduced to 25%, no matter what strategy is used to choose samples, a huge decline in the accuracy occurs. Then in our strategy, we adapt Logistic [20] as the classifier. With our strategy (PEM with sampling), we sample 1000 points for each class in test phase, the results are almost recovered while a fine-tune PEM could enhance the performance (from 0.943 to 0.988).

Then to further illustrate the advantages of our results, we visualize the 20 "selected" images selected from MiniImageNet dataset for the "dog" classes (Fig. 5). We observe that the selected images are much more diverse and representative than those selected randomly from the dataset (with random selection program), indicating such PEM-based distilled images can be used as a good "summary" of the dataset.

At last, we compare to the state of the art dataset distillation method [39]. We test our method on CIFAR-10 dataset. The model is identical to the ones used in [39], which can achieve about 80% test accuracy on CIFAR-10 in a fully supervised setting. The dataset distillation will synthesis 100 pictures (10

Table 1. Basic experiments of PEMDD.

Method	CUB-200	miniImageNet
Full data	1.000	0.938
Class-based random (10%)	0.378	0.258
Most remote (10%)	0.463	0.314
PEM	0.981	0.923
PEM+sampling	0.943	0.911
PEM+sampling+PEM	0.988	0.941

(a) Experiment results for basic classification task

Method	50	100	150	200
All random	0.091	0.132	0.158	0.165
Class-based random	0.092	0.143	0.174	0.220
K-means	0.105	0.184	0.223	0.347
Dataset distillation (random init)	–	0.368	–	–
Dataset distillation (fixed init)	–	0.540	–	–
PEMDD	0.247	0.519	0.614	0.719

(b) Experiment results compared with original dataset distillation algorithm

pictures for each class). The PEM based method, K-means and Random selection will select 50–200 pictures. The embedding dimension size of PEM is set to 10. All the result is shown in Table 1b.

4.3 Data Efficiency Application

In this section, we perform a series of experiments to test our strategy on different settings.

In Table 2a, we equally split the 10-th class. 300 samples are used to train the PEM and others are used as new samples. Other settings are same as previous one. In our setting, these two sets have different distribution on the 10-th class, therefore conventional method which builds on the i.i.d assumption may suffer from the performance decreasing. As shown in Table 2a, a simple fine-tune after adding new data, the PEM based method could get a good test accuracy performance (0.980).

Then, we remove the 8-th class in the training dataset and treat it as a totally new class. As shown in Table 2b, the PEM based with fine-tune could also give good result (0.981) in this setting.

Table 2. Data efficiency of PEMDD.

Method	CUB-200	miniImageNet
Full data	1.000	0.82
Class-based random (10%)	0.176	0.238
PEM+Sampling	0.516	0.610
PEM+Sampling+PEM	0.980	0.932

(a) Experiment results for adding new samples with same classes

Method	CUB-200	miniImageNet
Full data	1.000	0.938
Class-based random (10%)	0.168	0.185
PEM + Sampling	0.588	0.681
PEM + Sampling+PEM	0.981	0.940

(b) Experiment results for adding new classes

To illustrate the effectiveness of PEM in above three tasks, we visualize the feature on CUB-200 dataset with t-SNE [16] before and after the PEM optimization. In Fig. 4, (a), (c), (e) show the distribution condition before the PEM

Table 3. Experiment results of our few-shot learning solution.

Methods	miniImageNet		CUB	
	5way1shot	5way5shot	5way1shot	5way5shot
LEO [25]	63.80	77.59	–	–
Negative-Cosine [14]	62.33	80.94	72.66	89.40
TriNet [6]	58.12	76.92	69.61	84.10
E3BM [15]	63.80	80.29	–	–
LR with DC [42]	68.57	82.88	79.56	90.67
S2M2_R [17]	64.93	83.18	80.68	90.85
PEM-S	**58.48**	**82.75**	**72.81**	**90.55**

transformation. In (a), the distributions are hard to classify while in (b), features among classes are diverse and easy to classify. Similarly, in (c) and (e), when the new samples come in, the clusters become unstable and PEM could stabilize them again, as shown in (d) and (f). Overall, in the most stable state, all classes have a "safe" distance.

4.4 Few-Shot Application

In the experiment, a PEM transformation is learned based on few-shot training samples to get the stable state, then this transformation is used in the testing phase to transform the test set features into stable state. Our method, PEM-S, contains the up-sampling process mentioned before. The sampling number is 500. The experiment results are summarized in Table 3.

As shown in the Table 3, our results achieve SOTA in 5way5shot learning, though not has the best performance in 5way1shot learning. This is because our strategy partly rely on the inner class compactness and the 1-shot setting cannot provide such information for our PEM-S method.

4.5 Hyper-parameters

To show the effeteness and robustness of the proposed method, we run some Hyper-parameters studies in the baseline problems with mini-Imagenet dataset.

Properties of PE Function. The Properties of the PE function is mainly affected by the γ_{ij}. For simplicity, we fixed the value of τ_0 to 1, and run a comprehensive study of the selection τ_1. Figure 6a illustrates the average PE value $(1/N(N-1))$ of the PE of the whole stable distribution) at training dataset and test accuracy. It can be witnessed that appropriate values of τ_1 can make the system have lower PE and better test accuracy.

Embedding Size. We test our method with different embedding size. From Fig. 6b, one can found that the embedding size may slightly affect the PE values when > 7.

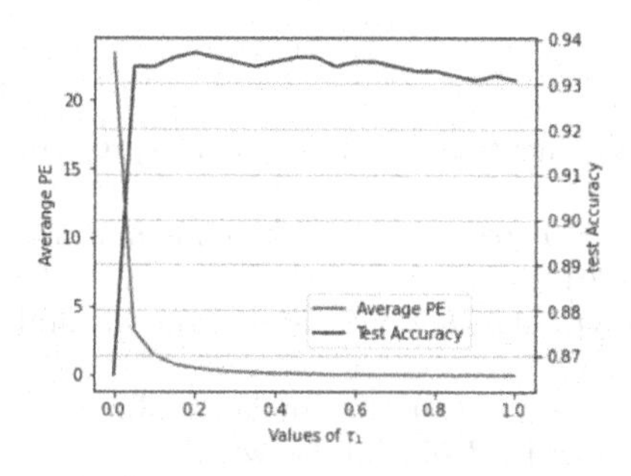

(a) The effect of different values of τ_1.

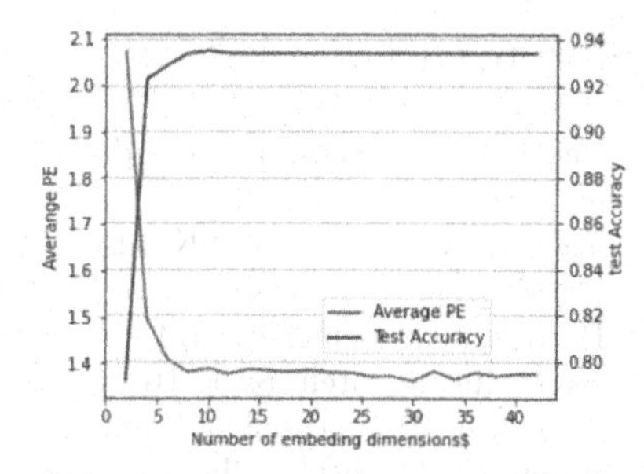

(b) The effect of the number of embedding dimensions.

Fig. 6. Hyper-parameters testing

Different Backbone Networks. Table 4 shows the consistent performance on different feature extractors, i.e., five convolutional layers (conv4), AlexNet, vgg16, resnet18, WRN28 (Baseline). It can be concluded that the PEM based method can achieves at almost 10× accuracy improvement for conv4, AlexNet and vgg16. Moreover, the WRN28 achieve the best performance. It is because of that WRS28 is a semi-supervised method that considers the main-fold information of the dataset.

Table 4. Experiment results with different backbones.

Backbones	Class-based random	PEM-based
conv4	0.089	0.858
AlexNet	0.098	0.913
vgg16	0.106	0.905
resnet18	0.151	0.921
WRN28 (Baseline)	0.258	0.941

5 Conclusion and Future Work

In this paper, we propose a PEM-based framework for DD and few-shot learning settings. PEM can help features to achieve a stable state in the new embedding space. In the new embedding space, the features will represent inner class compactness and inter class diversity, which is the foundation of UT based DD. Experiments results in multi-scenarios reveals the superiority of our PEM strategy. Our PEM-based framework shows that the limited data could be used to recover or even outperform the performance of original data while largely reducing the computation costs and storage costs. Future work will explore more application scenarios for distilled data. Moreover, statistical properties may also be used to generate samples, instead of choosing samples from the dataset. What's more, the PEM can be introduced as a first principle for machine learning problems instead of the conventional 'black box' models.

References

1. Ba, L.J., Caruana, R.: Do deep nets really need to be deep? arXiv preprint arXiv:1312.6184 (2013)
2. Bachem, O., Lucic, M., Krause, A.: Practical coreset constructions for machine learning. arXiv preprint arXiv:1703.06476 (2017)
3. Bezdek, J.C., Kuncheva, L.I.: Nearest prototype classifier designs: an experimental study. Int. J. Intell. Syst. **16**(12), 1445–1473 (2001)
4. Jain, J.M.C.: Textbook of Engineering Physics. Prentice-Hall of India Pvt. Limited, Indore (2009). https://books.google.com/books?id=DqZlU3RJTywC
5. Chen, W.Y., Liu, Y.C., Kira, Z., Wang, Y.C.F., Huang, J.B.: A closer look at few-shot classification. arXiv preprint arXiv:1904.04232 (2019)
6. Chen, Z., Fu, Y., Zhang, Y., Jiang, Y.G., Xue, X., Sigal, L.: Multi-level semantic feature augmentation for one-shot learning. IEEE Trans. Image Process. **28**(9), 4594–4605 (2019)
7. Cohn, D.A., Ghahramani, Z., Jordan, M.I.: Active learning with statistical models. J. Artif. Intell. Res. **4**, 129–145 (1996)
8. Dina, Z.E.A.: Force and potential energy, June 2019. https://chem.libretexts.org/@go/page/2063
9. Hariharan, B., Girshick, R.B.: Low-shot visual object recognition. CoRR abs/1606.02819 (2016). http://arxiv.org/abs/1606.02819
10. Hinton, G., Vinyals, O., Dean, J.: Distilling the knowledge in a neural network (2015)
11. Howard, A.G., et al.: MobileNets: efficient convolutional neural networks for mobile vision applications. arXiv preprint arXiv:1704.04861 (2017)
12. Kuhn, H.W.: The Hungarian method for the assignment problem. Naval Res. Logistics Q. **2**(1–2), 83–97 (1955)
13. Larrañaga, P., Lozano, J.A.: Estimation of Distribution Algorithms: A New Tool for Evolutionary Computation, vol. 2. Springer, Cham (2001). https://doi.org/10.1007/978-1-4615-1539-5
14. Liu, B., et al.: Negative margin matters: understanding margin in few-shot classification. In: Vedaldi, A., Bischof, H., Brox, T., Frahm, J.-M. (eds.) ECCV 2020. LNCS, vol. 12349, pp. 438–455. Springer, Cham (2020). https://doi.org/10.1007/978-3-030-58548-8_26
15. Liu, Y., Schiele, B., Sun, Q.: An ensemble of epoch-wise empirical Bayes for few-shot learning. In: Vedaldi, A., Bischof, H., Brox, T., Frahm, J.-M. (eds.) ECCV 2020. LNCS, vol. 12361, pp. 404–421. Springer, Cham (2020). https://doi.org/10.1007/978-3-030-58517-4_24
16. Van der Maaten, L., Hinton, G.: Visualizing data using T-SNE. J. Mach. Learn. Res. **9**(11), 2579–2605 (2008)
17. Mangla, P., Kumari, N., Sinha, A., Singh, M., Krishnamurthy, B., Balasubramanian, V.N.: Charting the right manifold: manifold mixup for few-shot learning. In: Proceedings of the IEEE/CVF Winter Conference on Applications of Computer Vision, pp. 2218–2227 (2020)
18. McCall, R.: Physics of the Human Body. Johns Hopkins University Press, Baltimore (2010). https://books.google.com/books?id=LSyC41h6CG8C
19. Mika, S., Ratsch, G., Weston, J., Scholkopf, B., Mullers, K.R.: Fisher discriminant analysis with kernels. In: Neural Networks for Signal Processing IX: Proceedings of the 1999 IEEE Signal Processing Society Workshop (Cat. no. 98th8468), pp. 41–48. IEEE (1999)

20. Pedregosa, F., et al.: Scikit-learn: machine learning in Python. J. Mach. Learn. Res. **12**, 2825–2830 (2011)
21. Peterson, L.E.: K-nearest neighbor. Scholarpedia **4**(2), 1883 (2009)
22. Ravi, S., Larochelle, H.: Optimization as a model for few-shot learning (2016)
23. Romero, A., Ballas, N., Kahou, S.E., Chassang, A., Gatta, C., Bengio, Y.: FitNets: hints for thin deep nets. arXiv preprint arXiv:1412.6550 (2014)
24. Rueda, M., Martínez, S., Martínez, H., Arcos, A.: Estimation of the distribution function with calibration methods. J. Stat. Plann. Infer. **137**(2), 435–448 (2007)
25. Rusu, A.A., et al.: Meta-learning with latent embedding optimization. arXiv preprint arXiv:1807.05960 (2018)
26. Salakhutdinov, R., Tenenbaum, J., Torralba, A.: One-shot learning with a hierarchical nonparametric Bayesian model. In: Proceedings of ICML Workshop on Unsupervised and Transfer Learning, pp. 195–206. JMLR Workshop and Conference Proceedings (2012)
27. Sener, O., Savarese, S.: Active learning for convolutional neural networks: a core-set approach. arXiv preprint arXiv:1708.00489 (2017)
28. Song, H., Diethe, T., Kull, M., Flach, P.: Distribution calibration for regression. In: International Conference on Machine Learning, pp. 5897–5906. PMLR (2019)
29. Strubell, E., Ganesh, A., McCallum, A.: Energy and policy considerations for deep learning in NLP (2019)
30. Sucholutsky, I., Schonlau, M.: Improving dataset distillation. CoRR abs/1910.02551 (2019). http://arxiv.org/abs/1910.02551
31. Sucholutsky, I., Schonlau, M.: 'Less than one'-shot learning: learning n classes from m < n samples (2020)
32. Tong, S., Koller, D.: Support vector machine active learning with applications to text classification. J. Mach. Learn. Res. **2**(Nov), 45–66 (2001)
33. Tsang, I.W., Kwok, J.T., Cheung, P.M., Cristianini, N.: Core vector machines: fast SVM training on very large data sets. J. Mach. Learn. Res. **6**(4), 363–392 (2005)
34. Tukey, J.W., et al.: Exploratory Data Analysis, vol. 2. Addison-Wesley, Reading (1977)
35. Vinyals, O., Blundell, C., Lillicrap, T., Kavukcuoglu, K., Wierstra, D.: Matching networks for one shot learning. arXiv preprint arXiv:1606.04080 (2016)
36. Wan, E.A., Van Der Merwe, R.: The unscented Kalman filter for nonlinear estimation. In: Proceedings of the IEEE 2000 Adaptive Systems for Signal Processing, Communications, and Control Symposium (Cat. No. 00EX373), pp. 153–158. IEEE (2000)
37. Wan, E.A., Van Der Merwe, R., Haykin, S.: The unscented Kalman filter. Kalman filtering and neural networks **5**(2007), 221–280 (2001)
38. Wang, T., Zhu, J., Torralba, A., Efros, A.A.: Dataset distillation. CoRR abs/1811.10959 (2018). http://arxiv.org/abs/1811.10959
39. Wang, T., Zhu, J.Y., Torralba, A., Efros, A.A.: Dataset distillation. arXiv preprint arXiv:1811.10959 (2018)
40. Welinder, P., et al.: Caltech-UCSD birds 200 (2010)
41. Xian, Y., Lorenz, T., Schiele, B., Akata, Z.: Feature generating networks for zero-shot learning. CoRR **abs/1712.00981** (2017). http://arxiv.org/abs/1712.00981
42. Yang, S., Liu, L., Xu, M.: Free lunch for few-shot learning: distribution calibration (2021)

Object Centric Point Sets Feature Learning with Matrix Decomposition

Zijia Wang(✉), Wenbin Yang, Zhisong Liu, Qiang Chen, Jiacheng Ni, and Zhen Jia

Dell Technologies OCTO Research Office, Shanghai, China
Zijia_Wang@dell.com

Abstract. A representation matching the invariance/equivariance characteristics must be learnt to rebuild a morphable 3D model from a single picture input. However, present approaches for dealing with 3D point clouds depend heavily on a huge quantity of labeled data, while unsupervised methods need a large number of parameters. This is not productive. In the field of 3D morphable model building, the encoding of input photos has received minimal consideration. In this paper, we design a unique framework that strictly adheres to the permutation invariance of input points. Matrix Decomposition-based Invariant (MDI) learning is a system that offers a unified architecture for unsupervised invariant point set feature learning. The key concept behind our technique is to derive invariance and equivariance qualities for a point set via a simple but effective matrix decomposition. MDI is incredibly efficient and effective while being basic. Empirically, its performance is comparable to or even surpasses the state of the art. In addition, we present a framework for manipulating avatars based on CLIP and TBGAN, and the results indicate that our learnt features may help the model achieve better manipulation outcomes.

Keywords: Object centric representation · 3D learning · Diffusion model

1 Introduction

Understanding objects is one of the core problems of computer vision, especially in the avatar generation process, learning an object-centric representation is important in many downstream tasks [14]. In machine learning, even a small attack [27] or image corruption [6] could produce a large accuracy decline. This is specifically important in 3D settings because a point cloud will be generated and can be varied based on different conditions. An object-centric representation is a graceful representation that can handle the distribution shift [4].

In this paper, we investigate deep learning architectures that are able to reason about three-dimensional geometric data, such as point clouds or meshes. In order to accomplish weight sharing and other kernel improvements, most convolutional architectures need extremely regular input data formats. Examples of such formats are picture grids and 3D voxels. Before feeding point clouds

Y. Zheng et al. (Eds.): ACCV 2022, LNCS 13848, pp. 256–268, 2023.
https://doi.org/10.1007/978-3-031-27066-6_18

or meshes to a deep learning architecture, the majority of researchers often convert such data to conventional 3D voxel grids or collections of pictures (for example, views) since point clouds and meshes do not have a regular format. This particular modification of the data representation, on the other hand, results in data that is needlessly large in volume, and it also introduces quantization artifacts, which have the potential to conceal the inherent invariances of the data.

To solve these problems, we propose a nearly non-parametric framework to learn a more useful representation and reconstruct the 3D avatar from the portraits. Concretely, our contributions to this paper are:

- A self-supervised learning framework to learn the canonical representation of the input. Two loss functions are introduced to make sure the learned representations are invariant/equivariant.
- Tensor decomposition for representation learning. The learned representation could satisfy the invariance/equivariance properties, which could be used in the 3D registration part which is the key component in 3D morphable model learning.
- A generalized application framework to deal with 3D images. We use the 3D morphable model reconstruction task as an example here, the results of the generated avatar show that our proposed algorithm can deal with the avatar manipulation well.

2 Related Work

2.1 Point Cloud Features

Current point cloud feature extraction algorithms are mostly handcrafted based on one specific task [18]. These features contain certain statistical properties that are invariant to certain transformations. Therefore, they can be categorized into intrinsic (local features) [1,2,23] and extrinsic (global features) [3,13,16,20,21]. However, it's also necessary to optimally combine these properties. Although [18] tried to perform the trade-off to find the best feature combination, it's not trivial to make the whole process explicit and efficient.

2.2 Deep 3D Representations

Currently, there exist many approaches for 3D feature learning like Volumetric CNNs [17,19,26] which utilize 3D convolutional neural networks to deal with voxelized shapes. However, data sparsity and computation cost of 3D convolution naturally limit the ability of the representations learned from these networks. Then FPNN [15] and Vote3D [25] proposed some metrics to solve the problem brought by data sparsity, but when these methods come to very large point clouds, their operations based on space volumes constrain them. These days, some new powerful methods like Multiview CNNs [19,22] and Feature-based DNNs [7,12] are proposed. However, the representative ability of the extracted features is still one of the key constraints of these metrics. Therefore, in this paper, we'll try to solve this problem.

3 Object-Centric Representation Learning

The key innovation of the proposed method is to utilize the simple but powerful matrix decomposition to generate invariance/equivariance properties for points set. In the Sect. 3.2, we first introduce the matrix decomposition module in our method, then in Sect. 3.3 we reveal our novel unsupervised pipeline. In the following, Sect. 3.4, we develop the decoder for input point sets reconstruction. At last, we discuss the avatar generation extension for the proposed method in the Sect. 3.6.

3.1 Problem Definition

We design a deep learning framework that directly consumes unordered point sets as inputs. A point cloud is represented as a set of 3D points [18]

$$\boldsymbol{P} = \{P_n | n = 1, ..., N\}, \tag{1}$$

where each point P_n is a vector of its (x, y, z) coordinate plus extra feature channels such as color, normal, etc. For simplicity and clarity, unless otherwise noted, we only use the (x, y, z) coordinate as our point's channels, as shown in Fig. 1.

The input point set is a subset of points from Euclidean space. It has three main properties: [18]

- Unordered. Unlike pixel arrays in images or voxel arrays in volumetric grids, the point cloud is a set of points.
- Interaction among points. The points are from a space with a distance metric. It means that points are not isolated, and neighboring points form a meaningful subset. Therefore, the model needs to be able to capture local structures from nearby points, and the interactions among local structures.
- Invariance under transformations. As a geometric object, the learned representation of the point set should be invariant to certain transformations. For example, rotating and translating points altogether should not modify the global point cloud category or the segmentation of the points.

3.2 Matrix Decomposition for Invariant and Equivariant Feature Learning

An overview of MD in the feature space is depicted in Fig. 3. The point cloud input will be transferred to a feature matrix $\boldsymbol{X} \in R^{M \times d}$, then we can decompose the matrix in the following manner:

$$\boldsymbol{X} = \boldsymbol{U}\boldsymbol{V} + \boldsymbol{E} \tag{2}$$

where $\boldsymbol{U} \in R^{M \times k}$ and $\boldsymbol{V} \in R^{k \times d}$ with $k < d$ are the decomposition factors; $\boldsymbol{E}$ is the residual. $\boldsymbol{U}$ can be considered as activation factor, which should be

Fig. 1. Some examples of 3d point sets. Different colors indicate different semantic parts of an object.

invariant and capture the most important information for dataset. While $\boldsymbol{V}$ can be considered as the template factor, which is equivariant to each input data sample. Therefore, $\boldsymbol{UV}$ is the low rank approximation of $\boldsymbol{X}$ [11]. As shown in Fig. 3, the MD is a white-box unsupervised decomposition, which utilized the low-rank properties of the Feature Matrix. In MD, it can be conducted online [10] (Fig. 2).

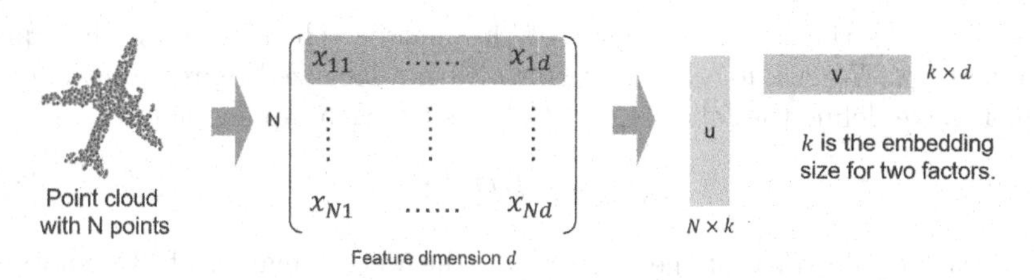

Fig. 2. The decomposition of the feature matrix. The invariant part $\boldsymbol{U}$ records the relative location, and the equivariant part $\boldsymbol{V}$ records the absolute location in world coordination.

3.3 Encoder

The encoder training pipeline of the proposed method is illustrated in the Fig. 3. Our network is trained by feeding **pairs of randomly rotated copies of the same shape**. The input point clouds are randomly generated from two random transformations $T_1, T_2 \in R^{(3)}$ (rotating and translating). Note that we train such a decomposition in a fully unsupervised fashion and that the network only ever sees randomly rotated point clouds.

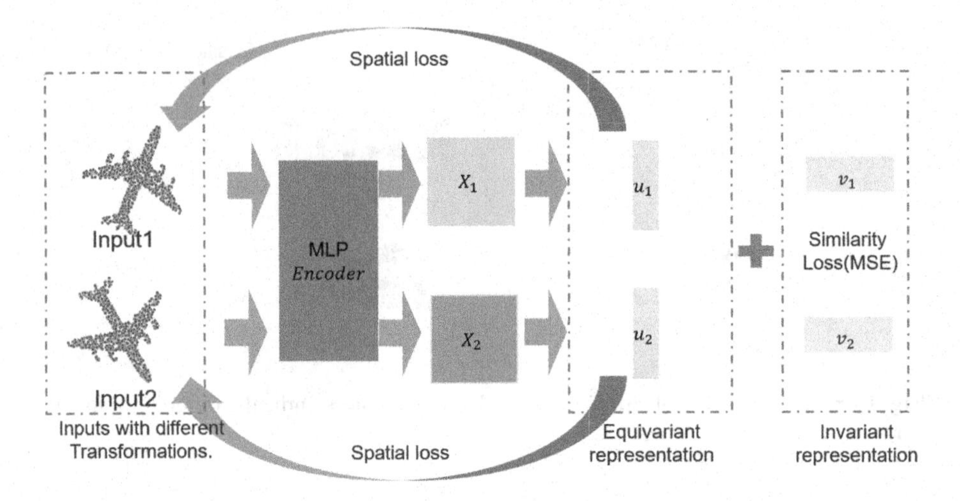

Fig. 3. The pipeline of the proposed method. The framework can take 2D- or 3D-points set as input.

As the two input point sets are of the same object, so the similarity loss can be expressed as

$$L_{sim} = ||\boldsymbol{V}_1 - \boldsymbol{V}_2||_F \tag{3}$$

where $||\cdot||_F$ is the Frobenius norm of the matrix. Then, to learn the equivalence features. We ask for the network to learn a localized representation of the geometry. we define the following spatial loss for each input point set.

$$L_{spa}(P) = tr(\boldsymbol{U}^T\boldsymbol{W}\boldsymbol{U}) \tag{4}$$

where $tr(\cdot)$ is the trace of the matrix; $\boldsymbol{W}$ the weight matrix of 3D points set. The $\boldsymbol{W}(m,n)$ is the weight between two points, and can be calculated as:

$$\boldsymbol{W}(m,n) = exp(\frac{||P_m - P_n||_2^2}{\sigma^2}) \tag{5}$$

where σ is the parameter to control the distance. For the sake of effectiveness, one can just use part of randomly selected points for reconstruction. The spatial loss considers the spatial relationship of the input 3D point set, as shown in the Fig. 7.

After training, we will choose the 3D points set with the minimal l_1 norm of $\boldsymbol{U}$ as the reference point set (Fig. 4).

3.4 Reconstruction Decoder

For downstream tasks:

- Classification: need the invariant representation.

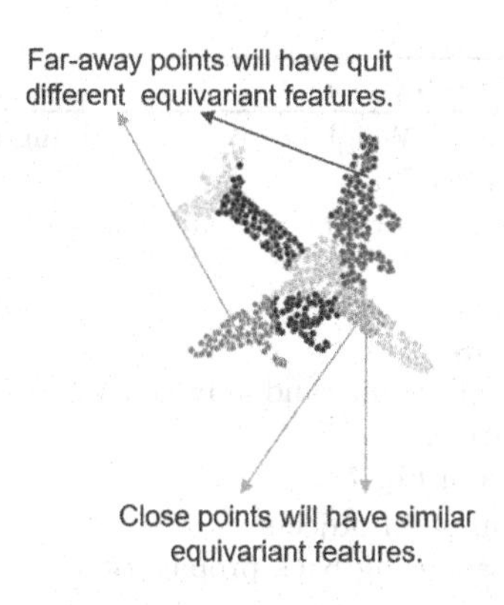
Far-away points will have quit different equivariant features.
Close points will have similar equivariant features.

Fig. 4. The spatial loss.

- Segmentation: need the equivariant representation.
- Point sets reconstruction: need the reconstructed matrix $\boldsymbol{UV}$ and an addictive decoder that transforms the latent feature into 3D point sets.

It is clear that, in applications like Avatar [8], one should reconstruct the lantern features to a 3D point set. In this paper, we design a decoder shown in the Fig. 5. The decoder MSE loss can be expressed as

$$L_{dec} = \frac{1}{N}\|\boldsymbol{P} - \hat{\boldsymbol{P}}\|_F \tag{6}$$

where $\hat{\boldsymbol{P}}$ is the output matrix of the decoder.

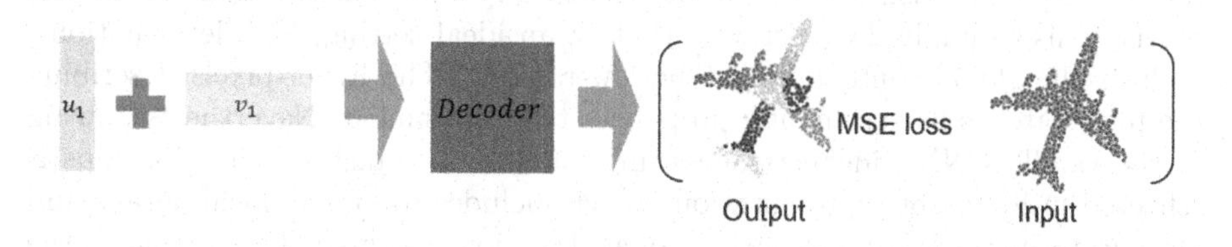
u_1
v_1
Decoder
MSE loss
Output
Input

Fig. 5. The decoder for point set reconstruction.

The decoder can be trained with/without the framework proposed in the Sect. 3.3. When a train with the encoder Sect. 3.3, the loss can be calculated as:

$$L = L_{sim} + \alpha \sum_{i}^{2} L_{spa}^{i} + \beta \sum_{i=1}^{2} L_{dec}^{i} \tag{7}$$

where α and β are weights to control the loss values. The joint training algorithm is summed in Algorithm 1.

Algorithm 1: MDI Training

Input : Dataset $\{\boldsymbol{P}\} \in \mathcal{P}$; Weight α and β, Training epoch number T
Output: Encoder and Decoder
Initialize: model parameters for Encoder and Decoder
for $t \leftarrow 1$ *to* T **do**
    for *each mini batch B* **do**
        for *each object* **do**
            compute inputs point randomly generated from two random transformations
            compute loss in Eq. 7
        compute the sum of all objects
        update parameters using back propagation
return

3.5 Theoretical Analysis

In this section, we intend to illustrate why the low-rank assumption is beneficial for modeling the global context of representations by providing an example. The low-rank assumption is advantageous because it illustrates the inductive bias that low-level representations include fewer high-level ideas than the scale of the representations. Consider a picture of a person walking on the road. The route will be described by a large number of hyperpixels retrieved using a CNN's backbone. Notabene que la carretera puede ser considerada como repeticiones de pequeos fragmentos de carretera, por lo que se puede representar la carretera mediante la modelación y It is mathematically equal to locating a limited set of bases D corresponding to various road patches and a coefficient matrix C that records the relationship between the elementary road patches and the hyper-pixels. This example demonstrates that in an ideal setting, high-level notions, such as the global context, might be low-ranking. The hyper-pixels describing the road patches have semantic properties that are similar. Nevertheless, owing to the vanilla CNN's ineffectiveness un modeling long-distance relationships is reflected in its learnt representation, which includes too many local details and inaccurate information and lacks global direction. Imagine the subject in the photograph wearing gloves. When we see the gloves patch in our community, we assume it defines gloves. When the broader context is considered, it becomes clear that this patch is a portion of a person. The semantic information is hierarchical, depending on the amount of comprehension desired.

The objective of this section is to enable networks to comprehend the context globally by means of the low-rank recovery formulation. Incorrect information, notably redundancies and incompletions, are modeled as a noise matrix. To highlight the global context, we split the representations into two parts, a low-rank global information matrix and a local equivariant matrix, using an optimization approach to recover the clean signal subspace, eliminate the noises, and improve the global information through the skip connection. The data might reveal how much global knowledge the networks need for a certain operation.

3.6 Measurement of Learned Features on Avatar Generation

Avatar generation can naturally be a good application for the testing of invariance and equivariance features. Figure 6 shows the framework of our avatar generation, which is based on the TBGAN proposed in [9]. Given a pre-trained TBGAN generator $\mathcal{G}$, let $z \in \mathcal{R}^d$ denote a d-dimensional random input vector sampled from a Gaussian distribution $\mathcal{N}\left(0, \sigma^2\right)$ and e originally denote a one-hot encoded facial expression vector initialized to zero to obtain a neutral expression. However, in this paper, the equivariant feature derived from the generated image from CLIP becomes e. Let $\mathbf{c} \in \mathcal{C}$ denote an intermediate layer vector obtained by partial forward propagation of $\mathbf{z}$ and e through the generator $\mathcal{G}$. Our method first generates a textured mesh by using the generated shape, normal, and texture UV maps via cylindrical projection. Then given a text prompt t such as 'happy human', $\mathbf{c}$ is optimized via gradient descent to find a direction $\Delta\mathbf{c}$, where $\mathcal{G}(\mathbf{c}+\Delta\mathbf{c})$ produces a manipulated textured mesh in which the target attribute specified by t is present or enhanced, while other attributes remain largely unaffected. In our work, we optimize the original intermediate latent vector $\mathbf{c}$ using gradient descent and work in the 4×4 dense layer s of the TBGAN generator.

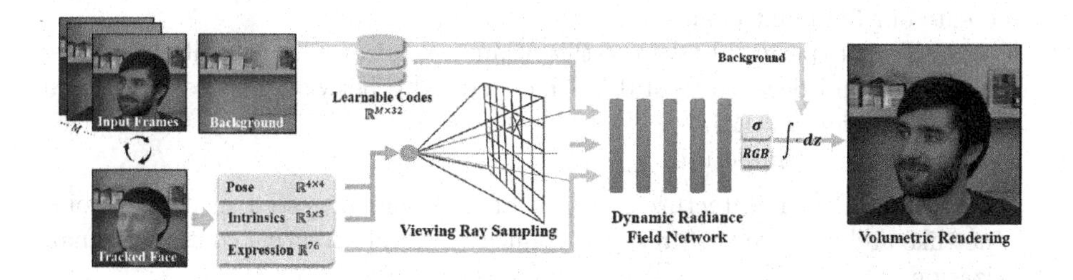

Fig. 6. The avatar generation framework for testing the learned feature extractor.

The optimized latent vector $c+\Delta c$ can then be fed into TBGAN to generate shape, normal, and texture UV maps, and finally a manipulated mesh with the target attributes. To perform meaningful manipulation of meshes without creating artifacts or changing irrelevant attributes, we use a combination of an equivariance loss, an identity loss, and an L2 loss as follows:

$$\underset{\Delta\mathbf{c}\in\mathcal{C}}{\arg\min} \quad \mathcal{L}_{\mathrm{eq}} + \lambda_{\mathrm{ID}}\mathcal{L}_{\mathrm{ID}} + \lambda_{\mathrm{L2}}\mathcal{L}_{\mathrm{L2}} \tag{8}$$

where λ_{ID} and λ_{L2} are the hyperparameters of $\mathcal{L}_{ID}$ and $\mathcal{L}_{\mathrm{L2}}$, respectively. While equivariance loss ensures that the user-specified attribute is present or enhanced, ID-loss and L2-loss leave other attributes unchanged, forcing disentangled changes. The identity loss $\mathcal{L}_{ID}$ minimizes the distance between the identity of the original renders and the manipulated renders:

$$\mathcal{L}_{\mathrm{ID}} = \|(\mathbf{U_{ori}} - \mathbf{U_{edi}})\|_2 \tag{9}$$

where $\mathbf{U_{ori}}$ is the invariant feature of the original image and the $\mathbf{U_{edi}}$ is the invariant feature for edited image. Similarly, the equivariance loss $\mathcal{L}_{eq}$ can be defined as:

$$\mathcal{L}_{\mathrm{eq}} = \|(\mathbf{V_{ori}} - \mathbf{V_{edi}})\|_2, \tag{10}$$

where $\mathbf{V_{ori}}$ and $\mathbf{V_{edi}}$ are the equivariant features of the original image and the edited image respectively. Finally, the L2 loss is used to prevent artifact generation and defined as:

$$\mathcal{L}_{\mathrm{L.2}} = \|\mathbf{c} - (\mathbf{c} + \Delta\mathbf{c})\|_2 \tag{11}$$

For TBGAN and renderer, we follow the same settings in [9].

4 Experiment Results

This section evaluates the proposed approach and compares it against State Of The Art methods. To evaluate our method, we rely on the ShapeNet (Core) dataset. We follow the category choices from AtlasNetV2, using the airplane and chair classes for single category experiments, while for multi-category experiments we use all the 13 classes in ShapeNet (Core) dataset. Unless noted otherwise, we randomly sample 1024 points from the object surface for each shape to create our 3D point clouds.

For all our experiments we use the Adam optimizer with an initial learning rate of 0.001 and decay rate of 0.1. Unless stated otherwise, we use $k = 10$ and feature dimension $d = 128$.

Our network architecture:

- **Encoder.** Our architecture is based on the one suggested in [24]: a point net-like architecture with residual connections and attentive context normalization.
- **Decoder.** Our decoder architecture is similar to AtlasNetV2 [5] (with trainable grids).

The last section in this part shows the avatar reconstruction results based on the framework shown in the Sect. 3.6.

4.1 Reconstruction Result

We evaluate the performance of our method for reconstruction against two baselines:

- AtlasNetV2 [5], the State Of The Art auto-encoder which utilizes a multi-head patch-based decoder;
- 3D-PointCapsNet [28], an auto-encoder for 3D point clouds that utilize a capsule architecture.

As shown in the Table 1 We achieve State Of The Art performance in both the aligned and unaligned settings.

We illustrate the reconstruction of 3D point clouds for all the methods. our method can provide semantically consistent decomposition, for example, the wings of the airplane have consistent colors.

Table 1. Reconstruction – Performance in terms of Chamfer distance.

	Aligned			Unaligned		
	Airplane	Chair	Multi	Airplane	Chair	Multi
3D-PointCapsNet	1.94	3.30	2.49	5.58	7.57	4.66
AtlasNetV2	1.28	2.36	2.14	2.80	3.98	3.08
MDI(Ours)	**0.93**	**2.01**	**1.66**	**1.05**	**3.75**	**2.20**

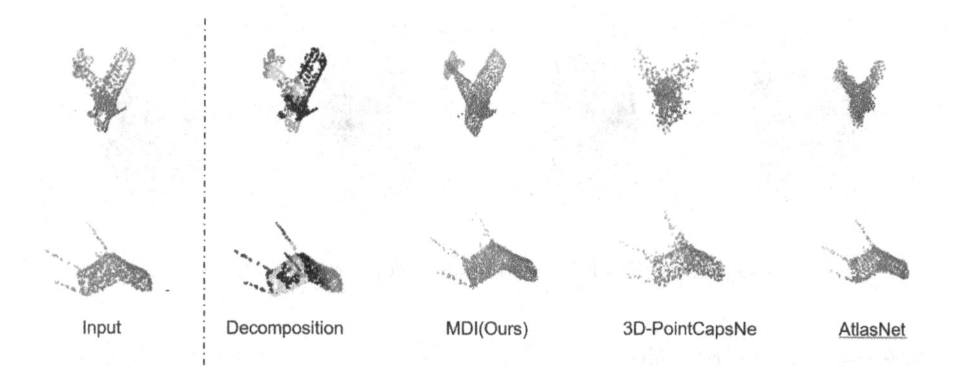

Fig. 7. Reconstruction results.

4.2 Classification Result

We compute the features from the auto-encoding methods compared in Sect. 4.1 - AtlasNetV2 [5], 3D-PointCapsNet [28], and our learned invariance features. We use them to perform 13-way classification with Support Vector Machine (SVM) and K-Means clustering. Our results are superior to the other SOTA method. We argue that the joint invariant and equivariant feature learning with MD is important to unsupervised learning. This is especially obvious for the unaligned part because of the advantages of our learned invariant and equivariant features. And for aligned ones, we can also achieve competitive results (Table 2).

Table 2. Classification – Top-1 accuracy (%)

	Aligned		Unaligned	
	SVM	K-Means	SVM	K-Means
AtlasNetV2	94.07	61.66	71.13	14.59
3D-PointCapsNet	93.81	65.87	64.85	17.12
MDI(Ours)	**93.78**	**71.42**	**86.58**	**49.93**

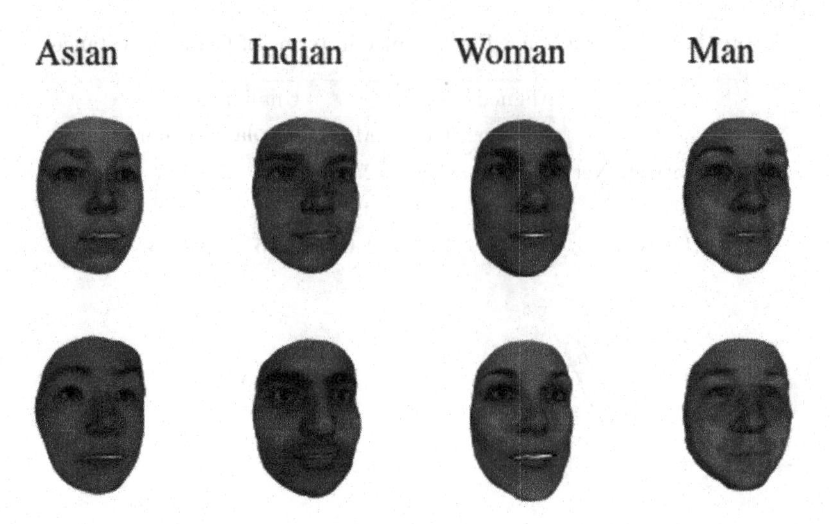

Fig. 8. Results of manipulation on equivariant features.

4.3 Avatar Generation

In this section, to show the power of our learned features, we follow the framework shown in the Fig. 6. By doing this, our method can be used to change their facial expressions such as 'smiling', 'angry', and 'surprised'. As can be seen in Fig. 8, our method can successfully manipulate a variety of complex emotions on various input meshes with almost no change to other attributes. By comparing with the results of TBGAN [9], the advantages of directly manipulating equivariant features are obvious.

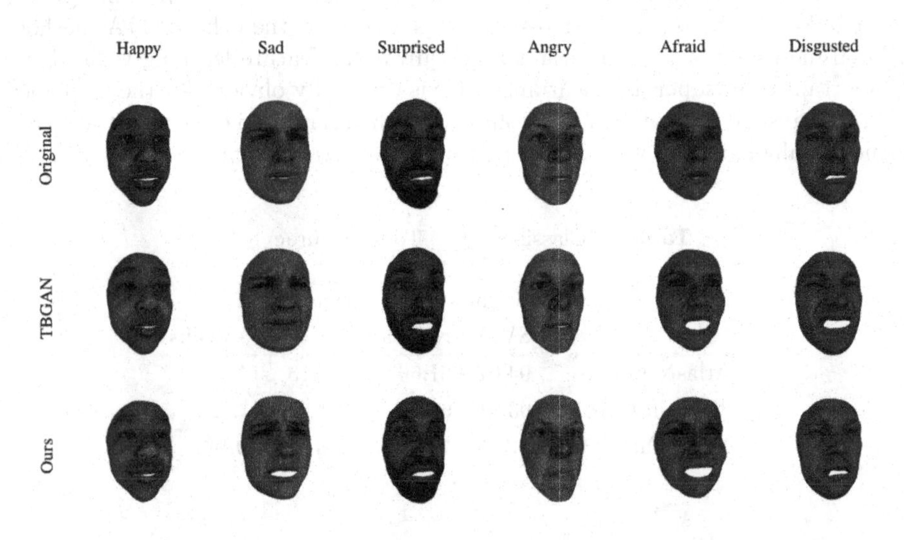

Fig. 9. Results of manipulation on invariant features.

By manipulating the invariant features, we slightly change the framework shown in Fig. 6. Concretely, the CLIP is used to generate the pictures with global features (invariant features), then the U is extracted from generated image. Then V is extracted from the original avatar images. The results also show that our method provides a global semantic understanding of more complex attributes such as 'man', 'woman', 'Asian', and 'Indian'. Figure 9 shows the results for manipulations on various randomly generated outputs, where we can see that our method can perform complex edits such as ethnicity and gender.

5 Conclusion

In this paper, we design a novel Matrix Decomposition-based Invariant (MDI) learning framework, which can provide a unified architecture for unsupervised invariant point sets feature learning.

Though simple, MDI is highly efficient and effective. Empirically, it shows strong performance on point sets reconstruction and unsupervised classification. Moreover, our framework will benefit other downstream like collaborative computing in avatar generation.

References

1. Aubry, M., Schlickewei, U., Cremers, D.: The wave kernel signature: a quantum mechanical approach to shape analysis. In: 2011 IEEE International Conference on Computer Vision Workshops (ICCV Workshops), pp. 1626–1633. IEEE (2011)
2. Bronstein, M.M., Kokkinos, I.: Scale-invariant heat kernel signatures for non-rigid shape recognition. In: 2010 IEEE Computer Society Conference on Computer Vision and Pattern Recognition, pp. 1704–1711. IEEE (2010)
3. Chen, D.Y., Tian, X.P., Shen, Y.T., Ouhyoung, M.: On visual similarity based 3D model retrieval. In: Computer Graphics Forum, vol. 22, pp. 223–232. Wiley Online Library (2003)
4. Creager, E., Jacobsen, J., Zemel, R.S.: Exchanging lessons between algorithmic fairness and domain generalization. CoRR **abs/2010.07249** (2020). https://arxiv.org/abs/2010.07249
5. Deprelle, T., Groueix, T., Fisher, M., Kim, V.G., Russell, B.C., Aubry, M.: Learning elementary structures for 3D shape generation and matching. arXiv preprint arXiv:1908.04725 (2019)
6. Duchi, J., Glynn, P., Namkoong, H.: Statistics of robust optimization: a generalized empirical likelihood approach. arXiv preprint arXiv:1610.03425 (2016)
7. Fang, Y., et al.: 3D deep shape descriptor. In: Proceedings of the IEEE Conference on Computer Vision and Pattern Recognition, pp. 2319–2328 (2015)
8. Gafni, G., Thies, J., Zollhofer, M., Nießner, M.: Dynamic neural radiance fields for monocular 4D facial avatar reconstruction. In: Proceedings of the IEEE/CVF Conference on Computer Vision and Pattern Recognition, pp. 8649–8658 (2021)
9. Gecer, B., et al.: Synthesizing coupled 3D face modalities by trunk-branch generative adversarial networks. In: Vedaldi, A., Bischof, H., Brox, T., Frahm, J.-M. (eds.) ECCV 2020. LNCS, vol. 12374, pp. 415–433. Springer, Cham (2020). https://doi.org/10.1007/978-3-030-58526-6_25

10. Geng, Z., Guo, M.H., Chen, H., Li, X., Wei, K., Lin, Z.: Is attention better than matrix decomposition? arXiv preprint arXiv:2109.04553 (2021)
11. Guan, N., Tao, D., Luo, Z., Shawe-Taylor, J.: MahNMF: Manhattan non-negative matrix factorization. arXiv preprint arXiv:1207.3438 (2012)
12. Guo, K., Zou, D., Chen, X.: 3D mesh labeling via deep convolutional neural networks. ACM Trans. Graph. (TOG) **35**(1), 1–12 (2015)
13. Johnson, A.E., Hebert, M.: Using spin images for efficient object recognition in cluttered 3D scenes. IEEE Trans. Pattern Anal. Mach. Intell. **21**(5), 433–449 (1999)
14. Li, N., Raza, M.A., Hu, W., Sun, Z., Fisher, R.: Object-centric representation learning with generative spatial-temporal factorization. In: Advances in Neural Information Processing Systems, vol. 34 (2021)
15. Li, Y., Pirk, S., Su, H., Qi, C.R., Guibas, L.J.: FPNN: field probing neural networks for 3D data. In: Advances in Neural Information Processing Systems, vol. 29 (2016)
16. Ling, H., Jacobs, D.W.: Shape classification using the inner-distance. IEEE Trans. Pattern Anal. Mach. Intell. **29**(2), 286–299 (2007)
17. Maturana, D., Scherer, S.: VoxNet: a 3D convolutional neural network for real-time object recognition. In: 2015 IEEE/RSJ International Conference on Intelligent Robots and Systems (IROS), pp. 922–928. IEEE (2015)
18. Qi, C.R., Su, H., Mo, K., Guibas, L.J.: Pointnet: deep learning on point sets for 3D classification and segmentation. In: Proceedings of the IEEE Conference on Computer Vision and Pattern Recognition, pp. 652–660 (2017)
19. Qi, C.R., Su, H., Nießner, M., Dai, A., Yan, M., Guibas, L.J.: Volumetric and multi-view CNNs for object classification on 3D data. In: Proceedings of the IEEE Conference on Computer Vision and Pattern Recognition, pp. 5648–5656 (2016)
20. Rusu, R.B., Blodow, N., Beetz, M.: Fast point feature histograms (FPFH) for 3D registration. In: 2009 IEEE International Conference on Robotics and Automation, pp. 3212–3217. IEEE (2009)
21. Rusu, R.B., Blodow, N., Marton, Z.C., Beetz, M.: Aligning point cloud views using persistent feature histograms. In: 2008 IEEE/RSJ International Conference on Intelligent Robots and Systems, pp. 3384–3391. IEEE (2008)
22. Su, H., Maji, S., Kalogerakis, E., Learned-Miller, E.: Multi-view convolutional neural networks for 3D shape recognition. In: Proceedings of the IEEE International Conference on Computer Vision, pp. 945–953 (2015)
23. Sun, J., Ovsjanikov, M., Guibas, L.: A concise and provably informative multi-scale signature based on heat diffusion. In: Computer Graphics Forum, vol. 28, pp. 1383–1392. Wiley Online Library (2009)
24. Sun, W., Tagliasacchi, A., Deng, B., Sabour, S., Yazdani, S., Hinton, G., Yi, K.M.: Canonical capsules: self-supervised capsules in canonical pose. In: Thirty-Fifth Conference on Neural Information Processing Systems (2021)
25. Wang, D.Z., Posner, I.: Voting for voting in online point cloud object detection. In: Robotics: Science and Systems, Rome, Italy, vol. 1, pp. 10–15 (2015)
26. Wu, Z., et al.: 3D shapenets: a deep representation for volumetric shapes. In: Proceedings of the IEEE Conference on Computer Vision and Pattern Recognition, pp. 1912–1920 (2015)
27. Zhang, B.H., Lemoine, B., Mitchell, M.: Mitigating unwanted biases with adversarial learning. In: Proceedings of the 2018 AAAI/ACM Conference on AI, Ethics, and Society, pp. 335–340 (2018)
28. Zhao, Y., Birdal, T., Deng, H., Tombari, F.: 3D point capsule networks. In: Proceedings of the IEEE/CVF Conference on Computer Vision and Pattern Recognition, pp. 1009–1018 (2019)

Aerial Image Segmentation via Noise Dispelling and Content Distilling

Yongqing Sun[1(✉)], Xiaomeng Wu[2], Yukihiro Bandoh[1], and Masaki Kitahara[1]

[1] NTT Computer and Data Science Laboratories, Tokyo, Japan
{yongqing.sun.fb,yukihiro.bandoh.pe,masaki.kitahara.ve}@hco.ntt.co.jp
[2] NTT Communication Science Laboratories, Kyoto, Japan
xiaomeng.wu.px@hco.ntt.co.jp

Abstract. Aerial image segmentation is an essential problem for land management which can be used for change detection and policy planning. However, traditional semantic segmentation methods focus on single-perspective images in road scenes, while aerial images are top-down views and objects are of a small size. Existing aerial segmentation methods tend to modify the network architectures proposed for traditional semantic segmentation problems, yet to the best of our knowledge, none of them focus on the noisy information present in the aerial images. In this work, we conduct an investigation on the effectiveness of each channels of the aerial image on the segmentation performance. Then, we propose a disentangle learning method to investigate the differences and similarities between channels and images, so that potential noisy information can be removed for higher segmentation accuracy.

Keywords: Aerial image segmentation · Disentangle learning · Semantic segmentation

1 Introduction

In the event of a large-scale disaster such as an earthquake or tsunami, it is necessary to quickly obtain information on a wide area in order to secure safe evacuation and rescue routes and to consider reconstruction measures. Aerial image segmentation is one of potential solutions. The result from aerial image segmentation can also be used to adjust governmental policy for subsidy, and to utilize the given resources for land management. In the past, this task was done manually and laboriously so it could only be done for a small number of photographs, which is not enough to capture the changes across large areas [5]. Moreover, such changes can be dramatic over time. Therefore, it is highly desirable to develop an automatic solution for the task.

A similar task known as semantic segmentation inputs an image and outputs a semantic segmentation map which indicates the class of an arbitrary pixel in the input image. For this task, various methods have been proposed using deep

Y. Zheng et al. (Eds.): ACCV 2022, LNCS 13848, pp. 269–279, 2023.
https://doi.org/10.1007/978-3-031-27066-6_19

learning. Among them, CNN [6,12,13] has been actively applied in research and has shown high performance. However, they heavily rely on manual annotation because they are fully supervised [1,3], yet remote-sensing data is not rich in training data, and it is not clear whether "deeper" learning is possible with a limited number of training samples.

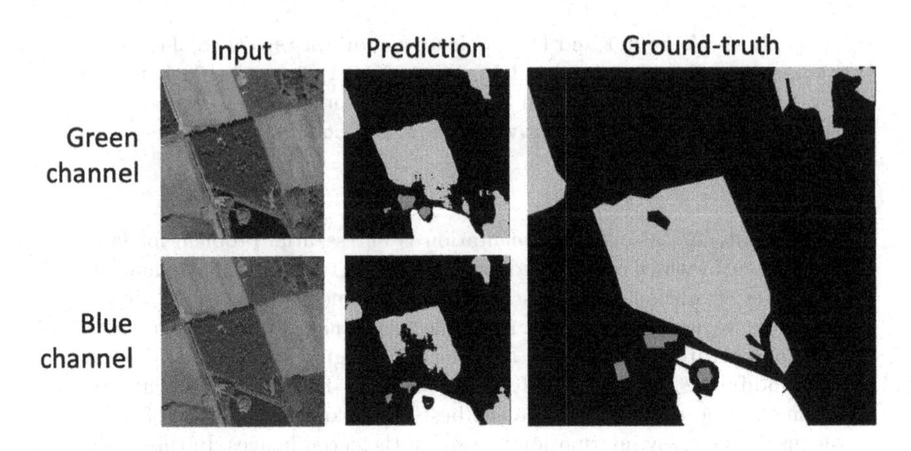

Fig. 1. U-Net results when trained on different channels show different effectiveness on different classes. The Green channel outperforms in the Woodland class (light gray) while the Blue channel outperforms in the Water class (white). (Color figure online)

Semantic segmentation of low-resolution images in aerial photography is a challenging task because their objects are tiny and observed from a top-down faraway viewpoint, which is entirely different compared to the traditional semantic segmentation task. Many aerial segmentation methods are based on network architectures proposed for the mainstream semantic segmentation task. Boguszewski et al. [2] employs DeepLabv3+ with the backbone being modified Xception71 and Dense Prediction Cell (DPC) for aerial segmentation. Khalel et al. [8] proposed 2-level U-Nets with data augmentation to refine the segmentation result on aerial images. Li et al. [9] proposed to add a group of cascaded convolution to U-Net to enhance the receptive field.

However, aerial images are generally captured from high altitude and available in low resolution, hence their illumination information is sensitive to noise. The situation is worsen with the presence of dense objects. Therefore, in order to achieve accurate segmentation and classification of aerial photographs, it is necessary to extract the essential semantic content in the images.

Disentangle learning indicates methods which allows encoding the input into separated features belonging to predefined sub-spaces. UNIT [11] and MUNIT [7] are popular unsupervised disentangle learning methods for image-to-image translation task, which can be deemed as the general task of many computer vision problems such as segmentation. They encode the image into a content code and a style code. Content code is the common feature that can be shared between 2 domains, e.g. the same pose shared by a cat image and a tiger image. Style code is the encoded distribution which gives the variant tastes to the content code and it cannot be shared between 2 domains, e.g. the difference in skin color and texture between cat and tiger images. However, UNIT only allows one-to-one translation while MUNIT allows one-to-many translation with random style codes. Inspired by this technique, we propose the content consistency loss to enforce the content code of different channels of the same image to be the same and style consistency loss to enforce the style code of the same channel of different images to be the same.

In this work, we conducted segmentation on each channel respectively and found that each channel has different segmentation effectiveness on objects as shown in Fig. 1. It is intuitively to think that different channels of the same image share the same content code, which is essential for aerial image segmentation. We also come up with a hypothesis that the same channels of different images share the same style code. The style code, which may reflect noise or specific illumination characteristic in channels, should not contribute to the segmentation result and has adverse effect on segmentation. Therefore, the content code should contain all the essential information needed for the segmentation task.

In our framework, there are 2 encoders and 3 generators. Each encoder and their corresponding generator is corresponding to an image channel. Taking the green channel as an example, the green channel encodes the green channel image into content code and style code, and the generator will generate back to the green channel image based on these two codes. We then define a reconstruction loss between two green channel images. The content consistency loss is added to ensure the content codes from green channel encoder and blue channel encoder to be the same for green channel images and blue channel images of the same image. The style consistency loss is also employed to ensure the style codes of the same channel is the same across different images. Finally, we use the content code of the two channels as the input to the segmentation generator followed with a semantic loss. The proposed method is evaluated on LandCover.ai dataset [2].

To summarize, our contribution is two-fold:

- We show that different channels have different focuses on different segmentation classes.
- We propose a disentangle learning framework that automatically remove the noisy information present in aerial images by mining the differences between channels of the same image and similarities of the same channel across different images.

The paper is organized as follows: Sect. 3 presents our proposed framework, Sect. 4 shows the experimental results and detailed discussion, then our work is concluded with Sect. 5.

2 Related Work

2.1 Traditional Semantic Segmentation

Semantic segmentation task inputs an image and outputs a semantic segmentation map which indicates the class of an arbitrary pixel in the input image. However, it does not tell if those pixels belong to different instances. Traditional semantic segmentation focus on a road scene dataset such as Cityscapes [4], where big objects are generally available. DeepLabv3+ [6] proposes to use atrous separable convolution for the encoder. U-net [15] proposes skips which connect low-level features and high-level features.

On the other hand, objects in aerial images are captured from top-down views and in a very tiny form. Therefore, aerial segmentation presents challenges demanding approaches different from traditional semantic segmentation.

2.2 Aerial Segmentation

Many aerial segmentation methods are based on network architecture proposed for the mainstream semantic segmentation task. A. Boguszewski et al. [2] employs DeepLabv3+ with the backbone being modified Xception71 and Dense Prediction Cell (DPC) for aerial segmentation. Khalel et al. [8] proposed 2-levels U-Nets with data augmentation to refine the segmentation result on aerial images. Li et al. [9] proposed to add a group of cascaded convolution to U-net to enhance the receptive field. However, none of them really focus on investigating the input aerial images for noise removal purpose.

We opt for U-net, which has a similar network architecture to FPN [10] for small object detection. The skip connections enhance semantic level of lower-level features and reduce information distortion at the input of high-level features. We do not employ FPN [10] directly as the size of the objects in the aerial images is already very small and that size does not vary much.

2.3 Disentangle Learning

Disentangle learning indicates methods which allows encoding the input into separate features belonging to predefined sub-spaces. UNIT [11] and MUNIT [7] are popular unsupervised disentangle learning methods for image-to-image translation task, which can be deemed as the general task of many computer vision problems such as segmentation. They encode the image into two content code and style code. Content code is the common feature can be shared between 2 domains, e.g. cats and tigers may share the same pose. Style code is the encoded distribution which gives the variant tastes to the content code and it cannot be shared between 2 domains, e.g. skin colors and patterns between cats and tigers are not the same. However, UNIT only allows one-to-one translation while MUNIT allows one-to-many translation with random style codes.

Yet, due to their unsupervised nature, none of them can utilize the advantage of doing disentangle learning on separate channels of the same image: Given an

RGB image, we know for sure that their content code must be the same. We also take our constraint hypothesis further, the style code of an interested channel, e.g. green channel, should be the same across different images, which is currently not endorsed by any of the existing methods.

FCN+MLP [14] up-samples the encoded features of each layers in Base FCN and then combine them to yield the final prediction. Khalel et al. [8] proposed 2-levels U-Nets with data augmentation.

3 Proposed Method

3.1 Framework

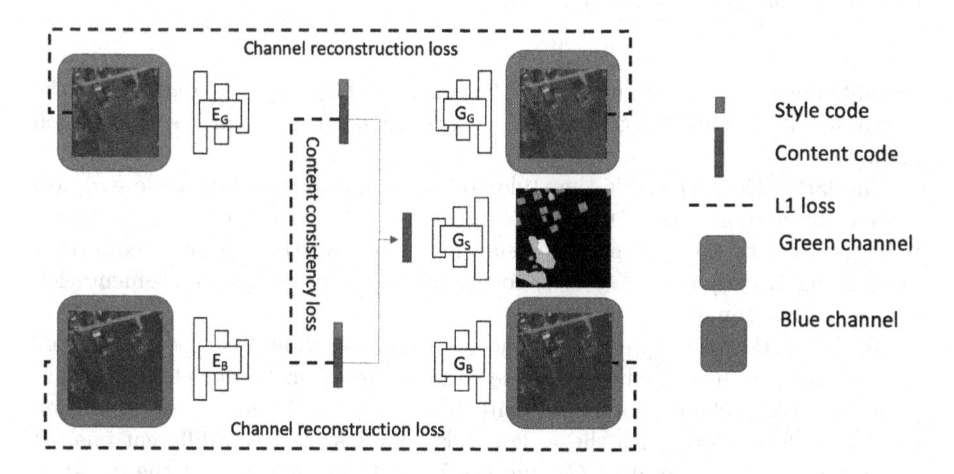

Fig. 2. (a) Style consistency loss

Let x_G, x_B be the G, B channels of an input image x. We exclude the R-channel from our framework as it is discovered to contain a lot of noisy information in Sect. 4.2.

We opt for U-net, which has a similar network architecture to FPN [10] for small object detection, as the network architecture of our encoders and generators. The skip connections enhance semantic level of lower-level features and reduce information distortion at the input of high-level features. We do not employ FPN [10] directly as the size of the objects in the aerial images is already very small and that size does not vary much.

As shown in Fig. 3, our network consists of two encoders and three generators. The two encoders are G-channel and B-channel encoder; the three generators are G-channel generator, B-channel generator, and segmentation map generator. Let E_G be the G-channel encoder, $E_G(x_G)$ yields a 256-channel feature map, whose 56 first channels act as the style code s_{x_G} and 200 remaining channels act as the

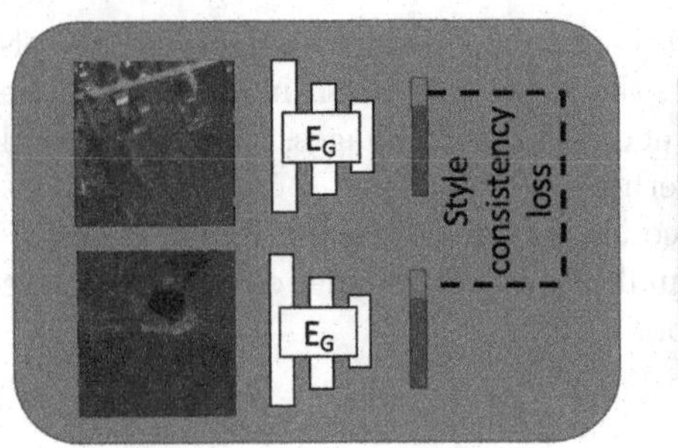

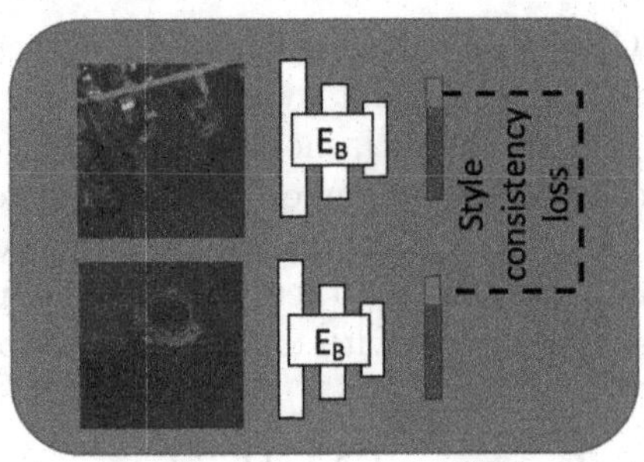

Fig. 3. (b) Content consistency loss and channel. An overview of our proposed framework described with Fig. (a) and (b). We employ U-net network architecture for our encoders E and generators G. Our network also uses a semantic segmentation loss, which is not shown for the sake of brevity.

content code c_{xG} as shown in Fig. 3. The generator $G_G(c_{xG}, s_{xG})$ reconstruct the G-channel input while the generator $G_S(c_{xG})$ predicts the semantic segmentation map.

Similarly, $E_B(x_B)$ yields the style code s_{xG} and the content code c_{xB}. We enforce the content code of G-channel c_{xG} and B-channel c_{xB} of the same image to be the same by the content consistency loss. This constraint is enforced further by training $G_S(c_{xG})$ and $G_S(c_{xB})$ to predict the same semantic segmentation map using the same network parameters G_S.

We argue that the style code of the same channel should be the same across different images, hence, while the style code is important for the channel reconstruction task, it should not contain any information useful for the segmentation task. Therefore, along with the style consistency loss between different images, the reconstruction generators G_G and G_B input the style code but the semantic segmentation map generator G_S does not.

3.2 Channel Reconstruction Loss

Let E_i and G_i be the i-channel encoder and i-channel reconstruction generator. Channel reconstruction loss enforces G_i to reconstruct the exact same channel which is the input of E_i.

$$L_r(x) = \sum_{i \in \{G,B\}} \|x_i - G_i(E_i(x_i))\|_1 \tag{1}$$

3.3 Content Consistency Loss

Let $(c_{x_i}, s_{x_i}) = E_i(x_i)$ be the content code c_{x_i} and style code s_{x_i} of the input i-channel. Content consistency loss enforces content codes of G-channel and B-channel of the same image to be the same.

$$L_c(x) = \|c_{x_G} - c_{x_B}\|_1 \tag{2}$$

One may concern that the content codes will converge to zero using this simple loss. However, the semantic segmentation loss will enforce them to be non-zero in order to generate a meaningful semantic segmentation map.

3.4 Style Consistency Loss

Style consistency loss, on the other hand, enforces the style codes of the same channel to be the same across different images. Let x and y be 2 different images, x_i and y_i be the i-channel of the two, style consistency loss can be computed as follows:

$$L_s(x, y) = \sum_{i \in \{G,B\}} \|s_{x_i} - s_{y_i}\|_1 \tag{3}$$

The style codes are also constrained to be non-zero by the channel reconstruction loss and content consistency loss. As the content codes are the same for all channels of the same image and the reconstructed channels should look differently, the style codes should be the main contributor of such differences, hence they are enforced to be non-zero.

If the batch size is larger than 2 then the style consistency loss will be computed on the first and second samples, second and third samples, etc. The style consistency loss is also applied on the first and last samples forming a cycle of style consistency.

3.5 Semantic Segmentation Loss

We use the weighted cross-entropy loss for the semantic segmentation task. Let M, N and K be the width, height and number of channels of the semantic segmentation map $z_i = G_S(x_i)$. Here, K is also the number of classes in the dataset and w_k is the loss weight of the class k. The larger w_k is, the more attention is spent on reducing the loss values from class k.

$$L_{ss}(x) = \tag{4}$$

$$-\sum_{i \in \{G,B\}} \sum_{m=1}^{M} \sum_{n=1}^{N} \sum_{k=1}^{K} w_k \log\left(\frac{e^{z_i(m,n,k)}}{\sum_{j=1}^{K} e^{z_i(m,n,j)}}\right) \tag{5}$$

3.6 Total Batch Loss

We set the task weight of the content consistency loss, style consistency loss as they are our fundamental constraint. We also set the task weight of the segmentation loss to 10 as it is our main task. Due to the nature of the style consistency loss, our total loss is provided in a form for the whole batch. Let $|B|$ be the batch size of a batch B and x_b be the b-th sample in the batch.

$$L(B) = \frac{10}{|B|}\left(L_s(x_1, x_{|B|}) + \sum_{b=2}^{|B|} L_s(x_{b-1}, x_b)\right) \tag{6}$$

$$+\sum_{b=1}^{|B|} (10 \times L_c(x_b) + L_r(x_b) + 10 \times L_{ss}(x_b)) \tag{7}$$

4 Experiment Result

4.1 LandCover.ai

To the best of our knowledge, LandCover.ai [2] is one of the benchmark aerial segmentation datasets that includes buildings, woodlands and water classes for our experimental purpose. They also has the background class, which involves objects and regions those do not belong to other 3 classes. It covers 216 km^2 of rural areas in Poland with the resolution being 25/50 cm/px. The training set has 7470 aerial images of size 512×512 pixels in the training set. The validation set and the test set have 1620 aerial images of the same size in each set.

We follow the previous work and use mIoU, which is the mean IoU of the four classes, as our evaluation metric.

$$IoU = \frac{TP}{TP + FN + FP} \tag{8}$$

Here, TP stands for True Positive, FN stands for False Negative, FP stands for False Positive.

The training dataset also suffers from serious class imbalance. To address this issue, we employ the weighted cross-entropy loss for the semantic segmentation task. The class distribution and our proposed class weights are shown in Table 1.

4.2 Investigation on RGB Channels

We investigate the effectiveness of each channels on the segmentation result by training U-net on each channel separately. The result is shown in Table 2. As expected, Green channel has the best performance in Woodlands as it is the dominant color in the Woodlands class, Blue channel has the best performance in Water as it is the dominant color in the Water class. However, Red channel's performance is very poor, especially on the Buildings class which it is supposed to prevail as the roof color might be red. We suspect this is due to the lighting condition in this dataset. Therefore, in this work and for this dataset, we will only consider G-channel and B-channel. Our framework can be easily extended to a 3-channel version.

Table 1. Class distribution in the training set of LandCover.ai and our used class weight.

	Buildings	Woodlands	Water	Background
Percentage	0.86%	33.21%	6.51%	59.42%
Class weight	0.8625	0.025	0.1125	0.0125

Table 2. mIoU of U-net trained on separate channels.

Channel	Buildings	Woodlands	Water	Background	Overall
Red	60.6%	78.26%	62.63%	64.75%	66.59%
Green	79.61%	**90.06%**	90.47%	**88.46%**	87.15%
Blue	**80.6%**	89.74%	**91.05%**	88.2%	**87.4%**
Green & Blue	79.09%	90.56%	91.33%	88.64%	87.41%

4.3 Aerial Segmentation Using Disentangle Learning on G-Channel and B-Channel

We initialize our network parameters with the pre-trained U-nets on G-channel and B-channel in the previous section where they are applicable. We train the network using Adam optimizer with weight decay set to 10^{-7}. As the optimization space of the network on this problem is very tricky, we first set the learning rate to 10^{-4} and then decrease in a phased manner to avoid overshooting.

As the training progresses, the content consistency loss and style consistency loss easily decrease down approximately to zero while the segmentation performance keeps increasing. This fact indicates that our hypothesis of style codes of the same channel across different images being the same is evidently reasonable.

We compared the proposed method with baseline implementation with DeepLabv3 in [2], which is non-augmentation version because we want to make clear performance comparison between two methods. The result is shown in Table 3, which demonstrate that our method can disentangle semantic content successfully across G and B channels and their corresponding style noises, and shows better performance on image segmentation.

Table 3. mIoU of the proposed method and baseline implementation with DeepLabV3+OS4 in [2] .OS denotes encoder output stride during training and evaluation.

Methods	Buildings	Woodlands	Water	Background	Overall
DeepLabV3	77.53%	91.05%	93.84%	93.02%	88.86%
Our methods	**79.47%**	**91.56%**	**94.33%**	92.64%	**89.5%**

5 Conclusion

We conduct an investigation on the effectiveness of R, G, B channels on the segmentation performance and find that each channel especially performs well for different classes due to their dominant color present in those classes. We propose a disentangle learning method to remove potential noisy information by setting 2 important constraints: channels of the same image should share the same content code and the same channel in different images should share the same style code. Our method demonstrates the effectiveness on aerial image segmentation. In the future work, We will continue to do investigation on disentanglement of more channels such as hyperspectral images.

References

1. Badrinarayanan, V., Kendall, A., Cipolla, R.: SegNet: a deep convolutional encoder-decoder architecture for image segmentation. IEEE Trans. Pattern Anal. Mach. Intell. **39**(12), 2481–2495 (2017)
2. Boguszewski, A., et al.: Landcover. AI: dataset for automatic mapping of buildings, woodlands and water from aerial imagery. arXiv preprint arXiv:2005.02264 (2020)
3. Chen, L.-C., Zhu, Y., Papandreou, G., Schroff, F., Adam, H.: Encoder-decoder with atrous separable convolution for semantic image segmentation. In: Ferrari, V., Hebert, M., Sminchisescu, C., Weiss, Y. (eds.) ECCV 2018. LNCS, vol. 11211, pp. 833–851. Springer, Cham (2018). https://doi.org/10.1007/978-3-030-01234-2_49
4. Cordts, M., et al.: The cityscapes dataset for semantic urban scene understanding. In: Proceedings of the IEEE Conference on Computer Vision and Pattern Recognition (CVPR) (2016)
5. Gerard, F., et al.: Land cover change in Europe between 1950 and 2000 determined employing aerial photography. Prog. Phys. Geogr. **34**(2), 183–205 (2010)
6. He, K., Gkioxari, G., Dollár, P., Girshick, R.: Mask r-CNN. In: Proceedings of the IEEE International Conference on Computer Vision, pp. 2961–2969 (2017)
7. Huang, X., Liu, M.-Y., Belongie, S., Kautz, J.: Multimodal unsupervised image-to-image translation. In: Ferrari, V., Hebert, M., Sminchisescu, C., Weiss, Y. (eds.) ECCV 2018. LNCS, vol. 11207, pp. 179–196. Springer, Cham (2018). https://doi.org/10.1007/978-3-030-01219-9_11
8. Khalel, A., El-Saban, M.: Automatic pixelwise object labeling for aerial imagery using stacked u-nets. arXiv preprint arXiv:1803.04953 (2018)
9. Li, X., Jiang, Y., Peng, H., Yin, S.: An aerial image segmentation approach based on enhanced multi-scale convolutional neural network. In: 2019 IEEE International Conference on Industrial Cyber Physical Systems (ICPS). pp. 47–52. IEEE (2019)
10. Lin, T.Y., Dollár, P., Girshick, R., He, K., Hariharan, B., Belongie, S.: Feature pyramid networks for object detection. In: Proceedings of the IEEE Conference on Computer Vision and Pattern Recognition, pp. 2117–2125 (2017)
11. Liu, M.Y., Breuel, T., Kautz, J.: Unsupervised image-to-image translation networks. Adv. Neural. Inf. Process. Syst. **30**, 700–708 (2017)
12. Liu, S., Qi, L., Qin, H., Shi, J., Jia, J.: Path aggregation network for instance segmentation. In: Proceedings of the IEEE Conference on Computer Vision and Pattern Recognition, pp. 8759–8768 (2018)

13. Long, J., Shelhamer, E., Darrell, T.: Fully convolutional networks for semantic segmentation. In: Proceedings of the IEEE Conference on Computer Vision and Pattern Recognition, pp. 3431–3440 (2015)
14. Maggiori, E., Tarabalka, Y., Charpiat, G., Alliez, P.: Can semantic labeling methods generalize to any city? the iNRIA aerial image labeling benchmark. In: 2017 IEEE International Geoscience and Remote Sensing Symposium (IGARSS), pp. 3226–3229. IEEE (2017)
15. Ronneberger, O., Fischer, P., Brox, T.: U-Net: convolutional networks for biomedical image segmentation. In: Navab, N., Hornegger, J., Wells, W.M., Frangi, A.F. (eds.) MICCAI 2015. LNCS, vol. 9351, pp. 234–241. Springer, Cham (2015). https://doi.org/10.1007/978-3-319-24574-4_28

Vision Transformers Theory and Applications

Temporal Cross-Attention for Action Recognition

Ryota Hashiguchi and Toru Tamaki(✉)

Nagoya Institute of Technology, Nagoya, Japan
r.hashiguchi.651@nitech.jp, tamaki.toru@nitech.ac.jp

Abstract. Feature shifts have been shown to be useful for action recognition with CNN-based models since Temporal Shift Module (TSM) was proposed. It is based on frame-wise feature extraction with late fusion, and layer features are shifted along the time direction for the temporal interaction. TokenShift, a recent model based on Vision Transformer (ViT), also uses the temporal feature shift mechanism, which, however, does not fully exploit the structure of Multi-head Self-Attention (MSA) in ViT. In this paper, we propose *Multi-head Self/Cross-Attention* (MSCA), which fully utilizes the attention structure. TokenShift is based on a frame-wise ViT with features temporally shifted with successive frames (at time $t+1$ and $t-1$). In contrast, the proposed MSCA replaces MSA in the frame-wise ViT, and some MSA heads attend to successive frames instead of the current frame. The computation cost is the same as the frame-wise ViT and TokenShift as it simply changes the target to which the attention is taken. There is a choice about which of key, query, and value are taken from the successive frames, then we experimentally compared these variants with Kinetics400. We also investigate other variants in which the proposed MSCA is used along the patch dimension of ViT, instead of the head dimension. Experimental results show that a variant, MSCA-KV, shows the best performance and is better than TokenShift by 0.1% and then ViT by 1.2%.

1 Introduction

Recognizing the actions of people in videos is an important topic in computer vision. After the emergence of Vision Transformer (ViT) [1] which has been shown to be effective for various image recognition tasks [2,3], research extending ViT to video has become active [4–8].

In recognition of video, unlike images, it is necessary to model the temporal information across frames of a given video clip. While most early works of action recognition were frame-wise CNN models followed by temporal aggregation [9, 10], 3D CNN models [11–13] were shown to be effective and well generalized when they were trained on large-scale datasets [14] and transferred to smaller datasets [15].

However, the computational cost of 3D convolution is usually high. To mitigate the issue of the trade-off between the ability of temporal modeling and the high computational cost, Temporal Shift Module (TSM) [16] was proposed by extending spatial feature shifting [17,18] to the temporal dimension. TSM uses

Y. Zheng et al. (Eds.): ACCV 2022, LNCS 13848, pp. 283–294, 2023.
https://doi.org/10.1007/978-3-031-27066-6_20

late fusion by applying 2D ResNet [19] to each frame, while it exploits the temporal interaction between adjacent frames by shifting the features in each layer of ResNet. The feature shift operation is computationally inexpensive, nevertheless it has been shown to perform as well as 3D CNN. Therefore, many related CNN models [20–22] have been proposed. TokenShift [23] is a TSM-like 2D ViT-based model that shifts a fraction of ViT features, showing that shifting only the class tokens is enough to achieve a good performance.

However, TokenShift doesn't fully exploit the advantage of the architecture of the attention mechanism of ViT, instead it simply shifts features like as in TSM. We propose a method that seamlessly utilizes the ViT structure to interact with features across adjacent frames. The original TSM or TokenShift uses a 2D model applied to each frame, while the output features of layers of the 2D model are then shifted from the current time t to one time step forward $t+1$ and backward $t-1$. In the encoder architecture of ViT, Multi-head Self-Attention (MSA) computes attentions between patches at a given frame t. The proposed method utilizes and modifies it for the temporal interaction; some patches at frame t attends to not only patches at the current frame t but also patches at the neighbor frames $t+1$ and $t-1$. We call this mechanisms *Multi-head Self/Cross-Attention* (MSCA). This allows the temporal interaction to be computed within the Transformer block without additional shift modules. Furthermore, the computational cost remains the same because some parts of self-attention computation is simply replaced with the temporal cross-attention. The contributions of this paper are as follows:

- We propose a new action recognition model with the proposed MSCA, which seamlessly combines the concepts of ViT and TSM. This avoids computing spatio-temporal attention directly in the 3D video volume.
- MSCA uses cross-attention to perform temporal interactions inside the Transformer block, unlike TokenShift that uses additional shifting modules. This allows for temporal interaction without increasing computational complexity nor changing model architecture.
- Experiments on Kinetics400 show that the proposed method improves performance compared to ViT and TokenShift.

2 Related Work

2.1 2D/3D CNN, and TSM

Early approaches of deep learning to action recognition are 2D-based methods [9,10] that apply 2D CNN to videos frame by frame, or Two-Stream methods [24] that use RGB frames and optical flow stacks. However, it is difficult to model long-term temporal information beyond several time steps, and this is where 3D-based methods [11–13] came in. These models simultaneously consider spatial and temporal information, but their computational cost is relatively high compared to 2D-based methods. Therefore, mixed models of 2D and 3D convolution [25–27] have been proposed by separating convolution in the spatial and temporal dimensions.

TSM [16] is an attempt to model temporal information for action recognition without increasing parameters and computational complexity while maintaining the architecture of 2D CNN-based methods. TSM shifts intermediate features between neighbor frames, allowing 2D CNNs of each frame to interact with other frames in the temporal direction. Many CNN-based variants have followed, such as Gated-Shift Network [20] that learns temporal features using only 2D CNN with the gate shift module, and RubiksNet [22] which uses three learnable shifts (spatial, vertical, and temporal) for end-to-end learning.

2.2 ViT-Based Video Models

ViT [1] performs very well in image recognition tasks, and its application to action recognition tasks has been actively studied [4–8,28,29]. It is known that the computational cost of self-attention in Transformer is $O(N^2)$ where N is the number of tokens, or the number of patches N for ViT. For videos, the number of frames T also contributes to the computational cost of spatio-temporal self-attention. A simple extension of ViT with full spatio-temporal attention gives a computational cost of $O(N^2T^2)$, since the number of tokens increases in proportion to the temporal extent. To alleviate this computational issue, TimeSformer [7] reduces the computational complexity to $O(T^2N + TN^2)$ by applying temporal and spatial attention separately, TokenLeaner [28] to $O(S^2T^2)$ by selecting $S(< N)$ patches that are important for the representation, and Space-time Mixing Attention [29] to $O(TN^2)$ by attending tokens in neighbor frames only.

TokenShift [23] is a TSM-like model that shifts a fraction of ViT features in each Transformer block. It computes the attention only in the current frame, hence the complexity is $O(TN^2)$ with fewer FLOPs than [29]. TokenShift shifts the class token only based on experimental results, however it doesn't fully exploit the attention mechanism. In contrast, the proposed method makes use of the attention mechanism for feature shift; we replace some of the self-attention modules with the proposed cross-attention modules that shift key, query, and value to neighbor frames, instead of naively shifting features.

3 Method

Figure 1(a) shows the original ViT encoder block that has the Multi-head Self-Attention (MSA) module. Figure 1(b) shows the encoder block of TokenShift. Two modules for shifting intermediate features are added to the original ViT block. These modules shift a portion of the class tokens in the temporal direction by one time step forward and backward, similar to TSM. Figure 1(c) shows the encoder block of the proposed MSCA. The difference is that MSA is replaced with the Multi-head Self/Cross-Attention (MSCA) module, and no additional modules exist.

3.1 TokenShift and ViT

In this section, we briefly review TokenShift [23] which is based on ViT [1].

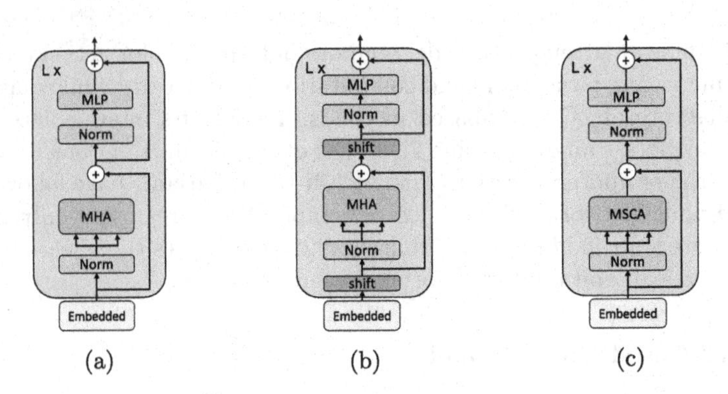

Fig. 1. Encoder blocks of (a) ViT, (b) TokenShift with two shift modules (in blue), and (c) the proposed method with MSCA (in green). (Color figure online)

Input Patch Embedding. Let an input video be $x \in \mathbb{R}^{T \times 3 \times H \times W}$, where T is the number of frames in the video clip, and H, W are the height and width of the frame. Each frame is divided into patches of size $P \times P$ pixels and transformed into a tensor $\hat{x} = [x_0^1, \ldots, x_0^N] \in \mathbb{R}^{T \times N \times d}$, where $x_0^i \in \mathbb{R}^{T \times d}$ denotes the i-th patch, $N = \frac{HW}{P^2}$ is the number of patches of dimension $d = 3P^2$.

The input patch x_0^i is then transformed by the embedding matrix $E \in \mathbb{R}^{d \times D}$ and the positional encoding E_{pos} as follows:

$$z_0 = [c_0, x_0^1 E, x_0^2 E, \ldots, x_0^N E] + E_{\text{pos}}, \tag{1}$$

where $c_0 \in \mathbb{R}^{T \times D}$ is the class token. This patch embedding $z_0 \in \mathbb{R}^{T \times (N+1) \times D}$ is the input to the first encoder block.

Encoder Block. Let z_ℓ be the input to the ℓ-th encoder block. The output z_ℓ of the block can be expressed as follows:

$$z'_\ell = \text{MSA}(\text{LN}(z_{\ell-1})) + z_{\ell-1} \tag{2}$$

$$z_\ell = \text{MLP}(\text{LN}(z'_\ell)) + z'_\ell, \tag{3}$$

where LN is the layer normalization, MSA is the multi-head self-attention, and MLP is the multi-layer perceptron. In the following, $z_{\ell,t,n,d}$ denotes the element at (t, n, d) in z_ℓ.

Shift Modules. TokenShift inserts two shift modules in the block as follows;

$$z'_{\ell-1} = \text{Shift}(z_{\ell-1}) \tag{4}$$

$$z''_\ell = \text{MSA}(\text{LN}(z'_{\ell-1})) + z'_{\ell-1} \tag{5}$$

$$z'''_\ell = \text{Shift}(z''_\ell) \tag{6}$$

$$z_\ell = \text{MLP}(\text{LN}(z'''_\ell)) + z'''_\ell. \tag{7}$$

The shift modules take the input $z_{\text{in}} \in R^{T\times(N+1)\times D}$ and compute the output z_{out} of the same size by shifting the part of z_{in} corresponding to the class tokens ($z_{\text{in},t,0,d}$, the first elements of the second dimension of z_{in}) while leaving the other parts untouched. This is implemented by the following assignments;

$$z_{\text{out},t,0,d} = \begin{cases} z_{\text{in},t-1,0,d}, & 1 < t \leq T, 1 \leq d < D_b \\ z_{\text{in},t+1,0,d}, & 1 \leq t < T, D_b \leq d < D_b + D_f \\ z_{\text{in},t,0,d}, & \forall t, D_b + D_f \leq d \leq D \end{cases} \tag{8}$$

$$z_{\text{out},t,n,d} = z_{\text{in},t,n,d}, \qquad \forall t, 1 \leq n < N, \forall d \tag{9}$$

The first equation shifts the class tokens; along the channel dimension D, the backward shift to $t-1$ is done for the first D_{b} channels, the forward shift to $t+1$ is done for the next D_{f}, and no shift for the rest channels. The second equation passes through features other than the class tokens.

3.2 MSCA

The proposed method replaces MSA with MSCA in the original block;

$$z'_\ell = \text{MSCA}(\text{LN}(z_{\ell-1})) + z_{\ell-1} \tag{10}$$

$$z_\ell = \text{MLP}(\text{LN}(z'_\ell)) + z'_\ell. \tag{11}$$

In the following, we first describe MSA, and then define MSCA.

MSA. The original MSA computes the key $K^{(t)}$, query $Q^{(t)}$, and value $V^{(t)}$ for the portion $z^{(t)} \in \mathbb{R}^{(N+1)\times D}$ of the input feature $z \in \mathbb{R}^{T\times(N+1)\times D}$ at time t, as follows;

$$K^{(t)}, Q^{(t)}, V^{(t)} = z^{(t)}[W_k, W_q, W_v], \tag{12}$$

where $W_k, W_q, W_v \in \mathbb{R}^{D\times D}$ are embedding matrices, and $z = [z^{(1)}, \dots, z^{(T)}]$. These are used to compute the i-th attention head;

$$\text{head}_i^{(t)} = a(Q_i^{(t)}, K_i^{(t)})V_i^{(t)} \in \mathbb{R}^{(N+1)\times D/h}, \tag{13}$$

at time t for $i = 1, \dots, h$, where the attention a is

$$a(Q, K) = \text{softmax}(QK^T/\sqrt{D}), \tag{14}$$

and $Q_i^{(t)} \in \mathbb{R}^{(N+1)\times D/h}$ is the part of $Q^{(t)}$ corresponding to i-th head

$$Q^{(t)} = [Q_1^{(t)}, \dots, Q_i^{(t)}, \dots, Q_h^{(t)}], \tag{15}$$

and $K_i^{(t)}, V_i^{(t)}$ are the same. These heads are finally stacked to form

$$\text{MSA}(z^{(t)}) = [\text{head}_1^{(t)}, \dots, \text{head}_h^{(t)}]. \tag{16}$$

In MSA, patches in t-th frame are attended from other patches of the same frame at time t, which means that there are no temporal interactions between frames.

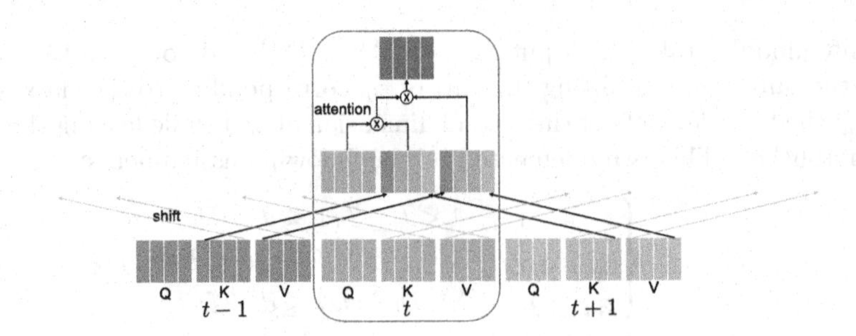

Fig. 2. Shit operation at $T = t$ in the MSCA-KV model

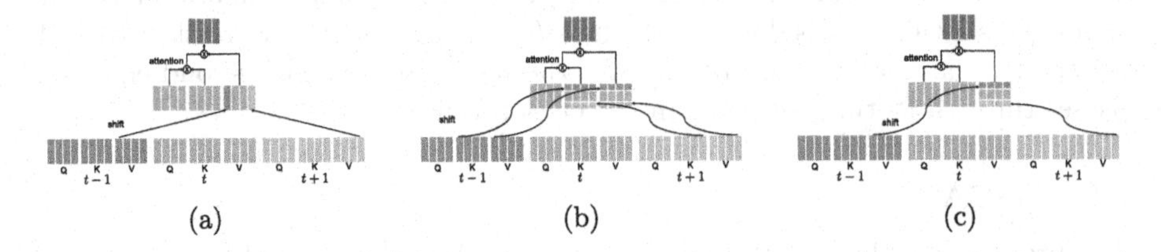

(a) (b) (c)

Fig. 3. Shit operations of (a) MSCA-V, (b) MSCA-pKV, and (c) MSCA-pV.

MSCA-KV. The proposed MSCA computes the attention across frames, that is, patches in t-th frame are also attended from patches in frames at time $t+1$ and $t-1$. This can be done with shift operations; after generating Q, K, V at each frame, these are exchanged with the neighbor frames.

There are choices of which of Q, K, or V to be shifted. A possible choice is to shift K and V, and the query in the current frame is attended by key-value pairs in other frames. This is expressed as follows;

$$\text{head}_i^{(t)} = \begin{cases} a(Q_i^{(t)}, K_i^{(t-1)})V_i^{(t-1)} & 1 \leq i < h_b \\ a(Q_i^{(t)}, K_i^{(t+1)})V_i^{(t+1)} & h_b \leq i < h_b + h_f \\ a(Q_i^{(t)}, K_i^{(t)})V_i^{(t)} & h_b + h_f \leq i \leq h. \end{cases} \tag{17}$$

Here, we have the first h_b heads with the backward shift, the next h_f heads with the forward shift, and the rest heads with no shift. We call this MSCA-KV, and Fig. 2 shows the diagram of the shift operation. First, queries, keys and values are computed at each frame, and then some of them are shifted before computing the attention. The solid arrows indicate the key-value shift from time $t-1$ and $t+1$. As in the same way, the key-value shift from the current frame t to $t-1$ and $t+1$ are shown in dotted arrows. There are shifts in all other frames at the same time in the same manner.

MSCA-V. Another choice of shift is shown in Fig. 3(a). Here, Q and K are not shifted, but only V is shifted as follows;

$$\text{head}_i^{(t)} = \begin{cases} a(Q_i^{(t)}, K_i^{(t)})V_i^{(t-1)} & 1 \leq i < h_b \\ a(Q_i^{(t)}, K_i^{(t)})V_i^{(t+1)} & h_b \leq i < h_b + h_f \\ a(Q_i^{(t)}, K_i^{(t)})V_i^{(t)} & h_b + h_f \leq i \leq h. \end{cases} \tag{18}$$

This might not be common because the key and value are now separated and taken from different frames. However, this makes sense for modeling temporal interactions because the values (which are mixed by the attention weights) come from different frames while the attention is computed in the current frame. We call this version MSCA-V. Therefore, in addition to shifting V in this way, there are seven possible combinations; Q, K, V, QK, KV, QV, and QKV. We compared these variants in the experiments. Note that MSCA-QKV is equivalent to a simple feature shifting because attended features are computed in each frame and then shifted.

MSCA-pKV. All of the above seven variants perform shift operations along the head (or channel) dimension D, but similar variants of shift are also possible for the patch dimension $N+1$.

The shapes of $K^{(t)}, Q^{(t)}, V^{(t)}$ are $\mathbb{R}^{(N+1)\times D}$, and the first dimension is for patches while the second is for heads. As in the same way the shift along the head dimension, we have a variation of shift operations along the patch dimension as shown in Fig. 3(b).

First, K and V are expressed as stacks of keys and values of patches at different frames as follows;

$$K^{(t)} = [K_0^{(t)}, K_1^{(t)}, \ldots, K_N^{(t)}] \tag{19}$$

$$V^{(t)} = [V_0^{(t)}, V_1^{(t)}, \ldots, V_N^{(t)}], \tag{20}$$

where $K_n^{(t)}, V_n^{(t)} \in \mathbb{R}^D$ are the key and value of patch n at time t.

Then, keys of some patches in the current frame are shifted to form K';

$$K_n'^{(t)} = \begin{cases} K_n^{(t-1)} & 0 \leq n < N_b \\ K_n^{(t+1)} & N_b \leq n < N_b + N_f \\ K_n^{(t)} & N_b + N_f \leq n \leq N, \end{cases} \tag{21}$$

and also V' in the same way. Finally, the i-th head is computed as follows

$$\text{head}_i^{(t)} = a(Q_i^{(t)}, K_i'^{(t)})V_i'^{(t)}. \tag{22}$$

We refer to this version as MSCA-pKV.

MSCA-pV. Like as MSCA-V, a variant of the shift in the patch direction with V only can be also considered as shown in Fig. 3(c), by the following shift;

$$\text{head}_i^{(t)} = a(Q_i^{(t)}, K_i^{(t)})V_i'^{(t)}. \tag{23}$$

As before, there are seven variants, and we call these MSCA-pV, and so on.

4 Experimental Results

4.1 Setup

Kinetics400 [14] was used to train and evaluate the proposed method. This dataset consists of a training set of 22k videos, a validation set of 18k videos, with 400 categories of human actions. Each video was collected from Youtube, and the portion corresponding to each category was cropped to a length of 10 s.

We used a 2D ViT pre-trained on ImageNet21k [30] with $h = 12$ heads, 12 encoder blocks, and patches of size $P = 16$ and $D = 768 = 3 \times 16 \times 16$ (these parameters are the same for TokenShift and MSCA models). For action recognition, ViT was applied to each frame and resulting frame-wise features were aggregated by temporal averaging (this is referred to as ViT in the experiment and in [23]). We compared this ViT, TokenShift and the proposed method. Note that we report the performance of TokenShift based on our reproduction using the author's code.[1]

For training, we used the same settings as in [23]. Input clips were of 8 frames with stride of 32 frames (starting frames were randomly chosen). Frames were flipped horizontally at a probability of 50%, and the short side was randomly resized in the range of [244, 330] pixels while maintaining the aspect ratio, then a random 224×224 pixel rectangle was cropped (therefore the number of patches is $N = 196 = 14 \times 14$). In addition, brightness change, saturation change, gamma correction, and hue correction were applied to frames, each at the probability of 10%. The number of epochs was set to 12, the optimizer to SDG with momentum of 0.9 and no weight decay. The initial learning rate was set to 0.1, and decayed by a factor of 10 at 10th epoch. The batch size was set to 42, and 21 batches were trained on each of two GPUs. Gradient updates were performed once every 10 iterations, so the effective batch size was 420.

We used the multi-view test [31]. From a validation video, one clip was sampled as in training, and this was repeated 10 times to sample 10 clips. Each clip was resized to 224 pixels on its short side while maintaining the aspect ratio, and cropped to 224×224 at the right, center, and left. The results of these 30 clips (views) were averaged to compute a single prediction score.

4.2 The Amount of Shift of MSCA-KV

We first investigate the effect of the number of heads to be shifted. Table 1 shows the performance of MSCA-KV. The best performance was obtained when only two heads (each for forward and backward) which corresponds to shifting $2/12 = 16.7\%$ of the channels. As the number of shifted heads increased, the performance decreased, suggesting that shifting a few heads is sufficient while most heads need not to be shifted. This observation coincides with the conclusions of TokenShift [23], which shows that the shift of the class token only is enough, and also TSM [16], which used the shift of $1/4 = 25\%$ of the channels (each 1/8 for forward and backward).

[1] https://github.com/VideoNetworks/TokShift-Transformer.

Table 1. The performance of MSCA-KV for the validation sets of Kinetics400. Note that the zero shift means a naive frame-wise ViT. The column "shift" means the percentage of the shifted dimensions to the total dimensions D.

Model	Heads	h_b, h_f	Shift %	Top-1	Top-5
ViT	0	0	0 (ViT)	75.65	92.19
MSCA-KV	2	1	16.7	**76.47**	**92.88**
	4	2	33.3	76.07	92.61
	6	3	50.0	75.66	92.30
	8	4	66.7	74.72	91.91

Table 2. The performance of MSCA variants shifting in the head direction.

Model	Top-1	Top-5
TokenShift	76.37	92.82
ViT	75.65	92.19
MSCA-Q	75.84	92.57
MSCA-K	75.76	92.13
MSCA-V	75.58	92.39
MSCA-QK	75.47	92.32
MSCA-KV	**76.47**	**92.88**
MSCA-QV	75.57	92.43
MSCA-QKV	75.78	92.37

Table 3. The performance with different numbers of blocks with MSCA modules of KV shift.

model	# MSCA	# MSA	Top-1	Top-5
ViT	0	12	75.65	92.19
	4	8	75.67	92.19
	8	4	76.40	92.77
MSCA-KV	12	0	**76.47**	**92.88**

In the following experiments, we used shifting two heads for all MSCA variations.

4.3 Comparison of MSCA Variations Shifting in the Head Direction

Table 2 shows the performance of the MSCA model variants with shift operations in the head direction. Among them, MSCA-KV performed the best, outperforming others by at least 0.5%. Other variants performed as same as the baseline ViT, indicating that the shift operation is not working effectively in such variations.

4.4 The Number of Encoder Blocks with MSCA

The proposed method replaces MSA modules in 12 encoder blocks in ViT with MSCA modules. However, it is not obvious that modules of all blocks need to be replaced. Table 3 shows the results when we replaced MSA modules with MSCA in some blocks near the end (or top) of the network. All 12 replacements correspond to MSCA-KV, and 0 (no MCSA) is ViT. Using four MSCA modules showed no improvement, while using 8 MSCA modules performed almost the same as MSCA-KV with all 12 MSCA modules. This is because the input video

Table 4. The performance of MSCA-pKV. The column "shift" means the percentage of the shifted patches to the total patches $N+1$.

Patches	N_b, N_b	Shift %	Top-1	Top-5
0	0	0 (ViT)	75.65	92.19
8	4	4.1	76.28	92.49
16	8	8.1	**76.35**	**92.91**
32	16	16.2	76.07	92.77
48	24	24.4	75.84	92.49
64	32	32.5	75.29	92.04

Table 5. The performance of MSCA models shifting in the patch direction.

Model	Top-1	Top-5
TokenShift	**76.37**	92.82
ViT	75.65	92.19
MSCA-pQ	75.75	92.10
MSCA-pK	75.69	92.24
MSCA-pV	75.84	92.43
MSCA-pQK	75.50	91.99
MSCA-pKV	76.35	**92.91**
MSCA-pQV	75.83	92.35
MSCA-pQKV	75.73	92.21

clip consists of 8 frames, and shifting more than 8 times with MSCA modules ensures that the temporal information from all frames is available for the entire network. Therefore, it would be necessary to use at least as many MSCA modules as the number of frames in the input clip.

4.5 Comparison of MSCA Variations Shifting in the Patch Direction

Table 4 shows the performance of MSCA-pKV with different amount of shift. The results indicate that, again, a small amount of shift is enough for a better performance and the best performance was obtained when 16 patches were shifted. The performance decreases for a smaller amount of shift, approaching the performance of ViT with no shift. In the experiments below, we used the shift of 16 patches.

Table 5 shows the performance of MSCA variants with shifting in the patch direction. Just like as MSCA-KV was the best for shifting in the head direction, MSCA-pKV has the best performance here, while it is just comparable to TokenShift. An obvious drawback of this approach is the first layer; the patches to be shifted were fixed to the first N_b and N_f patches in the order of the patch index, which is irrelevant to the content of the frame. This might be mitigated by not using MSCA modules in the first several layers.

5 Conclusions

In this paper, we proposed MSCA, a ViT-based action recognition model that replaces MSA modules in the encoder blocks. The MSCA modules compute the attention by shifting of the key, query, and value for temporal interaction between frames. Experimental results using Kinetics400 showed that the proposed method is effective for modeling spatio-temporal features and performs better than a naive ViT and TokenShift. Future work includes evaluations on other datasets and comparisons with similar methods such as Space-time Mixing Attention [29].

Acknowledgements. This work was supported in part by JSPS KAKENHI Grant Number JP22K12090.

References

1. Dosovitskiy, A., et al.: An image is worth 16×16 words: Transformers for image recognition at scale. In: International Conference on Learning Representations (2021)
2. Radford, A., et al.: Learning transferable visual models from natural language supervision. CoRR abs/2103.00020 (2021)
3. Ramesh, A., et al.: Zero-shot text-to-image generation. In Meila, M., Zhang, T., (eds.): Proceedings of the 38th International Conference on Machine Learning. Volume 139 of Proceedings of Machine Learning Research, PMLR pp. 8821–8831 (2021)
4. Arnab, A., Dehghani, M., Heigold, G., Sun, C., Lučić, M., Schmid, C.: Vivit: a video vision transformer. In: Proceedings of the IEEE/CVF International Conference on Computer Vision (ICCV), pp. 6836–6846 (2021)
5. Li, X., et al.: Vidtr: video transformer without convolutions. CoRR abs/2104.11746 (2021)
6. Girdhar, R., Carreira, J., Doersch, C., Zisserman, A.: Video action transformer network. In: Proceedings of the IEEE/CVF Conference on Computer Vision and Pattern Recognition (CVPR) (2019)
7. Bertasius, G., Wang, H., Torresani, L.: Is space-time attention all you need for video understanding? In: Proceedings of the International Conference on Machine Learning (ICML) (2021)
8. Sharir, G., Noy, A., Zelnik-Manor, L.: An image is worth 16×16 words, what is a video worth? CoRR abs/2103.13915 (2021)
9. Karpathy, A., Toderici, G., Shetty, S., Leung, T., Sukthankar, R., Fei-Fei, L.: Large-scale video classification with convolutional neural networks. In: Proceedings of the IEEE Conference on Computer Vision and Pattern Recognition (CVPR) (2014)
10. Donahue, J., et al.: Long-term recurrent convolutional networks for visual recognition and description. In: Proceedings of the IEEE Conference on Computer Vision and Pattern Recognition (CVPR) (2015)
11. Tran, D., Bourdev, L., Fergus, R., Torresani, L., Paluri, M.: Learning spatiotemporal features with 3d convolutional networks. In: Proceedings of the IEEE International Conference on Computer Vision (ICCV) (2015)
12. Carreira, J., Zisserman, A.: Quo vadis, action recognition? a new model and the kinetics dataset. In: Proceedings of the IEEE Conference on Computer Vision and Pattern Recognition (CVPR) (2017)
13. Hara, K., Kataoka, H., Satoh, Y.: Can spatiotemporal 3d cnns retrace the history of 2d CNNs and ImageNet? In: Proceedings of the IEEE Conference on Computer Vision and Pattern Recognition (CVPR), pp. 6546–6555 (2018)
14. Kay, W., et al.: The kinetics human action video dataset. CoRR abs/1705.06950 (2017)
15. Soomro, K., Zamir, A.R., Shah, M.: Ucf101: a dataset of 101 human actions classes from videos in the wild. CoRR abs/1212.0402 (2012)
16. Lin, J., Gan, C., Han, S.: Tsm: temporal shift module for efficient video understanding. In: Proceedings of the IEEE/CVF International Conference on Computer Vision (ICCV) (2019)

17. Chen, W., Xie, D., Zhang, Y., Pu, S.: All you need is a few shifts: designing efficient convolutional neural networks for image classification. In: Proceedings of the IEEE/CVF Conference on Computer Vision and Pattern Recognition (CVPR) (2019)
18. Wu, B., et al.: Shift: a zero flop, zero parameter alternative to spatial convolutions. In: Proceedings of the IEEE Conference on Computer Vision and Pattern Recognition (CVPR) (2018)
19. He, K., Zhang, X., Ren, S., Sun, J.: Deep residual learning for image recognition. In: Proceedings of the IEEE Conference on Computer Vision and Pattern Recognition (CVPR) (2016)
20. Sudhakaran, S., Escalera, S., Lanz, O.: Gate-shift networks for video action recognition. In: IEEE/CVF Conference on Computer Vision and Pattern Recognition (CVPR) (2020)
21. Chang, Y.L., Liu, Z.Y., Lee, K.Y., Hsu, W.: Learnable gated temporal shift module for deep video inpainting. In: BMVC (2019)
22. Fan, L., Buch, S., Wang, G., Cao, R., Zhu, Y., Niebles, J.C., Fei-Fei, L.: RubiksNet: learnable 3d-shift for efficient video action recognition. In: Vedaldi, A., Bischof, H., Brox, T., Frahm, J.-M. (eds.) ECCV 2020. LNCS, vol. 12364, pp. 505–521. Springer, Cham (2020). https://doi.org/10.1007/978-3-030-58529-7_30
23. Zhang, H., Hao, Y., Ngo, C.W.: In: Token Shift Transformer for Video Classification, pp. 917–925. New York, NY, USA, Association for Computing Machinery (2021)
24. Simonyan, K., Zisserman, A.: Two-stream convolutional networks for action recognition in videos. In Ghahramani, Z., Welling, M., Cortes, C., Lawrence, N., Weinberger, K.Q., (eds.) Advances in Neural Information Processing Systems. vol. 27, Curran Associates, Inc. (2014)
25. Qiu, Z., Yao, T., Mei, T.: Learning spatio-temporal representation with pseudo-3d residual networks. In: Proceedings of the IEEE International Conference on Computer Vision (ICCV) (2017)
26. Zhang, D., Dai, X., Wang, X., Wang, Y.F.: S3d: single shot multi-span detector via fully 3d convolutional network. In: Proceedings of the British Machine Vision Conference (BMVC) (2018)
27. Tran, D., Wang, H., Torresani, L., Ray, J., LeCun, Y., Paluri, M.: A closer look at spatiotemporal convolutions for action recognition. In: Proceedings of the IEEE Conference on Computer Vision and Pattern Recognition (CVPR) (2018)
28. Ryoo, M.S., Piergiovanni, A., Arnab, A., Dehghani, M., Angelova, A.: Tokenlearner: adaptive space-time tokenization for videos. In: Advances in Neural Information Processing Systems (NeurIPS) (2021)
29. Bulat, A., Perez-Rua, J.M., Sudhakaran, S., Martinez, B., Tzimiropoulos, G.: Space-time mixing attention for video transformer. In Beygelzimer, A., Dauphin, Y., Liang, P., Vaughan, J.W., (eds.) Advances in Neural Information Processing Systems (2021)
30. Deng, J., Dong, W., Socher, R., Li, L.J., Li, K., Fei-Fei, L.: Imagenet: a large-scale hierarchical image database. In: 2009 IEEE Conference on Computer Vision and Pattern Recognition, pp. 248–255 (2009)
31. Wang, X., Girshick, R., Gupta, A., He, K.: Non-local neural networks. In: Proceedings of the IEEE Conference on Computer Vision and Pattern Recognition (CVPR) (2018)

Transformer Based Motion In-Betweening

Pavithra Sridhar, V. Aananth, Madhav Aggarwal, and R. Leela Velusamy(✉)

National Institute of Technology - Tiruchirappalli, Tiruchirappalli, TN 620015, India
leela@nitt.edu

Abstract. In-Betweening is the process of drawing transition frames between temporally-sparse keyframes to create a smooth animation sequence. This work presents a novel transformer based in betweening technique that serves as a tool for 3D animators. We first show that this problem can be represented as a sequence to sequence problem and introduce TweenTransformers - a model that synthesizes high-quality animations using temporally-sparse keyframes as input constraints.

We evaluate the model's performance via two complementary methods - quantitative evaluation and qualitative evaluation. The model is compared quantitatively with the state-of-the-art models using LaFAN1, a high-quality animation dataset. Mean-squared metrics like L2P, L2Q, and NPSS are used for evaluation. Qualitatively, we provide two straightforward methods to assess the model's output. First, we implement a custom ThreeJs-based motion visualizer to render the ground truth, input, and output sequences side by side for comparison. The visualizer renders custom sequences by specifying skeletal positions at temporally-sparse keyframes in JSON format. Second, we build a motion generator to generate custom motion sequences using the model. Code can be found in https://github.com/Pavi114/motion-completion-using-transformers.

Keywords: Motion in-betweening · Kinematics · Transformer · LAFAN1

1 Introduction

Realistic and accurate animation generation is an important but challenging problem with many applications, including animating 3D characters in films, real-time character motion synthesis in Video Games, and Educational applications. One widely used method to generate animations is via motion inbetweening, commonly known as tweening. It generates intermediate frames called "inbetweens" between two temporally sparse keyframes to deliver an illusion of movement by smoothly transitioning from one position to another.

In traditional animation pipelines, animators manually draw motion frames between a set of still keyframes indicative of the most critical positions the body must be at during its motion sequence. Recent improvements include Motion Capture (MOCAP) technologies [9], and query-based methods [15,20]

Y. Zheng et al. (Eds.): ACCV 2022, LNCS 13848, pp. 295–307, 2023.
https://doi.org/10.1007/978-3-031-27066-6_21

to generate animations. However, MOCAP technology is expensive, and human-drawn animations are preferred. With the rise of computer-aided animation, deep learning-based algorithms have enabled the smooth generation of keyframes from sparse frames by learning from large-scale motion capture data. Existing models currently use Recurrent Neural Networks (RNNs) [7,10], Long Short Term Memory Networks (LSTMs) [8], and BERT-based models [3,4].

The complexity in generating character animations includes 1) replicating complex human behavior to create realistic characters, 2) predominantly used transition generation methods are either expensive or inefficient, 3) RNNs/LSTMs, though they can capture long-term dependencies, cannot be parallelized due to the sequential processing of input, resulting in longer training times, and 4) RNNs/LSTMs do not support transfer learning making it hard to use pre-trained models.

Inspired by the concept of self-attention to capture long-term dependencies, this paper proposes a transformer-based model to generate realistic animation sequences. Model generalization constitutes the main effort this framework puts into improving the performance of machine learning predictions. This would be analogous to large text transformer models like GPT-3 [2]. This work not only eases the effort put in by the animators but also helps researchers by unblocking transfer learning for the task of inbetweening, thus introducing a level of generalization into the model.

Overall, the contributions in this paper can be summarized as follows:

1. Represent motion in-betweening as a sequence to sequence problem where the input sequence consists of keyframes and the output sequence represents the complete and smoothed motion sequence.
2. Set a baseline for the input sequence by filling the frames between the keyframes with interpolated values.
3. Experiment with the efficiency and viability of using transformers to achieve sequence to sequence translation for human motion and compare them with the existing results.
4. Evaluate the model against other state-of-the-art models [4,8,17] for the same task using L2P, L2Q, and NPSS metrics.
5. Build a visualizer and a motion generator that qualitatively evaluates the output of the model in comparison to the ground truth and input sequences.

2 Related Work

The problem is analogous to machine translation, where sequence to sequence (seq2seq) architectures are prevalent [1,19,22]. "Encoder-only" models like BERT [3] is designed to learn the context of a word based on all its surroundings (left and right of the word), making them suitable for feature extraction, sentiment classification, or span prediction tasks but not for generative tasks like translation or sequence completion. The pre-training objectives used by encoder-decoder transformers like T5 [18] include a fill-in-the-blank task where the model predicts missing words within a corrupted piece of text that is analogous to in-betweening when motion sequences replace sentences.

Early works in human motion prediction include using Conditional Restricted Boltzmann Machines (RBMs) [21] to encode the sequence information in latent variables and predict using decoders. More recently, many RNN-based approaches like Encoder-Recurrent-Decoder (ERD) networks [5] propose separating spatial encoding and decoding from the temporal dependencies. Other recent approaches investigate new architectures like transformers [13] and loss functions to further improve prediction of human motion [6,12].

Initial approaches in motion in-betweening focused on generating missing frames by integrating keyframe information with spacetime models [24]. The following widely successful method for inbetweening adopted a probabilistic approach, framing it as a Maximum Aposterior Optimization problem (MAP) [14], dynamical Gaussian process model [23] or as Markov models with dynamic autoregressive forests [11]. The latest deep learning approaches include works by Holden et al. [10], and Harvey et al. [7] and helped RNNs dominate this field. The latest work using RNN focuses on augmenting a Long Short Term Memory(LSTM) based architecture with time-to-arrival embeddings and a scheduled target noise vector, allowing the system to be robust to target distortions [8]. Some recent work includes BERT-based encoder-only models [3,4] that predict the entire sequence in one pass and deep learning approaches for interpolation [16]. However, BERT-based models will be less effective than encoder-decoder models for generative tasks.

3 Methodology

The following sections detail the model architecture, **Tween Transformers**, to perform motion frame completion similar to sentence completion.

3.1 Tween Transformers (TWTR)

The architecture of Tween Transformers (TWTR) consists of four main components:

1. Input masking module
2. Input encoding neural network that encodes each motion sequence and converts the input to a set of sequential tokens
3. Transition generation network that includes a standard transformer comprising of encoder and decoder modules with feed-forward and multi-head attention networks.
4. Output decoding neural network that computes a sequence of character motion.

While the transition generation module learns the temporal dependencies, the input and output encoding networks aim to learn spatial dependencies between the different body joints for encoding and decoding motion sequences. Finally,

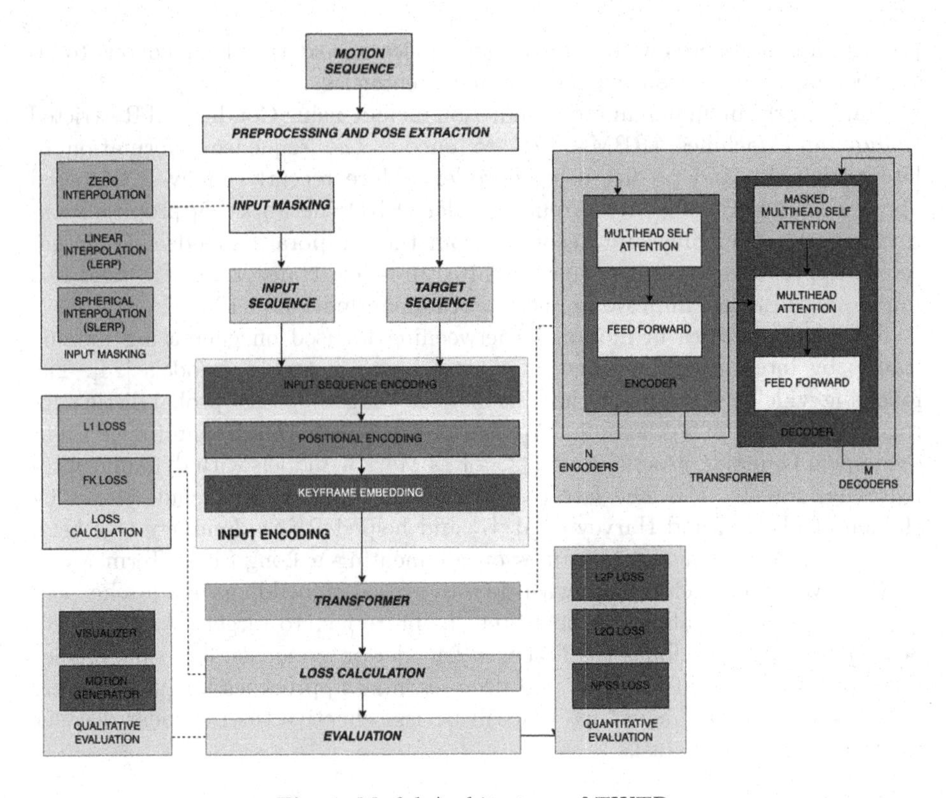

Fig. 1. Model Architecture of TWTR

the model also uses multiple losses, including forward kinematics loss, to improve the realism of the generated sequences. It is assumed that the input has both position (x, y, z) and orientation (q0, q1, q2, q3) variables. Therefore, a single pose can be defined with a root position coordinate $P \in R^3$ and a quaternion matrix $Q \in R^{J \times 4}$, where J represents the joint number of the input pose (here, 22). The following sections discuss the architecture of the model in detail, as indicated in Fig. 1.

Input Masking. There are multiple keyframe gaps k specified in the model configuration. The frames belonging to the keyframe gap are filled with interpolated values that are derived from the frames constituting the two ends of the keyframe gap. Two kinds of interpolations are carried out and compared. They are implemented in the following ways:

- Positions and rotations are linearly interpolated.
- Positions are linearly interpolated while rotations are spherically interpolated.

Input Encoding. As seen in Fig. 1, model encoding has three modules - Input Sequence Encoding, Positional Encoding, and Keyframe Embedding.

1. Input Sequence Encoding
   The input sequence encoder network is a set of three Linear encoders that are fully connected two-layer Feed-Forward Networks (FFN) with ReLU activations. The input sequence encoder takes in the global root position *root_p*, local quaternions q, and global root velocity *root_v* and outputs a set of "sequential tokens." The hidden size of the FFNs are 16, 8, and 8 for q, *root_p*, and *root_v* respectively. The output sizes of the FFNs are defined by the *embedding* hyperparameter. The outputs from the FFNs are concatenated to form the output of the input sequence encoding network. Equation (1) describes the Linear Encoder and Eq. (2) describes the Input Sequence Encoder.

$$\mathrm{L}(x) = \mathrm{Linear}(\mathrm{ReLU}(\mathrm{Linear}(x))) \tag{1}$$

$$\begin{aligned} \mathrm{I}(root\_p, root\_v, q) = \mathrm{L_p}(root\_p) \parallel \mathrm{L_v}(root\_v) \\ \parallel \mathrm{L_q}(q_1) \parallel ... \parallel \mathrm{L_q}(q_J) \end{aligned} \tag{2}$$

   where $root\_p \in R^3$, $root\_v \in R^3$, $q_i \in R^4$, I denotes the Input Sequence Encoder, and L denotes the Linear Encoder.
2. Positional Encoding: Positional encoding, a popular method introduced by Vaswani et al. [22], involves adding a set of predefined sinusoidal and cosine signals to introduce temporal knowledge to the transformer model. The positional encoding for source $\boldsymbol{Z_s} = [z_{tta,2i}]$ and target $\boldsymbol{Z_t} = [z_{tta,2i}]$ is computed using equation (3)

$$\begin{aligned} z_{tta,2i} &= \sin(\frac{tta}{basis^{2i/d}}) \\ z_{tta,2i+1} &= \cos(\frac{tta}{basis^{2i/d}}) \end{aligned} \tag{3}$$

   where *tta* is the number of timesteps until arrival and the *basis* component influences the rate of change in frequencies along the embedding dimension d. A basis of 10,000 is used.
3. Keyframe Embedding: Following previous works [4], the model incorporates additive keyframe embeddings. The keyframe embeddings E_{kf} classify the frames in the sequence into keyframes, unknown frames, and ignored frames. They're represented by learnable embedding vectors $\{\hat{e}_0, \hat{e}_1, \hat{e}_2\}$ respectively. The keyframe embeddings are represented by equation (4), where $e^t_{kf} \in \{\hat{e}_0, \hat{e}_1, \hat{e}_2\}$ and T is the sequence length. The embeddings are added to the input sequence, similar to positional encodings.

$$E_{kf} = [e^1_{kf}, e^2_{kf}, ..., e^T_{kf}] \tag{4}$$

Transformer. A transformer consists of multiple encoder and decoder layers. Each encoder consists of a multi-head self-attention layer (MHSA) and a feed-forward network (FFN) and each decoder consists of a masked multi-head self-attention layer (MMHSA), multi-head attention layer (MHA) and a feed-forward network. The attention function leveraged in the transformer maps a query and a set of key-value pairs - all vectors - to an output. The processing of a single attention head can be represented as follows:

$$Attention(Q, K, V) = Softmax(\frac{QK^T}{\sqrt{d_k}})V \tag{5}$$

where $Q = W_q A$ represents a query matrix, $K = W_k A$ represents a key matrix, and $V = W_v A$ represents a value matrix. W_q, W_k, and W_v are the corresponding weight matrices, and d_k represents the dimension of the key matrix. The Query matrix can be interpreted as the keyframe for which Attention is calculated. The Key and Value matrices represent the keyframes that are "attended to," i.e., how relevant that keyframe is to the query keyframe. In MMHSA, the target is masked before applying the attention mechanism. All the attention outputs are concatenated and sent to the FFN.

Output Decoding. The decoder takes in the concatenated "sequential tokens" outputted by the Input Sequence Encoder and outputs the global root position $root\_p$, local quaternions q, and global root velocity $root\_v$. To reverse engineer the spatial dependencies, each of the three FFNs, one for each output, comprises two linear layers with ReLU activation. The hidden size of the FFNs is the same as in the Input Sequence Encoder. The output sizes of the FFNs are defined by the original dimensions of the three parameters. Equation (6) describes the Output Decoder.

$$\mathrm{O}(x) = (\mathrm{L_p}(x[: d_p]), \mathrm{L_v}(x[d_p : d_p + d_v), Q) \tag{6}$$

$$Q = \begin{bmatrix} \mathrm{L_q}(x\,[\,d_p + d_v : d_p + d_v + d_q]) \\ \mathrm{L_q}(x\,[\,d_p + d_v + d_q : d_p + d_v + 2 \times d_q] \\ \cdots \\ \mathrm{L_q}(x\,[\,d_p + d_v + (J-1) \times d_q : d_p + d_v + J \times d_q] \end{bmatrix}$$

where d_p, d_v, and d_q are embedding dimensions for p, v, and q. $x[i : j]$ represents a tensor containing the values in x from the ith index to the $(j-1)^{th}$ index. J denotes the number of joints in the skeleton, $Q \in R^{J \times 4}$ denotes the tensor of stacked quaternions, O denotes the Output Decoder, and L denotes the Linear Encoder.

3.2 Loss Computation

Given a collection of predicted motion sequences and the ground truth, inbetweening loss is computed as the scaled sum of two individual losses - Reconstruction loss and Forward Kinematics (FK) loss.

$$L = \alpha_r L_R + \alpha_{fk} L_{FK} \tag{7}$$

where α_r and α_{FK} are constants to balance the disparity of individual losses. For training we use $\alpha_r = 100$ and $\alpha_{FK} = 1$.

Reconstruction Loss L_R. Reconstruction loss evaluates the ability of the model to "reconstruct" the target sequence from the input sequence. Reconstruction loss accounts for the difference in output and target quaternions values and is computed using an L1 norm. While Harvey et al. [8] compute and sum reconstruction losses for q, x, and *contacts*, they acknowledge that the most important component is q. Reconstruction loss is computed using equation (8).

$$L_R = \frac{1}{NT} \sum_{n=0}^{N-1} \sum_{t=0}^{T-1} \hat{q}_n^t - q_n^t \tag{8}$$

where $\hat{q}_n^t$ is the rotational quaternion of the predicted motion sequence n at time t. q refers to the ground truth quaternion. N refers to the number of sequences, and T refers to the length of each motion sequence.

Forward Kinematics Loss L_{FK}. Forward Kinematics loss compares the difference in the global positions of joints between the ground truth and the model's output. Forward Kinematics loss evaluates the ability of the model to "understand" the relationships between relative angles and global positions. Although the offsets of various joints in the skeleton are not provided to the model, it learns to respect human geometry and maintain correct posture by minimizing the Forward Kinematics loss. The Forward Kinematics loss is computed using equation (9).

$$L_{FK} = ||\hat{p}_{global} - p_{global}||_1 + ||\hat{q}_{global} - q_{global}||_1 \tag{9}$$

where $\hat{p}_{global}$ and $\hat{q}_{global}$ can be derived from the local coordinates using Forward Kinematics $FK(\hat{p}_{local}, \hat{q}_{local})$ and, similarly p_{global} and q_{global} can be derived from the local coordinates using Forward Kinematics $FK(p_{local}, q_{local})$.

3.3 Training

Following previous works [8,17], the entire dataset was split into windows of maximum length $T_{max} = 65$. To construct each batch, the number of start keyframes is set to 10 and the number of end keyframes to be 1. The number of in-between frames is sampled from the range $[5, 44]$ without replacement.

The weight associated with the number of in-between frames n_{in} is set to be inversely proportional to it, $w_{n_{in}} = \frac{1}{n_{in}}$. This prevents overfitting on the windows with a large number of in-between frames. Shorter windows are sampled more often as they are more abundant and hence harder to overfit. Therefore, the number of unique non-overlapping sequences of a given total length $10 + 1 + n_{in}$ is approximately inversely proportional to n_{in}. Finally, given the total sampled sequence length, the sequence start index is sampled uniformly at random in the range $[0, T_{max}(1 + 10 + n_{in})]$.

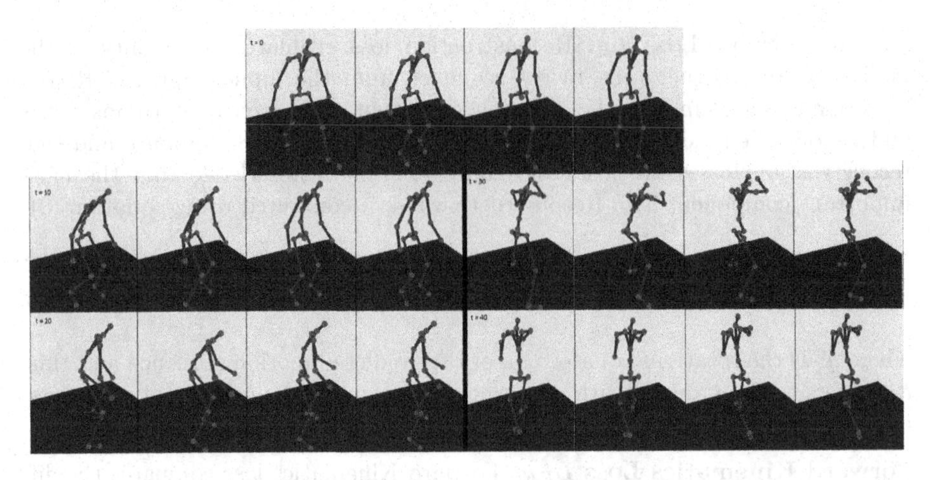

Fig. 2. Stills from the Ground Truth, LERP, Model output, and smoothed output sequences at different timestamps for the action "Aiming2" performed by subject "Subject5". Considering the frames at $t = 20$, it is clear that the output produced by our model resembles the ground truth more than the interpolated sequence.

4 Setup and Experimental Results

4.1 Evaluation Metrics

The model is evaluated against the L2P, L2Q, and NPSS metrics used in previous studies on the subject five sequences of the LAFAN1 dataset. The L2P defines the average L2 distances of the positions between the predicted motion sequence and the ground truth sequence. Equation 10 shows the L2P calculation. Similarly, the L2Q defines the average L2 distances of the global quaternions. A combination of local quaternions, positions, and motion sequence properties is used to compute these metrics. Equation 11 shows the L2Q calculation.

$$L2P = \frac{1}{NT} \sum_{n=0}^{N-1} \sum_{t=0}^{T-1} \hat{p}_n^t - {p_n}^t \tag{10}$$

$$L2Q = \frac{1}{NT} \sum_{n=0}^{N-1} \sum_{t=0}^{T-1} \hat{q}_n^t - {q_n}^t \tag{11}$$

where $\hat{q}$ is the rotational quaternion of the predicted motion sequence n at time t. q refers to the ground truth quaternion. Similarly, $\hat{p}$ refers to the position of the predicted motion sequence p refers to the ground truth position. N refers to the number of sequences, and T refers to the length of each motion sequence.

Normalized Power Spectrum Similarity (NPSS) is an approach comparing angular frequencies with the ground truth. It is an Earth Mover Distance (EMD) based metric over the power spectrum which uses the squared magnitude spectrum values of the Discrete Fourier Transform coefficients. Equation (12) computes the NPSS metric.

Fig. 3. Still from the motion generator

$$NPSS = \frac{\sum_{i=0}^{N-1} \sum_{j=0}^{T-1} w_{i,j} * emd_{i,j}}{\sum_{i=0}^{N-1} \sum_{j=0}^{T-1} w_{i,j}} \tag{12}$$

where $emd_{i,j}$ refers to the EMD distance, and $w_{i,j}$ refers to the weights.

Harvey et al. [8] state that the L2P metric is a better metric than any angular loss for assessing the visual quality of transitions with global displacements as it helps us weigh the positions of the bones and joints. Hence, they argue that L2P is a much more critical metric than L2Q and NPSS.

4.2 Dataset

The publicly available Ubisoft La Forge Animation (LaFAN1) Dataset was used for all the experiments. Introduced by Harvey et al. [8] in Ubisoft, LaFAN1 consists of general motion capture clips in high definition. The motion sequences are in BVH format. The LaFAN1 dataset comprises five subjects, 77 sequences, and 496,672 motion frames at 30fps for a total of 4.6 h. There are around 15 themes, from everyday actions like walking, sprinting, and falling to uncommon actions like crawling, aiming, and a few sports movements. Similar to other works [4,8,17], all subject five sequences were used for testing and benchmarking, with the remaining used for training.

4.3 Data Preprocessing

First, the local position and orientation values from the BVH files provided in the LaFAN1 dataset [7] are extracted. 22 joints are considered for the skeleton model. Forward Kinematics was used to compute the absolute positions of each joint from the relative positions (relative to hip) given in the dataset. Positions are modeled as standard matrices, and orientations are modeled using quaternions.

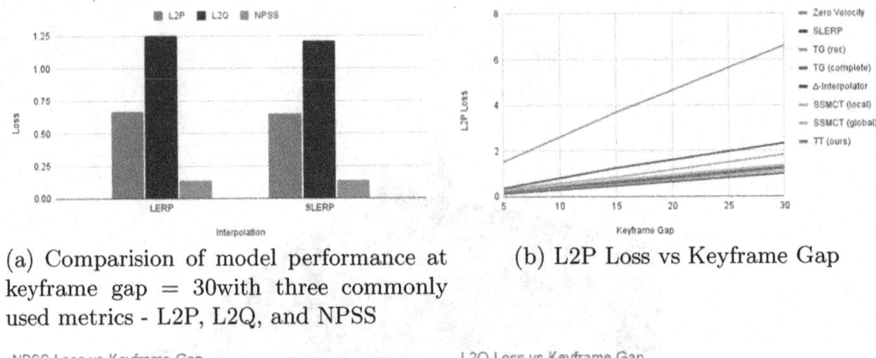

(a) Comparision of model performance at keyframe gap = 30with three commonly used metrics - L2P, L2Q, and NPSS

(b) L2P Loss vs Keyframe Gap

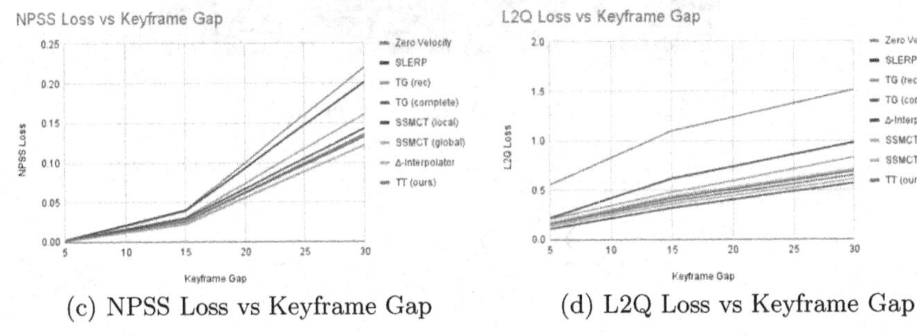

(c) NPSS Loss vs Keyframe Gap

(d) L2Q Loss vs Keyframe Gap

Fig. 4. .

Further, global position and root velocity are computed from local positions using Forward kinematics.

4.4 Hyperparameters

Most hyperparameters from previous baselines are retained in order to show the relative improvement in performance using Transformers. This study presents a novel hyperparameter comparison using different interpolation techniques - Linear and Spherical, to compare the performance of several baseline studies. A batch size of 64 for 100 epochs was used. Adam optimizer with a learning rate of 10^{-4} along with a constant dropout of 0.2 was utilised. Keyframe gaps of 5, 15, and 30 were tested to compare the performance of the transformer over higher frame gaps.

4.5 Visualizer and Motion Generator

To qualitatively evaluate the model, a visualizer was built using Node and ThreeJs that juxtaposes the ground truth, interpolated sequence, output sequence, and a smoothed output sequence of the transformer model. The model's output is

Table 1. The Tween Transformer model is compared with baseline Motion In-Betweening methods using L2P, L2Q, and NPSS metrics for various sequence lengths. The Interpolation based methods are included as part of the study. TT (Ours) refers to the Tween Transformer model.

Length	L2Q			L2P			NPSS		
	5	15	30	5	15	30	5	15	30
Zero Velocity	0.56	1.10	1.51	1.52	3.69	6.60	0.0053	0.0522	0.2318
SLERP	0.22	0.62	0.98	0.37	1.25	2.32	0.0023	0.0391	0.2013
TG_{rec}	0.21	0.48	0.83	0.32	0.85	1.82	0.0025	0.0304	0.1608
$TG_{complete}$	0.17	0.42	0.69	0.23	0.65	1.28	0.0020	0.0258	0.1328
$SSMCT_{local}$	0.17	0.44	0.71	0.23	0.74	1.37	0.0019	0.0291	0.143
$SSMCT_{Global}$	0.14	0.36	0.61	0.22	0.56	1.1	0.0016	0.0234	0.1222
Δ-Interpolator	**0.11**	**0.32**	**0.57**	**0.13**	**0.47**	**1.00**	**0.0014**	**0.0217**	**0.1217**
TT (Ours)	0.16	0.39	0.65	0.21	0.59	1.21	0.0019	0.0261	0.1358

stored in JSON format and rendered using a custom web-based visualizer. The visualizer was built from scratch using Typescript, NodeJs, Express, and ThreeJs. Figure 2 shows a sample output of the model generated using the visualizer. Further, motion generator was built using Python, Flask, Node, and ThreeJs using the visualizer module as a base. The motion generator allows a user to modify keyframes in a given motion sequence and generate inbetween frames for the same. The plugin consists of a backend Flask server that uses an instance of our model to generate the inbetween frames. Figure 3 shows a still from the motion generator where the stick modelis animating a generated custom motion sequence.

4.6 Inferences

As expected, SLERP performs better than LERP. However, it is observed that the performance at 30fps is almost comparable, as seen in Fig. 4a. This is because the spherical motion becomes almost linear for very short timescales. As seen in Table 1, it is inferred that the Tween Transformer model outperforms the interpolation model and performs closely with the baseline models. From Figs. 4b, 4d, and 4c, it is seen that Tween Transformers follow a similar trend to that of other models. Experiments show that training is crucial; Moving Average Smoothing is observed to have minimal effect on the output sequence as the model trains.

5 Conclusion

This work presents the Tween Transformer, a novel, robust, transformer-based motion in-betweening technique that serves as a tool for 3D animators and overcomes the challenges faced by existing RNN-based models [8,17], including sequential training, capturing long-term dependencies, and transfer learning. The generic model treats the application of in-betweening as a sequence to sequence problem and solves it using a transformer-based encoder-decoder architecture. It unboxes the potential of robust Transformer-based models for motion in-betweening applications. To conclude, the results encourage the application of low-resource cost-efficient models and enable further developments with the scope of transfer learning on the generalized implementation.

References

1. Bahdanau, D., Cho, K., Bengio, Y.: Neural machine translation by jointly learning to align and translate. In: Bengio, Y., LeCun, Y. (eds.) 3rd International Conference on Learning Representations, ICLR 2015, San Diego, CA, USA, May 7-9, 2015, Conference Track Proceedings (2015). http://arxiv.org/abs/1409.0473
2. Brown, T., et al.: Language models are few-shot learners. In: Larochelle, H., Ranzato, M., Hadsell, R., Balcan, M.F., Lin, H. (eds.) Advances in Neural Information Processing Systems. vol. 33, pp. 1877–1901. Curran Associates, Inc. (2020). https://proceedings.neurips.cc/paper/2020/file/1457c0d6bfcb4967418bfb8ac142f64a-Paper.pdf
3. Devlin, J., Chang, M.W., Lee, K., Toutanova, K.: BERT: pre-training of deep bidirectional transformers for language understanding. In: Proceedings of the 2019 Conference of the North American Chapter of the Association for Computational Linguistics: Human Language Technologies, Volume 1 (Long and Short Papers), pp. 4171–4186. Association for Computational Linguistics, Minneapolis, Minnesota, June 2019. https://doi.org/10.18653/v1/N19-1423, https://aclanthology.org/N19-1423
4. Duan, Y., Shi, T., Zou, Z., Lin, Y., Qian, Z., Zhang, B., Yuan, Y.: Single-shot motion completion with transformer. arXiv preprint arXiv:2103.00776 (2021)
5. Fragkiadaki, K., Levine, S., Malik, J.: Recurrent network models for kinematic tracking. CoRR **abs/1508.00271** (2015), http://arxiv.org/abs/1508.00271
6. Gopalakrishnan, A., Mali, A.A., Kifer, D., Giles, C.L., II, A.G.O.: A neural temporal model for human motion prediction. CoRR **abs/1809.03036** (2018). http://arxiv.org/abs/1809.03036
7. Harvey, F.G., Pal, C.: Recurrent transition networks for character locomotion. In: SIGGRAPH Asia 2018 Technical Briefs. SA 2018, Association for Computing Machinery, New York, NY, USA (2018). https://doi.org/10.1145/3283254.3283277, https://doi.org/10.1145/3283254.3283277
8. Harvey, F.G., Yurick, M., Nowrouzezahrai, D., Pal, C.: Robust motion in-betweening. ACM Trans. Graph. **39**(4) (2020). https://doi.org/10.1145/3386569.3392480
9. Holden, D.: Robust solving of optical motion capture data by denoising. ACM Trans. Graph. **37**(4) (2018). https://doi.org/10.1145/3197517.3201302, https://doi.org/10.1145/3197517.3201302
10. Holden, D., Saito, J., Komura, T.: A deep learning framework for character motion synthesis and editing. ACM Trans. Graph. **35**(4) (2016). https://doi.org/10.1145/2897824.2925975, https://doi.org/10.1145/2897824.2925975
11. Lehrmann, A.M., Gehler, P.V., Nowozin, S.: Efficient nonlinear markov models for human motion. In: Proceedings of the IEEE Conference on Computer Vision and Pattern Recognition (CVPR), June 2014

12. Liu, Z., et al.: Towards natural and accurate future motion prediction of humans and animals. In: 2019 IEEE/CVF Conference on Computer Vision and Pattern Recognition (CVPR), pp. 9996–10004 (2019). https://doi.org/10.1109/CVPR.2019.01024
13. Martínez-González, Á., Villamizar, M., Odobez, J.: Pose transformers (POTR): human motion prediction with non-autoregressive transformers. CoRR **abs/2109.07531** (2021), https://arxiv.org/abs/2109.07531
14. Min, J., Chen, Y.L., Chai, J.: Interactive generation of human animation with deformable motion models. ACM Trans. Graph. **29**(1) (2009). https://doi.org/10.1145/1640443.1640452
15. Müller, M., Röder, T., Clausen, M.: Efficient content-based retrieval of motion capture data. ACM Trans. Graph. **24**(3), 677–685 (2005). https://doi.org/10.1145/1073204.1073247
16. Oreshkin, B.N., Valkanas, A., Harvey, F.G., Ménard, L.S., Bocquelet, F., Coates, M.J.: Motion Inbetweening via Deep Δ-Interpolator. arXiv e-prints, January 2022 arXiv:2201.06701
17. Oreshkin, B.N., Valkanas, A., Harvey, F.G., Ménard, L.S., Bocquelet, F., Coates, M.J.: Motion inbetweening via deep -interpolator (2022). https://doi.org/10.48550/ARXIV.2201.06701, https://arxiv.org/abs/2201.06701
18. Prafulla, D., Sastry, G., McCandlish, S.: Enct5: fine-tuning t5 encoder for discriminative tasks (2021)
19. Ren, M., Kiros, R., Zemel, R.S.: Exploring models and data for image question answering. In: Proceedings of the 28th International Conference on Neural Information Processing Systems, NIPS 2015, vol. 2, pp. 2953–2961. MIT Press, Cambridge, MA, USA (2015)
20. Tanuwijaya, S., Ohno, Y.: Tf–df indexing for mocap data segments in measuring relevance based on textual search queries. Vis. Comput. **26**(6–8), 1091–1100 (2010). https://doi.org/10.1007/s00371-010-0463-9
21. Taylor, G.W., Hinton, G.E.: Factored conditional restricted boltzmann machines for modeling motion style. In: Proceedings of the 26th Annual International Conference on Machine Learning, ICML 2009, pp. 1025–1032. Association for Computing Machinery, New York, NY, USA (2009). https://doi.org/10.1145/1553374.1553505, https://doi.org/10.1145/1553374.1553505
22. Vaswani, A., et al.: Attention is all you need. In: Proceedings of the 31st International Conference on Neural Information Processing Systems, NIPS 2017, pp. 6000–6010. Curran Associates Inc., Red Hook, NY, USA (2017)
23. Wang, J.M., Fleet, D.J., Hertzmann, A.: Gaussian process dynamical models for human motion. IEEE Trans. Pattern Anal. Mach. Intell. **30**(2), 283–298 (2008). https://doi.org/10.1109/TPAMI.2007.1167
24. Witkin, A., Kass, M.: Spacetime constraints. In: Proceedings of the 15th Annual Conference on Computer Graphics and Interactive Techniques, SIGGRAPH 1988, pp. 159–168. Association for Computing Machinery, New York, NY, USA (1988). https://doi.org/10.1145/54852.378507, https://doi.org/10.1145/54852.378507

Convolutional Point Transformer

Chaitanya Kaul[1(✉)], Joshua Mitton[1], Hang Dai[2], and Roderick Murray-Smith[1]

[1] School of Computing Science, University of Glasgow, Glasgow G12 8RZ, Scotland
chaitanya.kaul@glasgow.ac.uk

[2] Mohamed bin Zayed University of Artificial Intelligence, Abu Dhabi, UAE

Abstract. We present CpT: Convolutional point Transformer – a novel neural network layer for dealing with the unstructured nature of 3D point cloud data. CpT is an improvement over existing MLP and convolution layers for point cloud processing, as well as existing 3D point cloud processing transformer layers. It achieves this feat due to its effectiveness in creating a novel and robust attention-based point set embedding through a convolutional projection layer crafted for processing dynamically local point set neighbourhoods. The resultant point set embedding is robust to the permutations of the input points. Our novel layer builds over local neighbourhoods of points obtained via a dynamic graph computation at each layer of the network's structure. It is fully differentiable and can be stacked just like convolutional layers to learn intrinsic properties of the points. Further, we propose a novel Adaptive Global Feature layer that learns to aggregate features from different representations into a better global representation of the point cloud. We evaluate our models on standard benchmark ModelNet40 classification and ShapeNet part segmentation datasets to show that our layer can serve as an effective addition for various point cloud processing tasks while effortlessly integrating into existing point cloud processing architectures to provide significant performance boosts.

1 Introduction

3D data takes many forms. Meshes, voxels, point clouds, multi view 2D images, RGB-D are all forms of 3D data representations. Amongst these, the simplest raw form that 3D data can exist in is a discretized representation of a continuous surface. This can be visualized as a set of points (in $\mathbb{R}^3$) sampled from a continuous surface. Adding point connectivity information to such 3D point cloud representations creates representations known as 3D meshes. Deep learning progress on processing 3D data was initially slow primarily due to the fact that early deep learning required a structured input data representation as a prerequisite. Thus, raw sensor data was first converted into grid-like representations such as voxels, multi view images, or data from RGB-D sensors was used to interpret geometric information. However, such data is computationally expensive to process, and in many cases, it is not the true representation of the 3D structure that may be required for solving the task. Modern applications of point clouds require a large amount of data processing. This makes searching for salient features in

Y. Zheng et al. (Eds.): ACCV 2022, LNCS 13848, pp. 308–324, 2023.
https://doi.org/10.1007/978-3-031-27066-6_22

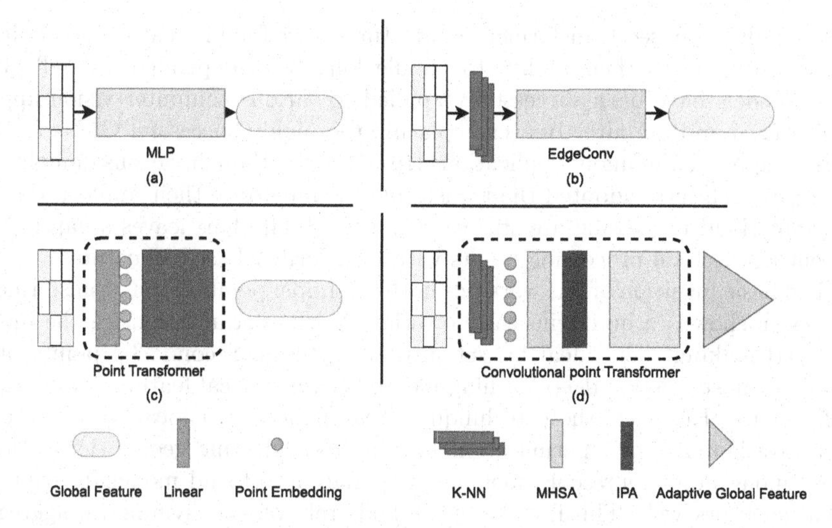

Fig. 1. Overview of the structure of (a) PointNet, (b) DGCNN, (c) Point Transformer, and, (d) Convolutional point Transformer illustrating the differences between these point based convolution layers. Global Feature denotes max or average pooling operations, MHSA denotes multi-head self attention, IPA denotes InterPoint Attention.

their representations a tedious task. Processing 3D meshes (along with added 3D point cloud information), requires dealing with their own sets of complexities and combinatorial irregularities, but such structures are generally efficient to process due to knowledge of point set connectivity. This motivation has sparked interest in processing 3D points directly at a point level instead of converting the points to intermediate representations. Lack of any general structure in the arrangement of points serves as a challenge in the ability to process them. This is due to the fact that point clouds are essentially set representations of a continuous surface in a 3D space. The seminal works on processing points proposed approaches to deal with this lack of order by constructing set based operations to ingest 3D points directly. They then created a symmetric mapping of such set based representations in a high dimensional space [22,42]. This representation was further processed by symmetry preserving operations to create a feature representation of the point cloud, before passing them through multi layer perceptrons for solving the task. Representations created on a per-point basis are generally not robust as there is no concept of locality in the data representations used to create them. Various methods have been introduced in literature to add the notion of locality to point set processing [10,11,16,23,33,38].

Recently proposed transformer architectures have found great success in language tasks due to their ability to handle long term dependencies well [31]. These models have been successfully applied to various computer vision applications [2,3] and are already even replacing the highly successful CNNs as the de facto approach in many applications [19,35]. Even though various domains of vision research have adopted the transformer architecture, their applications to 3D point cloud processing are still very limited [4,44]. This leaves a gap in the current research on processing points with transformer based structures.

The basic intuition of our work lies in three simple points. First, transformers have been shown to be intrinsically invariant to the permutations in the input data [41] making them ideal for set processing tasks. Second, 3D points processed by most existing deep learning methods exploit local features to improve performance. However, these techniques treat points at a local scale to keep them invariant to input permutations leading to neglecting geometric relationships among point representations, and the inability of said models to capture global concepts well. Third, existing methods rely on one symmetric aggregation function (e.g. max pooling) to aggregate the features that point operators (such as MLPs and 1D Convolutions) learn. This function is likely to leave out important feature information about the point set. To address these points, we propose *CpT: Convolutional point Transformer*. CpT differs from existing point cloud processing methods due to the following reasons:

i) CpT uses a K-nearest neighbour graph at every layer of the model. Such a dynamic graph computation [33] requires a method to handle the input data, in order to create a data embedding that can be fed into a transformer layer. Towards this end, we propose a novel Point Embedding Module that first constructs a dynamic point cloud graph at every stage of the network, and creates a point embedding to feed into a transformer layer for its processing.
ii) Transformers employ multi-head self attention to create contextual embeddings. Such an attention mechanism works to enhance the learning of features in the data. However, it has been shown that adding sample wise attention to data can help improve the performance of a transformer model even further [6,8,24,27]. To this end, we propose to add an InterPoint Attention Module (Fig. 2(c)) to the Transformer which learns to enhance the output by learning to relate each point in the input to every other point. This helps capture better geometric relationships between the points and aids in better learning of the local and global concepts in the data. The resultant transformer block, i.e., the CpT layer (shown in Fig. 2(a)) is a combination of multi-head self attention operation, followed by an InterPoint Attention module. The Q, K, V attention projections in this block are convolutional in nature to facilitate learning spatial context.
iii) We propose a novel Global Feature Aggregation layer (Fig. 2(b)) that learns to aggregate features from multiple symmetric global representations of the points through a spatial attention mechanism. Our main novelty here lies in the adaptive nature of this layer's information aggregation as it scales

the attention output through multiplication with a vector β and learns to weight the attention aggregated output over the training of the model.

We evaluate CpT on 3-point cloud processing tasks – classification of ModelNet40 CAD models and estimating their normals, and segmentation on the ShapeNet part dataset. We perform extensive experiments and multiple studies and ablations to show the robustness of our proposed model. Our results show that CpT can serve as an accurate and effective backbone for various point cloud processing tasks.

2 Related Literature

Processing Point Clouds Using Deep Learning. The lack of a grid-like structure to points makes applying convolutions directly on them a tedious task. Due to this reason, all previous works in processing 3D points using deep learning required the points to be first converted to a structured representation like voxels, depth maps, multiple 2D views of an object, etc., and then process the resultant representation with conventional deep models [21,29,45]. This process has massive computational overheads which result from first converting the data to the grid-like representation and then training large deep learning models on it. The first methods that proposed to treat points as set embeddings in a 3D space were PointNet [22] and Deep Sets [42]. They took the approaches of creating data embeddings that preserve point set symmetry using permutation invariance and permutation equivariance respectively. This drastically reduced the computational overhead as the models used to train the data worked directly on the raw data points, and the models used to train on such data were not very large scale, and yet accurate. These models however, suffered from the drawback of only looking globally at the points. This meant that the higher dimension embeddings created by such methods were not robust to occlusions as the representations only took the particular input point into consideration while creating its representation. This drawback was tackled in works where local neighbourhoods of points were taken into consideration to compute per point representations. PointNet++ [23] used farthest point sampling to estimate the locality of points first, before processing them via weight-shared multilayer perceptrons (MLPs). ECC Nets [26] proposed the Edge Conv operation over a local neighbourhoods of points. Dynamic Graph CNNs [33] created local neighbourhoods by computing a graphical representation of the points before every shared-weighted MLP layer to create a sense of locality in the points. Parameterized versions of convolution operations for points have also been proposed such as Spider CNN [38], and PointCNN [16]. Previous work with kernel density functions to weight point neighbourhoods [36] treated convolutional kernels as non linear functions of the 3D points containing both weight and density estimations. The notion of looking both locally and globally at points was proposed in SAWNet [10]. Other notable approaches that process 3D points as a graph either densely connect local neighbourhoods [43], operate on a superpoint graph to create contextual relationships [13], or use spectral graph convolutions to process the points [32].

Attention Mechanisms in Vision. Various attention mechanisms exist in the deep learning literature, with one of the first works introducing them for natural language understanding being [20]. The first attention mechanisms applied to vision were based on self attention via the squeeze-and-excite operation of SE Nets [7]. This attention mechanism provided channel-wise context for feature maps and was widely successful in increasing the accuracy over non attention based methods on ImageNet. It was extended in [1] to non-local networks, and in [25] to fully convolutional networks for medical imaging applications. One of the first works that combined channel-wise self attention with spatial attention for images [9,34] first created attention maps across the entire feature map, followed by a feature map re-calibration step to only propagate the most important features in the networks forward. FatNet [11] successfully applied this concept to point clouds by first applying spatial attention over feature groups of local regions of points, followed by feature weighting using a squeeze-and-excite operation.

Transformers in 2D and 3D Vision. A concurrent line of upcoming work applies the dot product attention to input data for its efficient processing. The Vision Transformer (ViT) [3] was the first real application of the Transformer to vision. It employed a Transformer encoder to extract features from image patches of size 16×16. An embedding layer converted the patches into a transformer-friendly representation and added positional embedding to them. This general structure formed the basis of initial transformer based research in vision, but required large amounts of data to train. DETR [2] was the first detection model created by processing the features of a Convolutional Neural Network by a transformer encoder, before adding a bipartite loss to the end to deal with a set based output. The Convolutional Vision Transformer (CvT) [35] improved over the ViT by combining convolutions with Transformers via a novel convolutional token embedding and a convolutional projection layer. The recently proposed Compact Convolutional Transformer (CCT) [5] improve over the ViT and CvT models by showing that with the right size and tokenization, transformers can perform head-to-head with state of the art CNN models on small scale datasets. The Swin Transformer [19] proposed a hierarchical transformer network whose representations are computed with shifted windows.

The application of the transformer to point cloud processing is limited. Existing works include the Point Transformer [44] which applies a single head of self attention on the points to create permutation invariant representations of local patches obtained via K-NN and furthest point sampling, and the Point Cloud Transformer [4] which applies a novel offset attention block to 3D point clouds. Over work extends over the existing literature to create a novel transformer block that performs feature wise as well as pointwise attended features of the input for its accurate and effective processing.

3 Convolutional Point Transformer

Notation. We denote a set of N-points with a D-dimensional embedding in the lth CpT layer as, $\mathbf{X}^l = \{\mathbf{x}_1^l, \ldots, \mathbf{x}_N^l\}$, where $\mathbf{x}_i^l \in \mathbb{R}^D$. For K nearest neighbours of point i ($i \in \{1 \ldots N\}$), we define the set of all nearest neighbours for that point to be $\Delta\mathbf{X}_i^l = \{\mathbf{x}_{j_1}^l - \mathbf{x}_i^l, \ldots, \mathbf{x}_{j_K}^l - \mathbf{x}_i^l\}$, where $j_k : j_k \in \{1, \ldots, N\} \wedge (j_k \neq i)$.

The Convolutional point Transformer (CpT) layer is shown in Fig. 2. Our main contributions are the Point Embedding Module (Sect. 3.1), the InterPoint Attention Module with a convolutional attention projection (Sect. 3.2) and the Adaptive Global Feature module (Sect. 3.3). We use CpT in 3 different network architectures - PointNet [22], DGCNN [33], Point Transformer [44]. We do this by replacing the point convolution layers in PointNet, the EdgeConv layer in DGCNN and the attention layer in Point Transformer with CpT. The information flow in all these architectures is similar. When an input point cloud of size $N \times 3$ is passed through the architecture, a graph of the points is computed via finding its K-Nearest Neighbours based on Euclidean distance. This representation is then passed through a point embedding layer that maps the input data into a representation implicitly inclusive of the nearest neighbours of the points. This is done via a 2D convolution operation whose degree of overlap across points can be controlled through the length of the stride. A multi-head attention operation is then applied to this embedded representation which is followed InterPoint Attention. The multi-head attention can be seen as learning relevant features of a points embedding as a function of its K nearest neighbours. Such an attention mechanism learns to attend to the features of the points rather than the points themselves (column-wise matrix attention), i.e. for a set of points in a batch, it learns to weight individual feature transformations. The InterPoint Attention on the other hand can be interpreted as learning the relationships between different the points themselves, within a batch (a row-wise matrix attention operating per point embedding, rather than per individual feature of the points). Each attention block is followed by a residual addition and a combination of Layer-Normalization and 2 1D Convolutions. These operations form one CpT layer. We update the graph following these operations and pass it into the next CpT layer. The output of the final CpT layer in all network architectures is fed through our Adaptive Global Feature module. It progressively learns to attend to features of max pooled and average pooled global representations of the features during training, to create a richer final feature representation.

3.1 Point Embedding Module

The point embedding module takes a k-NN graph as an input. The general structure of the input to this module is denoted by $(B, f, N, \Delta\mathbf{X}_i^l) \in \mathbb{R}^{(B \times f \times N \times \Delta\mathbf{X}_i^l)}$, where B and f are the batch size and input features respectively. This layer is essentially a mapping function F_θ that maps the input into an embedding E_θ. This operation is denoted by,

$$\mathbf{y} = \mathbf{F}_\theta(I(B, f, N, \Delta\mathbf{X}_i^l)), \mathbf{y} \in \mathbb{R}^{(B \times f \times \mathbf{E}_\theta)}.$$

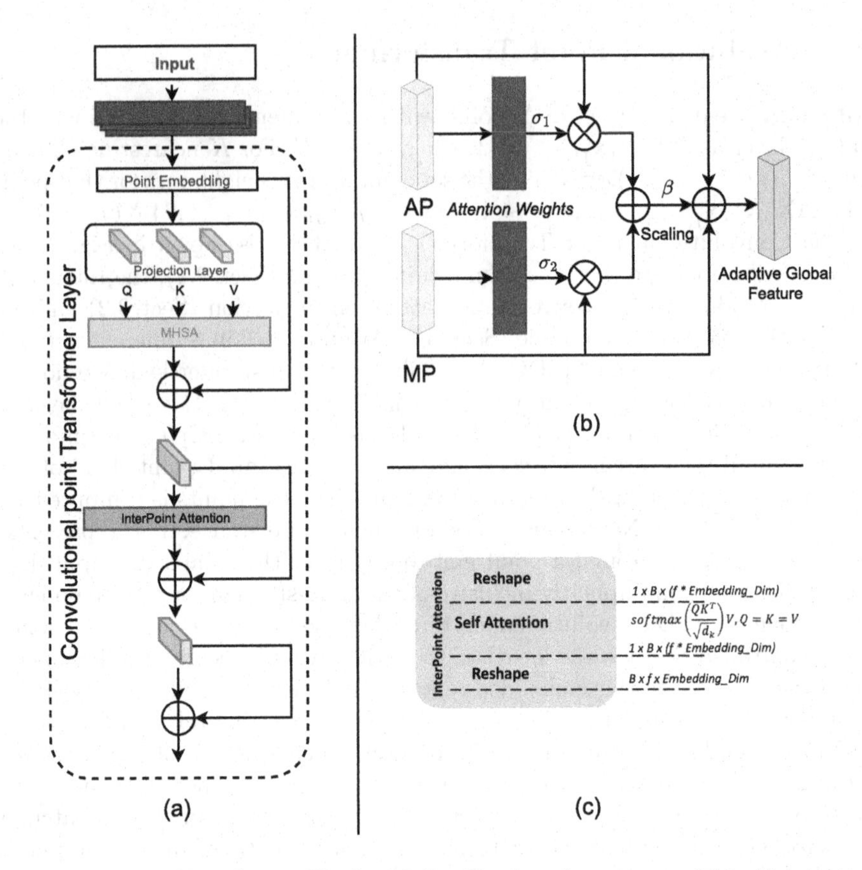

Fig. 2. The CpT Transformer Block with Dot Product Attention and InterPoint Attention are shown in (a). (c) shows the inner workings of the InterPoint Attention module. Flow of information is sequential from top to bottom. (b) shows the Adaptive Global Feature module. AP is average pooling and MP is max pooling. The Flow of information is from left to right.

There are many choices available for the mapping function $\mathbf{F}_\theta$. We use a convolution operation with a fixed size padding and stride. This allows us to train this module in a end-to-end setting along with the rest of the network as it is fully differentiable and can be plugged anywhere in the architecture.

3.2 Convolutional Point Transformer Layer

Our Convolutional point Transformer layer uses a combination of multi-head self attention and InterPoint attention to propagate the most important, salient features of the input through the network. The CpT layer leverages spatial context and moves beyond a fully connected projection, by using a convolution layer to sample the attention matrices. Instead of creating a more complicated network design, we leverage the ability of convolutions to learn relevant feature sets for

3D points. Hence, we replace the original fully connected layers in the attention block with depthwise convolution layers which forms our convolutional projection to obtain the attention matrices. The general structure of the CpT layer is shown in Fig. 2(a). The functionality of the rest of the transformer block is similar to ViT [3] where normalization and feedforward layers are added after every attention block and residual mappings are used to enhance the feature learning. The notable difference from ViT is that the feedforward layers are replaced with 1D convolution operations.

Formally, the projection operation of the CpT layer is denoted by the following operation,

$$z_i^{q/k/v} = Convolution(z_i, p, s),$$

where $z^{q/k/v}$ is the input for the $Q/K/V$ attention matrices for the i-th layer and z_i is the input to the convolution block. The convolution operation is implemented as a depthwise separable convolution operation with tuples of kernel size p and stride s. We now formalize the flow of data through the entire CpT block for a batch size B as,

$$out_i^a = \gamma(MHSA(z_i^{q/k/v})) + z_i^{q/k/v},$$

where out_i^a is the output of the multi-head attention block, $MHSA$ is the multi-head attention given by $MHSA(\cdot) = softmax(\frac{QK^T}{\sqrt{d}}V)$. $\sqrt{d}$ is used to scale the attention weights to avoid gradient instabilities. γ is the layer normalization operation. The addition denotes a residual connection. This output is further processed as in a vanilla transformer in the following way,

$$out_i^b = \gamma(FF(out_i^a)) + out_i^a.$$

FF here denotes the feedforward 1D convolution layers. The InterPoint attention is then computed on this output as,

$$out_i^c = \gamma(IPA(out_i^b)_{i=1}^B) + out_i^b,$$

which is then processed in a similar manner by layer normalization and feedforward layers as,

$$out_i^d = \gamma(FF(out_i^b) + out_i^b.$$

Following this step, the graph is recomputed and the process restarts for the $l+1$-th layer.

3.3 Adaptive Global Feature

The Adaptive Global feature layer (Fig. 2(b)) takes the output of the last CpT layer in a network and learns to create a robust attention-based global representation from it. The output of the final CpT layer are passed through max pooling and average pooling operations. Then attention weights σ_1 and σ_2 are learnt to weight these global representations before combining them via an aggregation function. The output of this aggregation function is further scaled via a vector β. β is initialized as a vector of all zeros, and the same shape as the pooled feature outputs. It gradually learns to assign weight to the attended features as the training progresses. Let the pooling operations, P be defined as the set, $P = \{AP, MP\}$. The output of the Adaptive Global Feature layer is denoted by,

$$AGF = \beta(\sum_{i=1}^{2} \sigma_i \cdot P) + AP + MP$$

4 Experiments and Results

We evaluate our model on two different datasets for point cloud classification, part segmentation and surface normal estimation tasks. For classification and surface normal estimation, we use the benchmark ModelNet40 dataset [37] and for object part segmentation, we use the ShapeNet Part dataset [40].

4.1 Implementation Details

Unless stated otherwise, all our models are trained in PyTorch on a batch size of 32 for 250 epochs. SGD with an initial learning rate of 0.1 and momentum 0.9 is used. We use a cosine annealing based learning rate scheduler. The momentum for batch normalization is 0.9. Batch Normalization decay is not used. Dropout, wherever used, is used with a rate of 0.5. Custom learning rate schedules are used for the segmentation tasks after initial experimentation. For the classification and 3D indoor scene segmentation tasks, we compute dynamic graphs using 20 nearest neighbours, while for the part segmentation task, we use 40 nearest neighbours. We use NVIDIA A6000 GPUs for our experiments.

4.2 Classification with ModelNet40

The ModelNet40 dataset [37] contains meshes of 3D CAD models. A total of 12,311 models are available belonging to 40 categories, split into a training-test set of 9,843–2,468 respectively. We use the official splits provided for all our experiments and datasets to keep a fair comparison. In terms of data pre-processing, we follow the same steps as [22]. We uniformly sample 1024 points from the mesh surface and rescale the point cloud to fit a unit sphere. Data augmentation is used during the training process. We perturb the points with random jitter and scalings during the augmentation process.

It can be seen from the results that CpT outperforms existing classification methods, including the Point Transformer [44]. CpT, when replacing the point convolution in PointNet and the EdgeConv in DGCNN, can be seen to provide significant performance boosts. In the Static Graph approach, the K-NN graph is only computed once (in the first CpT layer) and the same graph (and it's features) is used by the subsequent layers in the network. Here, even when we do

Table 1. Classification results on the ModelNet40 dataset. CpT (PointNet*) and CpT (DGCNN*) denote the backbone architecture the CpT layer was placed in. All trained architectures use our Adaptive Global Feature module.

Method	Class accuracy (%)	Instance accuracy (%)
3D ShapeNets [40]	77.3	84.7
VoxNet [21]	83.0	85.9
PointNet [22]	86.0	89.2
PointNet++ [23]	–	90.7
SpiderCNN [38]	–	90.5
PointWeb [43]	89.4	92.3
PointCNN [16]	88.1	92.2
Point2Sequence [17]	90.4	92.6
ECC [26]	83.2	87.4
DGCNN [33]	90.2	92.2
FatNet [11]	90.6	93.2
KPConv [30]	–	92.9
SetTransformer [14]	–	90.4
PCT [4]	–	93.2
PointTransformer [44]	90.6	93.7
CpT (PointNet*)	88.1	90.9
CpT (Static Graph)	90.3	92.1
CpT (DGCNN*)	**90.9**	**93.9**

not recompute the graph at every layer, CpT outperforms Dynamic Edge Conditioned Filters (ECC) [26] and performs at par with DGCNN. Dynamic graph computations before every layer help CpT surpass the accuracy of all existing graph and non graph based approaches, including outperforming existing transformer based approaches [4,44].

4.3 Segmentation Results with ShapeNet Part

The ShapeNet Part dataset [40] contains 16,881 3D shapes from 16 object categories. A total of 50 object parts are available to segment. We sample 2048 points from each shape and follow the official training-testing splits for our experiments. Our results on the dataset are summarized in Table 2, while the visualizations produced by our model are shown in Fig. 3. We train two CpT models for this task, with a DGCNN backbone, and a Point Transformer backbone. We note that CpT (DGCNN*) outperforms DGCNN by 0.9% on the IoU metric while CpT (PointTransformer*) out performs the point transformer by 0.2%.

Fig. 3. Visualizing the segmentation results from the ShapeNet Part Dataset. Ground truth images are in the top row and their corresponding segmentation maps in the bottom row.

Table 2. Results on the ShapeNet Part Segmentation dataset. CpT (DGCNN*) and CpT (PointTransformer*) denote the backbone architecture the CpT layer was placed in.

Method	IoU
Kd-Net [12]	82.3
SO-Net [15]	84.6
PointNet++ [23]	85.1
SpiderCNN [38]	85.3
SPLATNet [28]	85.4
PointCNN [16]	86.1
PointNet [22]	83.7
DGCNN [33]	85.1
FatNet [11]	85.5
PointASNL [39]	86.1
RSCNN [18]	86.2
KPConv [30]	86.4
PCT [4]	86.4
PointTransformer [44]	86.6
CpT (DGCNN*)	85.9
CpT (PointTransformer*)	**86.8**

4.4 Normal Estimation

We estimate normals of the ModelNet40 dataset where each point cloud has a corresponding normal label. Normals are crucial in understand the shape and underlying geometry of 3D objects. We train 2 backbone architectures for this task – CpT (PointNet*) and CpT (DGCNN*). Our results are summarized in Table 3. It can be seen that, when replacing the point convolution in PointNet, and the EdgeConv in DGCNN with our CpT layer, we get significant performance boosts over the baseline models.

Table 3. Results on the normal estimation task. Results on the PointNet and DGCNN backbones are reported. The reported error is the average cosine distance error. Lower is better.

Method	Error
PointNet [22]	0.47
CpT (PointNet*)	**0.31**
DGCNN [33]	0.29
CpT (DGCNN*)	**0.15**

5 Ablation Study

In this section, we detail experiments of our extensive studies to shed light at the inner workings of the CpT layer. We conducted a series of ablation studies to highlight how the different building blocks of CpT come together. The DGCNN backbone is used for these experiments.

Static v/s Dynamic Graphs. We trained two CpT models for these experiments. The K-NN graph computation is only done before the first CpT Layer in the Static CpT model. The results for this experiment are shown in Table 1. Computing the graph before each transformer layer helps boost CpT's instance accuracy by 1.8%.

Global Representations vs Graph Representations. We compared dynamic graph computation with feeding point clouds directly into the CpT layer. This equates to removing the K nearest neighbour step from the layer in Fig. 2(a). As the points were directly fed into CpT, we also replaced the Point Embedding Module with a direct parameterized relational embedding This relation was learnt using a convolution operation. The rest of the layer structure remained unchanged. The resultant model trained on global representations performed well. The results are summarized in Table 4. PointNet [22] is added to the table for reference as it also takes a global point cloud representation as an input. The CpT with dynamic graph computation outperforms both methods, but it is interesting to note that the CpT model that works directly on the entire point cloud manages a higher class accuracy than PointNet.

Table 4. Comparison of different input representations.

Model	Class accuracy (%)
PointNet [22]	86.2
CpT (No locality)	87.6
CpT (Dynamic graph)	90.6

Number of Nearest Neighbours for Dynamic Graph Computation. We also experiment with the number of nearest neighbours used to construct the graph. CpT layers implemented with 20 nearest neighbour graphs performs the best. This is due to the fact that for large distances beyond a particular threshold, the euclidean distance starts to fail to approximate the geodesic distance. This leads to capturing points that may not lie in the true neighbourhood of the point while estimating its local representation.

Table 5. CpT with different number of nearest neighbours.

k	Class accuracy (%)
10	89.6
20	90.6
30	90.3
40	90.4

CpT Robustness. We test the robustness of CpT towards classifying sparse datasets by sampling point clouds at various resolutions from the ModelNet40 dataset and evaluating CpT's performance on them. We sample 1024, 768, 512, 256 and 128 points for each CAD model in the dataset, keeping the input point clouds' resolution small. Figure 4(a) shows the robustness of CpT compared to PointNet and DGCNN. CpT is not tied to observing input data through receptive fields like CNNs in order to construct its feature space. Even at very small resolutions of the point cloud, the multi-head attention and InterPoint attention learns to model long range dependencies in the input points well to create a local and global understanding of the shape. This leads to a high performance even when there are a small sample of points available.

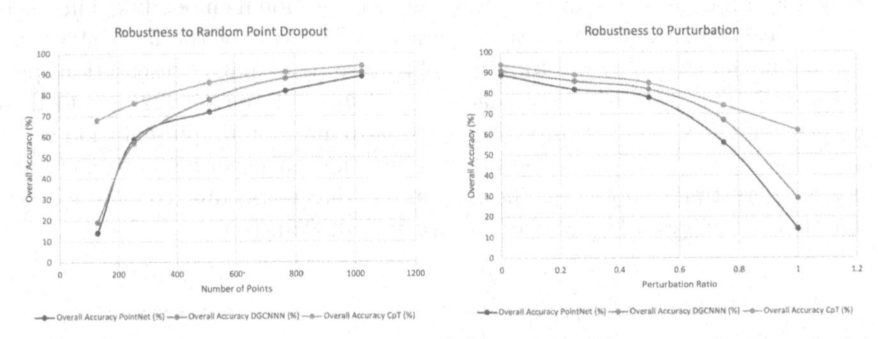

(a) Effect of randomly dropping out points on performance of CpT.

(b) Effect of adding perturbations to points on the performance of CpT.

Fig. 4. Comparing the robustness of CpT to PointNet [22] and DGCNN [33]

To observe the effect of perturbations on CpT's performance, we perturb each point in the point clouds independently using Gaussian noise. Our results (summarized in Fig. 4(b)) show that our network still manages to remain robust to the added noise even during high amounts of perturbations. The X-axis of the graph in Fig. 4(b)) shows the amount of standard deviation of the Gaussian noise which is linearly increased to perturb the points by a larger amount.

6 Conclusions and Future Work

In this paper, we proposed CpT: Convolutional point Transformer. We showed how transformers can be effectively used to process 3D points with the help of dynamic graph computations at each intermediate network layer. The main contributions of our work include, first, a Point Embedding Module capable of taking a dynamic graph as an input and transforming it into a transformer friendly data representation. Second, the InterPoint Attention Module which uses self attention to facilitate cross talk between the points in an arbitrary batch. Through this work, we have shown for the first time that different self attention methods can be efficiently used inside a single transformer layer for 3D point cloud processing ([44] only uses one form of self attention while [4] only uses an offset attention operator). We also proposed an Adaptive Global Feature module that compliments CpT by learning to attend to the most important features inside different global representations of point cloud features. CpT outperforms most existing convolutional and transformer based approaches for point cloud processing on a variety of benchmark tasks. To improve performance, future directions of this research lie in learning to sample points uniformly along the manifold of the 3D point cloud to preserve its local shape. This can lead to learning better local representations of the data and in turn, creating models with improved accuracy. Our results already show that CpT is capable of taking local context and processing it effectively with global information present in the points. We have shown how CpT can be easily integrated into various existing architectures that are used for processing 3D points. We believe that CpT can serve as an effective backbone for future point cloud processing tasks and be extended to various applications.

Acknowledgements. Chaitanya Kaul and Roderick Murray-Smith acknowledge funding from the QuantIC project funded by the EPSRC Quantum Technology Programme (grant EP/MO1326X/1) and the iCAIRD project, funded by Innovate UK (project number 104690). Joshua Mitton is supported by a University of Glasgow Lord Kelvin Adam Smith Studentship. Roderick Murray-Smith acknowledges funding support from EPSRC grant EP/R018634/1, Closed-loop Data Science.

References

1. Cao, Y., Xu, J., Lin, S., Wei, F., Hu, H.: GCNet: non-local networks meet squeeze-excitation networks and beyond. In: Proceedings of the IEEE/CVF International Conference on Computer Vision Workshops, pp. 1971–1980 (2019)
2. Carion, N., Massa, F., Synnaeve, G., Usunier, N., Kirillov, A., Zagoruyko, S.: End-to-end object detection with transformers. In: Vedaldi, A., Bischof, H., Brox, T., Frahm, J.-M. (eds.) ECCV 2020. LNCS, vol. 12346, pp. 213–229. Springer, Cham (2020). https://doi.org/10.1007/978-3-030-58452-8_13
3. Dosovitskiy, A., et al.: An image is worth 16 × 16 words: transformers for image recognition at scale. In: International Conference on Learning Representations (2020)
4. Guo, M.-H., Cai, J.-X., Liu, Z.-N., Mu, T.-J., Martin, R.R., Hu, S.-M.: PCT: point cloud transformer. Compu. Visual Media **7**(2), 187–199 (2021). https://doi.org/10.1007/s41095-021-0229-5
5. Hassani, A., Walton, S., Shah, N., Abuduweili, A., Li, J., Shi, H.: escaping the big data paradigm with Compact Transformers. arXiv preprint arXiv:2104.05704 (2021)
6. Ho, J., Kalchbrenner, N., Weissenborn, D., Salimans, T.: Axial attention in multi-dimensional transformers. arXiv preprint arXiv:1912.12180 (2019)
7. Hu, J., Shen, L., Sun, G.: Squeeze-and-excitation networks. In: Proceedings of the IEEE Conference on Computer Vision and Pattern Recognition, pp. 7132–7141 (2018)
8. Iida, H., Thai, D., Manjunatha, V., Iyyer, M.: Tabbie: pretrained representations of tabular data. arXiv preprint arXiv:2105.02584 (2021)
9. Kaul, C., Manandhar, S., Pears, N.: FocusNet: an attention-based fully convolutional network for medical image segmentation. In: 2019 IEEE 16th International Symposium on Biomedical Imaging (ISBI 2019), pp. 455–458. IEEE (2019)
10. Kaul, C., Pears, N., Manandhar, S.: SAWNet: a spatially aware deep neural network for 3D point cloud processing. arXiv preprint arXiv:1905.07650 (2019)
11. Kaul, C., Pears, N., Manandhar, S.: FatNet: A feature-attentive network for 3D point cloud processing. In: 2020 25th International Conference on Pattern Recognition (ICPR). pp. 7211–7218. IEEE (2021)
12. Klokov, R., Lempitsky, V.: Escape from cells: Deep KD-networks for the recognition of 3d point cloud models. In: Proceedings of the IEEE International Conference on Computer Vision, pp. 863–872 (2017)
13. Landrieu, L., Simonovsky, M.: Large-scale point cloud semantic segmentation with superpoint graphs. In: 2018 IEEE/CVF Conference on Computer Vision and Pattern Recognition, pp. 4558–4567 (2018). https://doi.org/10.1109/CVPR.2018.00479
14. Lee, J., Lee, Y., Kim, J., Kosiorek, A., Choi, S., Teh, Y.W.: Set transformer: a framework for attention-based permutation-invariant neural networks. In: International Conference on Machine Learning, pp. 3744–3753. PMLR (2019)
15. Li, J., Chen, B.M., Lee, G.H.: SO-Net: self-organizing network for point cloud analysis. In: 2018 IEEE/CVF Conference on Computer Vision and Pattern Recognition, pp. 9397–9406. IEEE (2018)
16. Li, Y., Bu, R., Sun, M., Wu, W., Di, X., Chen, B.: PointCNN: convolution on $\mathcal{X}$-transformed points. Adv. Neural. Inf. Process. Syst. **31**, 820–830 (2018)

17. Liu, X., Han, Z., Liu, Y.S., Zwicker, M.: Point2sequence: learning the shape representation of 3D point clouds with an attention-based sequence to sequence network. In: Proceedings of the AAAI Conference on Artificial Intelligence, vol. 33, pp. 8778–8785 (2019)
18. Liu, Y., Fan, B., Xiang, S., Pan, C.: Relation-shape convolutional neural network for point cloud analysis. In: Proceedings of the IEEE/CVF Conference on Computer Vision and Pattern Recognition, pp. 8895–8904 (2019)
19. Liu, Z., et al.: Swin transformer: hierarchical vision transformer using shifted windows. arXiv preprint arXiv:2103.14030 (2021)
20. Luong, M.T., Pham, H., Manning, C.D.: Effective approaches to attention-based neural machine translation. In: Proceedings of the 2015 Conference on Empirical Methods in Natural Language Processing, pp. 1412–1421 (2015)
21. Maturana, D., Scherer, S.: VoxNet: a 3D convolutional neural network for real-time object recognition. In: 2015 IEEE/RSJ International Conference on Intelligent Robots and Systems (IROS), pp. 922–928 (2015). https://doi.org/10.1109/IROS.2015.7353481
22. Qi, C.R., Su, H., Mo, K., Guibas, L.J.: PointNet: deep learning on point sets for 3D classification and segmentation. In: Proceedings of the IEEE Conference on Computer Vision and Pattern Recognition, pp. 652–660 (2017)
23. Qi, C.R., Yi, L., Su, H., Guibas, L.J.: Pointnet++: deep hierarchical feature learning on point sets in a metric space. In: 30th Proceedings Conference on Advances in Neural Information Processing Systems (2017)
24. Rao, R.M., et al.: MSA transformer. In: Meila, M., Zhang, T. (eds.) Proceedings of the 38th International Conference on Machine Learning. Proceedings of Machine Learning Research, vol. 139, pp. 8844–8856. PMLR (18–24 July 2021). https://proceedings.mlr.press/v139/rao21a.html
25. Roy, A.G., Navab, N., Wachinger, C.: Concurrent spatial and channel squeeze & excitation in fully convolutional networks. In: Frangi, A.F., Schnabel, J.A., Davatzikos, C., Alberola-López, C., Fichtinger, G. (eds.) MICCAI 2018. LNCS, vol. 11070, pp. 421–429. Springer, Cham (2018). https://doi.org/10.1007/978-3-030-00928-1_48
26. Simonovsky, M., Komodakis, N.: Dynamic edge-conditioned filters in convolutional neural networks on graphs. In: Proceedings of the IEEE Conference on Computer Vision and Pattern Recognition, pp. 3693–3702 (2017)
27. Somepalli, G., Goldblum, M., Schwarzschild, A., Bruss, C.B., Goldstein, T.: Saint: improved neural networks for tabular data via row attention and contrastive pre-training. arXiv preprint arXiv:2106.01342 (2021)
28. Su, H., et al.: SplatNet: sparse lattice networks for point cloud processing. In: Proceedings of the IEEE Conference on Computer Vision and Pattern Recognition, pp. 2530–2539 (2018)
29. Su, H., Maji, S., Kalogerakis, E., Learned-Miller, E.: Multi-view convolutional neural networks for 3D shape recognition. In: Proceedings of the IEEE International Conference on Computer Vision, pp. 945–953 (2015)
30. Thomas, H., Qi, C.R., Deschaud, J.E., Marcotegui, B., Goulette, F., Guibas, L.J.: KPConv: flexible and deformable convolution for point clouds. In: Proceedings of the IEEE/CVF International Conference on Computer Vision, pp. 6411–6420 (2019)
31. Vaswani, A., et al.: Attention is all you need. In: Advances in Neural Information Processing Systems, pp. 5998–6008 (2017)

32. Wang, C., Samari, B., Siddiqi, K.: Local spectral graph convolution for point set feature learning. In: Ferrari, V., Hebert, M., Sminchisescu, C., Weiss, Y. (eds.) ECCV 2018. LNCS, vol. 11208, pp. 56–71. Springer, Cham (2018). https://doi.org/10.1007/978-3-030-01225-0_4
33. Wang, Y., Sun, Y., Liu, Z., Sarma, S.E., Bronstein, M.M., Solomon, J.M.: Dynamic graph CNN for learning on point clouds. ACM Trans. Graphics **38**(5), 1–12 (2019)
34. Woo, S., Park, J., Lee, J.-Y., Kweon, I.S.: CBAM: convolutional block attention module. In: Ferrari, V., Hebert, M., Sminchisescu, C., Weiss, Y. (eds.) ECCV 2018. LNCS, vol. 11211, pp. 3–19. Springer, Cham (2018). https://doi.org/10.1007/978-3-030-01234-2_1
35. Wu, H., et al.: CVT: introducing convolutions to vision transformers. arXiv preprint arXiv:2103.15808 (2021)
36. Wu, W., Qi, Z., Fuxin, L.: PointConv: Deep convolutional networks on 3D point clouds. In: Proceedings of the IEEE/CVF Conference on Computer Vision and Pattern Recognition, pp. 9621–9630 (2019)
37. Wu, Z., et al.: 3D ShapeNets: a deep representation for volumetric shapes. In: Proceedings of the IEEE Conference on Computer Vision and Pattern Recognition, pp. 1912–1920 (2015)
38. Xu, Y., Fan, T., Xu, M., Zeng, L., Qiao, Yu.: SpiderCNN: deep learning on point sets with parameterized convolutional filters. In: Ferrari, V., Hebert, M., Sminchisescu, C., Weiss, Y. (eds.) ECCV 2018. LNCS, vol. 11212, pp. 90–105. Springer, Cham (2018). https://doi.org/10.1007/978-3-030-01237-3_6
39. Yan, X., Zheng, C., Li, Z., Wang, S., Cui, S.: PointASNL: robust point clouds processing using nonlocal neural networks with adaptive sampling. In: Proceedings of the IEEE/CVF Conference on Computer Vision and Pattern Recognition, pp. 5589–5598 (2020)
40. Yi, L., et al.: A scalable active framework for region annotation in 3D shape collections. ACM Trans. Graph. **35**(6) (2016). https://doi.org/10.1145/2980179.2980238
41. Yun, C., Bhojanapalli, S., Rawat, A.S., Reddi, S., Kumar, S.: Are transformers universal approximators of sequence-to-sequence functions? In: International Conference on Learning Representations (2019)
42. Zaheer, M., Kottur, S., Ravanbakhsh, S., Poczos, B., Salakhutdinov, R.R., Smola, A.J.: Deep sets. In: 30th Proceedings of Conference on Advances in Neural Information Processing Systems (2017)
43. Zhao, H., Jiang, L., Fu, C.W., Jia, J.: PointWeb: enhancing local neighborhood features for point cloud processing. In: 2019 IEEE/CVF Conference on Computer Vision and Pattern Recognition (CVPR), pp. 5560–5568 (2019). https://doi.org/10.1109/CVPR.2019.00571
44. Zhao, H., Jiang, L., Jia, J., Torr, P.H., Koltun, V.: Point transformer. In: Proceedings of the IEEE/CVF International Conference on Computer Vision, pp. 16259–16268 (2021)
45. Zhou, Y., Tuzel, O.: VoxelNet: end-to-end learning for point cloud based 3D object detection. In: Proceedings of the IEEE Conference on Computer Vision and Pattern Recognition, pp. 4490–4499 (2018)

Cross-Attention Transformer for Video Interpolation

Hannah Halin Kim(✉), Shuzhi Yu, Shuai Yuan, and Carlo Tomasi

Duke University, Durham, NC 27708, USA
{hannah,shuzhiyu,shuai,tomasi}@cs.duke.edu

Abstract. We propose TAIN (Transformers and Attention for video INterpolation), a residual neural network for video interpolation, which aims to interpolate an intermediate frame given two consecutive image frames around it. We first present a novel vision transformer module, named Cross-Similarity (CS), to globally aggregate input image features with similar appearance as those of the predicted interpolated frame. These CS features are then used to refine the interpolated prediction. To account for occlusions in the CS features, we propose an Image Attention (IA) module to allow the network to focus on CS features from one frame over those of the other. TAIN outperforms existing methods that do not require flow estimation and performs comparably to flow-based methods while being computationally efficient in terms of inference time on Vimeo90k, UCF101, and SNU-FILM benchmarks.

1 Introduction

Video interpolation [1–7] aims to generate new frames between consecutive image frames in a given video. This task has many practical applications, ranging from frame rate up-conversion [8] for human perception [9], video editing for object or color propagation [10], slow motion generation [11], and video compression [12].

Recent work on video interpolation starts by estimating optical flow in both temporal directions between the input frames. Some systems [3–5,13–15] use flow predictions output by an off-the-shelf pre-trained estimator such as FlowNet [16] or PWC-Net [17], while others estimate flow as part of their own pipeline [7,11, 18–24]. The resulting bi-directional flow vectors are interpolated to infer the flow between the input frames and the intermediate frame to be generated. These inferred flows are then used to warp the input images towards the new one.

While these flow-based methods achieve promising results, they also come with various issues as follows. First, they are computationally expensive and rely heavily on the quality of the flow estimates (see Fig. 1). Specifically, state-of-the art flow estimators [16,17,25] require to compute at least two four-dimensional cost volumes at all pixel positions. Second, they are known to suffer at occlusions, where flow is undefined; near motion boundaries, where flow is discontinuous; and in the presence of large motions [26,27]. For instance, RAFT [25] achieves an

Y. Zheng et al. (Eds.): ACCV 2022, LNCS 13848, pp. 325–342, 2023.
https://doi.org/10.1007/978-3-031-27066-6_23

	Network	Inference Time	PSNR
No Flow	EDSC	**19.37 ± 0.06**	34.84
	CAIN	25.21 ± 0.45	34.65
	TAIN (ours)	32.59 ± 0.82	**35.02**
Flow	DAIN	168.56 ± 0.33	34.71
	BMBC	333.24 ± 6.60	35.01
	ABME	**116.66 ± 1.07**	36.18
	VFIformer	476.06 ± 11.92	**36.50**

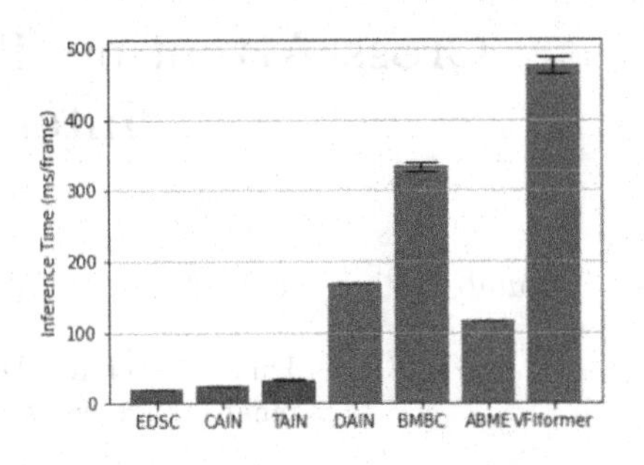

Fig. 1. Inference times (in milliseconds per frame prediction) of existing video interpolation methods and TAIN (ours) on a single P100 GPU. The inference times are computed as average and standard deviation over 300 inferences using two random images of size 256 × 256 as input. As a reference, we also list performance (PSNR) on Vimeo-90k [7]. Flow-based methods (red) have much higher inference times compared to TAIN (blue) and other non-flow-based methods (green). Bold values show the best performance in each panel, and underlined bold values show the best performance across both panels. Best viewed in color. (Color figure online)

End-Point Error (EPE) of 1.4 pixels on Sintel [28], but the EPE rises to 6.5 within 5 pixels from a motion boundary and to 4.7 in occlusion regions. Warping input images or features with these flow estimates will lead to poor predictions in these regions. This is especially damaging for video interpolation as motion boundaries and occlusions result from motion [26], and are therefore important regions to consider for motion compensation. Further, these flow estimators are by-and-large trained on synthetic datasets [28,29] to avoid the cost and difficulties of annotating real video [30], and sometimes fail to capture some of the challenges observed in real data, to which video interpolation is typically applied.

Apart from flow inputs, many of these methods also utilize additional inputs and networks to improve performance, which adds to their computational complexity. These include depth maps [5], occlusion maps [5,7,11,14,31], multiple input frames [32] for better flow estimation, image classifiers (*e.g.*, VGG [33], ResNet [34]) pretrained on ImageNet [35] for contextual features [1–3,11,31], adversarial networks for realistic estimations [31], and event cameras [36–39] to detect local brightness changes across frames. Instead, we achieve competitive performance with a single network that uses two consecutive input frames captured with commonly available devices.

The proposed TAIN system (Transformers and Attention for video INterpolation) is a residual neural network that requires no estimation of optical flow. Inspired by the recent success of vision transformers [40–43], we employ a novel transformer-based module named Cross Similarity (CS) in TAIN. Video interpolation requires the network to match corresponding points across frames, and our CS transformer achieves this through cross-attention. Vision Transformers typically use self-attention to correlate each feature of an image to every other feature *in the same image*. The resulting similarity maps are then used to collect relevant features from other image locations. Our CS module instead compares features *across frames*, namely, between an input frame and the current intermediate frame prediction. High values in this cross-frame similarity map indicate features with similar appearance. These maps are then used to aggregate features from the appropriate input image to yield CS features, and these are used in turn to refine the frame prediction. While there exists recent work on video interpolation that utilize vision transformers [22,32], our CS is specifically designed for video interpolation as it computes cross-frame similarity rather than within-frame self-similarity.

To account for occlusions or motion boundaries, we use CS scores in an Image Attention (IA) module based on spatial attention [44]. Given a feature in the tentative frame prediction, its maximum similarity score from our CS module will indicate whether or not a similar feature exists in either input frame. This maximum score will be high if such a feature exists and low if it does not. If the feature in the current frame prediction is occluded in one of the input images, the corresponding score will likely be low. Features at positions that straddle a motion boundary in the frame prediction often have a low maximum score as well. This is because boundary features contain information about pixels from both side of the boundary, and the particular mixture of information will change if the two sides move differently. The CS features aggregated from these low maximum similarity scores will not help in refining the current frame prediction, and our IA module learns to suppress them.

Thanks to CS transformer and IA module, TAIN improves or performs comparably to existing methods on various benchmarks, especially compared to those that do not require flow estimation. Our contributions are as follows:

- A novel Cross Similarity module based on vision transformer that aggregates features of the input image frames that have similar appearance to predicted-frame features. These aggregated features help refine the frame prediction.
- A novel Image Attention module that gives the predictor the ability to weigh features in one input frame over those in the other. This information is shown to be especially helpful near occlusions and motion boundaries.
- State-of-the-art performance on Vimeo-90k, UCF101, and SNU-FILM among methods that do not require optical flow estimation.

2 Related Work

2.1 Video Interpolation

Deep learning has rapidly improved the performance of video interpolation in recent work. Long *et al.* [45] are the first to use a CNN for video interpolation. In their system, an encoder-decoder network predicts the intermediate frame directly from two image frames using inter-frame correspondences. Subsequent work can be categorized largely into kernel-, flow-, and attention-based.

Kernel-Based approaches compute the new frame with convolutions over local patches, and use CNNs to estimate spatially-adaptive convolutional kernels [2,6,46]. These methods use large kernel sizes to accommodate large motion, and large amounts of memory are required as a result when frames have high resolution. Niklaus *et. al* propose to use separable convolutional kernels to reduce memory requirements and further improve results. Since kernel-based methods cannot handle motion larger than the pre-defined kernel size, EDSC [2] estimates not only adaptive kernels, but also offsets, masks, and biases to retrieve information from non-local neighborhoods. EDSC achieves the current state-of-the-art performance among methods that do not require optical flow.

Flow-Based approaches to video interpolation rely on bi-directional flow estimates using off-the-shelf pre-trained flow estimators (e.g. FlowNetS, PWC-Net) [3–5,13–15], or estimate flow as part of their pipeline [7,11,18–24]. Most of these methods assume linear motion between frames and use the estimated flow to warp input images and their features to the intermediate time step for prediction. In order to avoid making this linear motion assumption, Gui *et al.* [18] do not predict flow between two input frames but instead attempt to produce flow directly between the intermediate frame being predicted and the two input frames. Park *et al.* [23] do compute a tentative intermediate frame from the given flow estimates, but they re-estimate flow between the tentative predicted frame and the input images. These new estimates are then used for a final estimation of the intermediate frame. Multiple optical flow maps have also been used to account for complex motion patterns and mitigate the resulting prediction artifacts [15,21]. While these flow-based methods show promising results, they are computationally expensive. We do not require flow estimates in our approach.

Some approaches [5,14,31] integrate **both flow-based and kernel-based methods** by combining optical flow warping with learned adaptive local kernels. These methods perform robustly in the presence of large motions and are not limited by the assumption of a fixed motion range. They use small kernels that require less memory but are still expensive.

Recently, work by Choi *et al.* [1] proposes a residual network called CAIN that interpolates video through **attention mechanism** [44] without explicit computation of kernels or optical flow to curb model complexity and computational cost. The main idea behind their design is to distribute the information in a feature map into multiple channels through PixelShuffle, and extract motion information by processing the channels through a Channel Attention module. In our work, we extend CAIN with a novel vision transformer module and spatial attention module, still without requiring flow estimates or adaptive kernels.

2.2 Vision Transformers

Transformers [42,43] have shown success in both computer vision [40,41,47] and natural language processing [42,48] thanks to their ability to model long-range dependencies. Self-attention [42,48] has shown the most success among various modules in the transformer architecture, and derives query, key, and value vectors from the same image. Due to their content-adaptive nature, transformers have also been applied to video interpolation. Shi *et. al* [32] consider four frame inputs and propose to use a self-attention transformer based on SWIN [49] to capture long-range dependencies across both space and time. Lu *et. al* [22] propose a self-attention transformer along with a flow estimator to model long-range pixel correlation for video interpolation. Their work currently achieves the state-of-the-art performance on various video interpolation benchmarks. Different from existing work, we do not use self-attention in our work, but instead use cross-attention. In self-attention, query, key, and value are different projections of the same feature. We use cross-attention, where query and key are the same projections (shared weights) of different features. Specifically, our transformer module selects and refines features based on the similarities between the interpolated frame (query) and the two input frames (keys) while handling the occlusions that occur in the two input frames. On the other hand, existing transformer-based methods compare the features of the input frames without any explicit consideration of the interpolated frame or occlusions.

3 Method

The proposed TAIN method for video interpolation aims to predict frame $I_t \in \mathbb{R}^{h \times w \times 3}$ at time $t = 0.5$, given two consecutive images $I_0, I_1 \in \mathbb{R}^{h \times w \times 3}$ at times 0 and 1. We do not require any computation of flow, adaptive convolution kernel parameters, or warping, but instead utilize cross-similarity transformer and spatial attention mechanism. Specifically, a novel vision transformer module called the Cross Similarity (CS) module globally aggregates features from input images I_0 and I_1 that are similar in appearance to those in the current prediction $\hat{I}_t$ of frame I_t (Sect. 3.2). These aggregated features are then used to refine the prediction $\hat{I}_t$, and the output from each residual group is a new refinement. To account for occlusions of the interpolated features in the aggregated CS features, we propose an Image Attention (IA) module to enable the network to prefer CS features from one frame over those of the other (Sect. 3.3). See Fig. 2 for an overview of the network. Before describing our network we summarize CAIN [1], on which our work improves.

3.1 CAIN

CAIN [1] is one of the top performers for video interpolation and does not require estimation of flow, adaptive convolution kernels, or warping. Instead, CAIN utilizes PixelShuffle [50] and a channel attention module. PixelShuffle [50] rearranges the layout of an image or a feature map without any loss of information. To down-shuffle, activation values are merely rearranged by reducing each of the two spatial dimensions by a factor of s and increasing the channel dimension by a factor of s^2. Up-shuffling refers to the inverse operation. This parameter-free

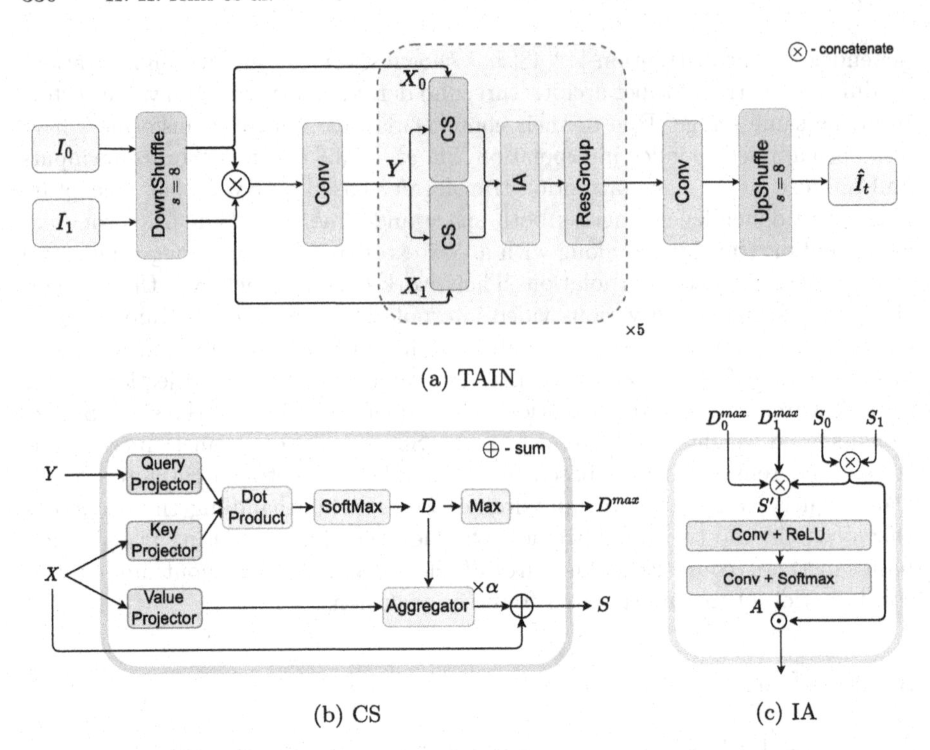

Fig. 2. (a) Overview of the proposed TAIN network for video interpolation. The two consecutive frames I_0 and I_1 are down-shuffled and concatenated along the channel dimension. They are then processed with five residual groups (ResGroups) with a (b) CS transformer and an (c) IA module before being up-shuffled back to the original resolution to yield the final prediction $\hat{I}_t$. The output from each ResGroup is a refinement of the output from the previous block.

operation allows CAIN to increase the receptive field size of the network's convolutional layers without losing any information. CAIN first down-shuffles ($s = 8$) the two input images and concatenates them along the channel dimension before feeding them to a network with five ResGroups (groups of residual blocks). With the increased number of channels, each of the blocks includes a channel attention module that learns to pay attention to certain channels to gather motion information. The size of the features remains $h/8 \times w/8 \times 192$ throughout CAIN, and the final output map is up-shuffled back to the original resolution of $h \times w \times 3$.

3.2 Cross Similarity (CS) Module

All points in the predicted frame I_t appear either in I_0 or I_1 or in both, except in rare cases where a point appears for a very short time between the two consecutive time frames. These ephemeral apparitions cannot be inferred from I_0 or I_1 and are ignored here. For the remaining points, we want to find features of I_0

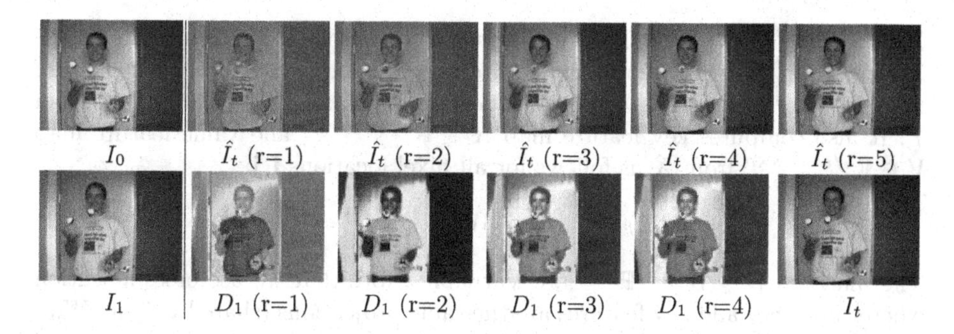

Fig. 3. Visualization of intermediate predictions and their cross similarity maps D_1 with I_1 across all five ResGroups ($r = 1 \dots 5$) using an example from Middlebury [51] dataset. The first column shows the two input frames, I_0 (top) and I_1 (bottom). The next four columns show the predicted intermediate frame $\hat{I}_t$ after each of the first four residual blocks with a query point highlighted in blue (top), and its corresponding similarity map D_1 from our CS module with the point of highest similarity highlighted in red (bottom). The values in the similarity maps D_1 show large scores (closer to white) whenever query and key features are similar in appearance. The last column shows the final output from TAIN (top), which is not used in any CS module, and the ground-truth intermediate frame I_t (bottom) as a reference. (Color figure online)

or I_1 that are similar in appearance to the features in the predicted intermediate frame $\hat{I}_t$, and use them to refine $\hat{I}_t$. We use a transformer to achieve this, where we compare each feature from $\hat{I}_t$ with features from I_0 and I_1, and use similar features to refine $\hat{I}_t$.

While transformers typically use self-similarity [42,48], wherein query, key, and value are based on similarities within the same feature array, we extend this notion through the concept of cross-image similarity. Specifically, we use features from the input images I_0 or I_1 (first column in Fig. 3) as keys and values, and features from the current frame prediction $\hat{I}_t$ (the remaining images in the first row of Fig. 3), as queries. Given a query feature normalized to have unit Euclidean norm (*e.g.*, features for the blue points on the ball in the first row of Fig. 3), our CS module compares it to all the similarly-normalized key features through a dot product, which are then Softmax-normalized to obtain the corresponding similarity maps D (*e.g.*, images in the second row of Fig. 3). The similarity matrix D is used to find the location of the largest similarity score (*e.g.*, red dots in the second row of Fig. 3), where we retrieve our value features (equation (3) later on), which are projections of the input image features, to yield aggregated input features S based on similarity. These aggregated features are then used to refine the intermediate prediction of $\hat{I}_t$.

Mathematical Formulation. Let $X \in \mathbb{R}^{h/s \times w/s \times d}$ denote the (down-shuffled) feature map from one of the input images $I_0, I_1 \in \mathbb{R}^{h \times w \times 3}$ and let

$Y \in \mathbb{R}^{h/s \times w/s \times d}$ denote the feature map of the predicted intermediate frame $\hat{I}_t \in \mathbb{R}^{h \times w \times 3}$. Let $M^{\mathbf{i}}$ represent feature at pixel location $\mathbf{i}$ in a given feature map M. Our CS module computes query feature map $Q \in \mathbb{R}^{h/s \times w/s \times d}$ from Y. It also computes key feature map $K \in \mathbb{R}^{h/s \times w/s \times d}$ and value feature map $V \in \mathbb{R}^{h/s \times w/s \times d}$ from X as follows for all pixel locations $\mathbf{i}, \mathbf{j}$:

$$Q^{\mathbf{i}} = W_{qk} Y^{\mathbf{i}}, \quad K^{\mathbf{j}} = W_{qk} X^{\mathbf{j}}, \quad V^{\mathbf{j}} = W_v X^{\mathbf{j}} . \tag{1}$$

The matrices $W_{qk}, W_v \in \mathbb{R}^{d \times d}$ are learnable. Note that we use cross-attention where the query and key features are the same projections (shared weights W_{qk}) of different features. This is different from the self-attention transformers used in existing work [22,32], where the query, key, and value features are different projections (no shared weights) of the same features.

Each query feature $Q^{\mathbf{i}}$ is compared with all the key features $K^{\mathbf{j}}$ to compute a similarity matrix $D^{\mathbf{i}} \in \mathbb{R}^{h/s \times w/s}$ that captures their similarity:

$$D^{\mathbf{i}} = sim(Q^{\mathbf{i}}, K^{\mathbf{j}}) = \frac{\exp({Q^{\mathbf{i}}}^T K^{\mathbf{j}} / \sqrt{d})}{\sum_{\mathbf{j}} \exp({Q^{\mathbf{i}}}^T K^{\mathbf{j}} / \sqrt{d})} . \tag{2}$$

Using this similarity matrix $D^{\mathbf{i}}$, the CS module finally computes the aggregated similarity feature $S^{\mathbf{i}} \in \mathbb{R}^{h/s \times w/s \times d}$ by taking the value feature $V^{\mathbf{i}_{\max}}$ at location $\mathbf{i}_{\max}$ corresponding to the maximum similarity score for each query feature $Q^{\mathbf{i}}$ and adding the result to $Y^{\mathbf{i}}$

$$S^{\mathbf{i}} = Y^{\mathbf{i}} + \alpha V^{\mathbf{i}_{\max}} , \tag{3}$$

where α is a learnable scalar parameter initialized to zero.

3.3 Image Attention (IA) Module

We propose an Image Attention (IA) module based on spatial attention [44] to weigh our two CS features S_0 and S_1 computed from the input frames as shown above. The IA module enables TAIN to prioritize or suppress features from one input image over those from the other for a given spatial location. This is useful especially at occlusions, where a feature from I_t appears in one input image (likely yielding high similarity scores) but not in the other (likely yielding low similarity scores). This also helps on motion boundaries, where features encode information from both sides of the boundary. As the two sides move relative to each other, the specific mixture of features changes. The IA module compares the two CS feature maps S_0 and S_1 to construct two IA weight maps $A_0, A_1 \in [0,1]^{h/s \times w/s \times 1}$ where $A_0 + A_1 = 1^{h/s \times w/s \times 1}$ as shown below. These weight maps are multiplied with the corresponding CS maps S_0 and S_1 before they are concatenated and fed to the next ResGroup. See Fig. 4.

Mathematical Formulation. Let $S' \in \mathbb{R}^{h/s \times w/s \times 2(d+1)}$ be the concatenation of S_0, S_1, and the two maximum similarity maps D_0 and D_1 along the

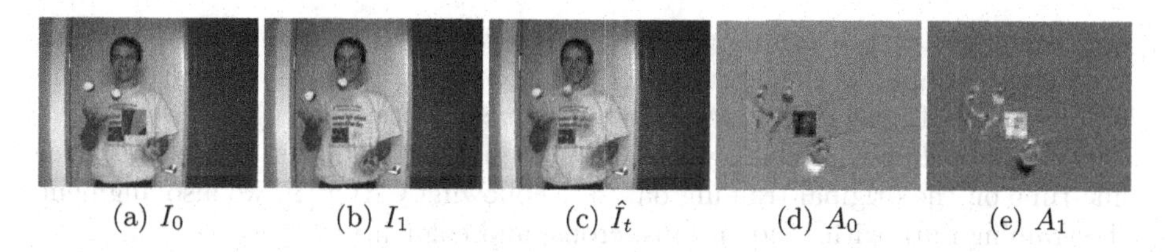

(a) I_0 (b) I_1 (c) $\hat{I}_t$ (d) A_0 (e) A_1

Fig. 4. Visualization of Image Attention maps (d) A_0 and (e) A_1 using an example from Middlebury [51] (White is large and black is small). For visualization purposes only, we add a square patch to (a) I_0 on the person's chest, and show that the IA module assigns a higher weight to the CS features from I_1 on the person's chest in (c) $\hat{I}_t$ than to those from I_0, which the patch occludes. (Color figure online)

channel dimension. Our IA module first computes image attention weights $A \in [0,1]^{h/s \times w/s \times 2}$ by applying two 1×1 convolutional layers with ReLU and Softmax on S' as shown in Fig. 2c. More formally, we compute A as:

$$A = \sigma(W_2 * (\rho(W_1 * S'))) \tag{4}$$

where $\sigma(\cdot)$ denotes the Softmax function, $\rho(\cdot)$ denotes the ReLU function, and W_1 and W_2 are the weights of the two 1×1 convolution layers. This image attention weight map A can be seen as a concatenation of maps $A_0, A_1 \in [0,1]^{h/s \times w/s \times 1}$ where $A_0 + A_1 = 1$ which are used to weigh the aggregated similarity features S_0 and S_1 to obtain the weighted features $\tilde{S_0}$ and $\tilde{S_1}$:

$$\tilde{S_0} = A_0 \odot S_0 \quad \text{and} \quad \tilde{S_1} = A_1 \odot S_1 , \tag{5}$$

where $\odot$ is the element-wise product. Note that this is different from the channel attention module from CAIN [1]. The channel attention module computes a $(2d)$-dimensional vector weight over the channel dimension, while the IA module computes a $h/s \times w/s$ weight over the spatial dimension.

3.4 TAIN Architecture and Training Details

Figure 2a shows an overview of the TAIN network for video interpolation. TAIN extends CAIN by applying the proposed CS transformer and IA module after each of the intermediate ResGroups. Each CS transformer obtains query features Q from the features Y from the previous ResGroup, and key and value features from one of the two input image features X_0 or X_1. We remove the residual connections around each ResGroup so that the output from each ResGroup is a new refinement of the output from the previous ResGroup. Figure 3 shows sample predictions after each of the five ResGroups that the query features are based on. With the IA module, the CS features S_0 and S_1 are weighted based on their maximum similarity scores from D_0 and D_1. The two weighted CS features, $\tilde{S_0}$ and $\tilde{S_1}$, are then used to refine the prediction in the next ResGroup.

In order to train the network for cases of occlusions and large motion, we add synthetic moving occlusion patches to the training data. Specifically, we first randomly crop a square patch of size between 21×21 and 61×61 from a different sample in the training set. This cropped patch is pasted onto the input and label images and translated in a linear motion across the frames. We apply this occluder patch augmentation to first pre-train on TAIN and then fine-tune on the original training dataset. Following CAIN [1], we also augment the training data with random flips, crops, and color jitter.

Following the literature [1,3,6,7,11,31], we train our network using the L_1 loss on the difference between the predicted $\hat{I}_t$ and true I_t intermediate frames: $\|\hat{I}_t - I_t\|_1$. As commonly done in the flow literature [52,53], we also include an L_1 loss on the difference between the gradients of the predicted and true intermediate frames: $\|\nabla\hat{I}_t - \nabla I_t\|_1$. We use the weighted sum of the two losses as our final training loss: $\mathcal{L} = \|\hat{I}_t - I_t\|_1 + \gamma\|\nabla\hat{I}_t - \nabla I_t\|_1$, where $\gamma = 0.1$.

Another common loss used in the literature [1–3,11,31] is perceptual loss: $\|\phi\left(\hat{I}_t\right) - \phi\left(I_t\right)\|_2^2$, where $\phi\left(\cdot\right)$ is a feature from a ImageNet pretrained VGG-19. As this loss depends on another network, adding to the computational complexity, we do not use this loss and still show performance improvements.

We implement TAIN in PyTorch [54][1] and train our network with a learning rate of 10^{-4} through Adam [55] optimizer.

4 Datasets and Performance Metrics

As customary [1], we train our model on Vimeo90K [7], and evaluate it on four benchmark datasets for video interpolation, *i.e.*, Vimeo90K [7], UCF101 [56], SNU-FILM [1], and Middlebury [51]. Vimeo90K [7] consists of 51,312 triplets with a resolution of 256×448 partitioned into a training set and a testing set. UCF101 [56] contains human action videos of resolution 256×256. For the evaluation of video interpolation, Liu *et al.* [19] constructed a test set by selecting 379 triplets from UCF101. The SNU Frame Interpolation with Large Motion (SNU-FILM) [1] dataset contains videos with a wide range of motion sizes for evaluation of video interpolation methods. The dataset is stratified into four settings, Easy, Medium, Hard, and Extreme, based on the temporal gap between the frames. Middlebury [51] includes 12 sequences of images for evaluation. The images are combination of synthetic and real images that are often used as evaluation for video interpolation.

Following the literature, we use Peak Signal-to-Noise Ratio (PSNR) and Structural Similarity Index (SSIM) [57] to measure performance. For the Middlebury dataset [51], we use Interpolation Error (IE).

5 Results

We first compare TAIN with the current state-of-the-art methods on video interpolation with two frame inputs in Table 1. Top panel of Table 1 lists kernel-based

[1] Code is available at https://github.com/hannahhalin/TAIN.

Table 1. Comparison to the existing methods across Vimeo90k, UCF101, SNU, and Middlebury (M.B.) datasets. Top panel shows kernel-based methods (K), second panel shows attention-based methods (Att.), third panel shows methods based on both kernels and flow (K+F), and bottom panel shows flow-based methods. Higher is better for PSNR and SSIM, and lower is better for IE. Bold values show the best performance in each panel, and underlined values show the best performance across all panels. TAIN (ours) outperforms existing methods based on kernel and attention, and performs comparably to those based on flow on Vimeo90k, UCF101, and SNU datasets.

	Method	Vimeo90k		UCF101		SNU-easy		SNU-extreme		M.B.
		PSNR	SSIM	PSNR	SSIM	PSNR	SSIM	PSNR	SSIM	IE
K	SepConv	33.79	97.02	34.78	96.69	39.41	99.00	24.31	**84.48**	2.27
	EDSC	**34.84**	**97.47**	**35.13**	**96.84**	**40.01**	**99.04**	**24.39**	84.26	**2.02**
Att.	CAIN	34.65	97.29	34.91	96.88	39.89	99.00	24.78	85.07	**2.28**
	TAIN (Ours)	**35.02**	**97.51**	**35.21**	**96.92**	**40.21**	**99.05**	**24.80**	**85.25**	2.35
K+F	AdaCoF	34.47	97.30	34.90	96.80	**39.80**	99.00	24.31	84.39	2.24
	MEMC	34.29	97.39	34.96	96.82	–	–	–	–	2.12
	DAIN	**34.71**	**97.56**	**34.99**	**96.83**	39.73	**99.02**	**25.09**	**85.84**	**2.04**
Flow	TOFlow	33.73	96.82	34.58	96.67	39.08	98.90	23.39	83.10	2.15
	CyclicGen	32.09	94.90	35.11	96.84	37.72	98.40	22.70	80.83	–
	BMBC	35.01	97.64	35.15	96.89	39.90	99.03	23.92	84.33	2.04
	ABME	36.18	98.05	35.38	96.98	39.59	99.01	25.42	86.39	2.01
	VFIformer	**36.50**	**98.16**	**35.43**	**97.00**	**40.13**	**99.07**	**25.43**	**86.43**	**1.82**

methods, *i.e.*, SepConv [2] and EDSC [2], second panel lists attention-based methods, *i.e.*, CAIN [1] and TAIN (ours), third panel lists methods based on both kernels and flow estimations, *i.e.*, AdaCoF [31], MEMC [14], and DAIN [5], and bottom panel lists flow-based methods, *i.e.*, TOFlow [7], CyclicGen [58], BMBC [4], ABME [23], and VFIformer [22]. Bold values show the highest performance in each panel while underlined values show the highest performance across all panels. TAIN outperforms existing kernel-based methods across all benchmarks. Comparing to the flow-based methods, TAIN outperforms all listed methods except for ABME and VFIformer. However, as shown in Fig. 1, the inference times of AMBE and VFIformer are about 4 and 15 times longer, respectively, than those of TAIN. DAIN and BMBC take around 5 times longer than TAIN to inference while performing comparably to TAIN.

Figure 5 visualizes examples of predictions from TAIN and the best performing methods in each panel of Table 1. Compared with kernel- and attention-based methods, TAIN is able to find the correct location of the moving object, *e.g.* shovel and hula hoop in red box, while keeping the fine details, *e.g.* shaft of the shovel, letters on the plane, and the hula hoop.

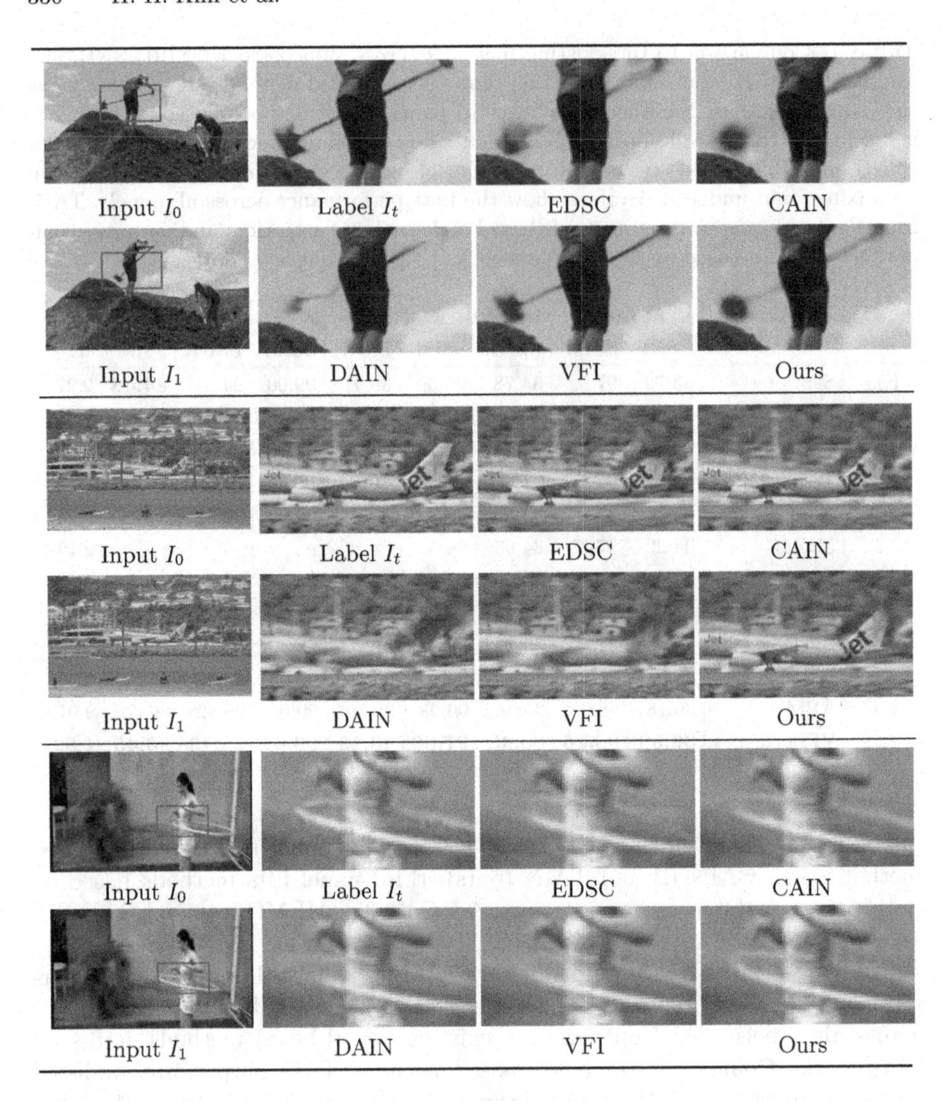

Fig. 5. Visualization of our proposed method and its comparison to the current state-of-the-art methods [1,2,5,22] on examples from Vimeo90k and UCF101. (Color figure online)

Inference Time. Figure 1 compares the inference time of TAIN and other state-of-the-art methods listed in Table 1. To measure time, we create two random images of size 256×256, the same size as those of UCF101 dataset, and evaluate it 300 times using a P100 GPU and report their average and standard deviation. As mentioned above, TAIN achieves competitive performance as the existing flow-based methods while taking a fraction of time for inference.

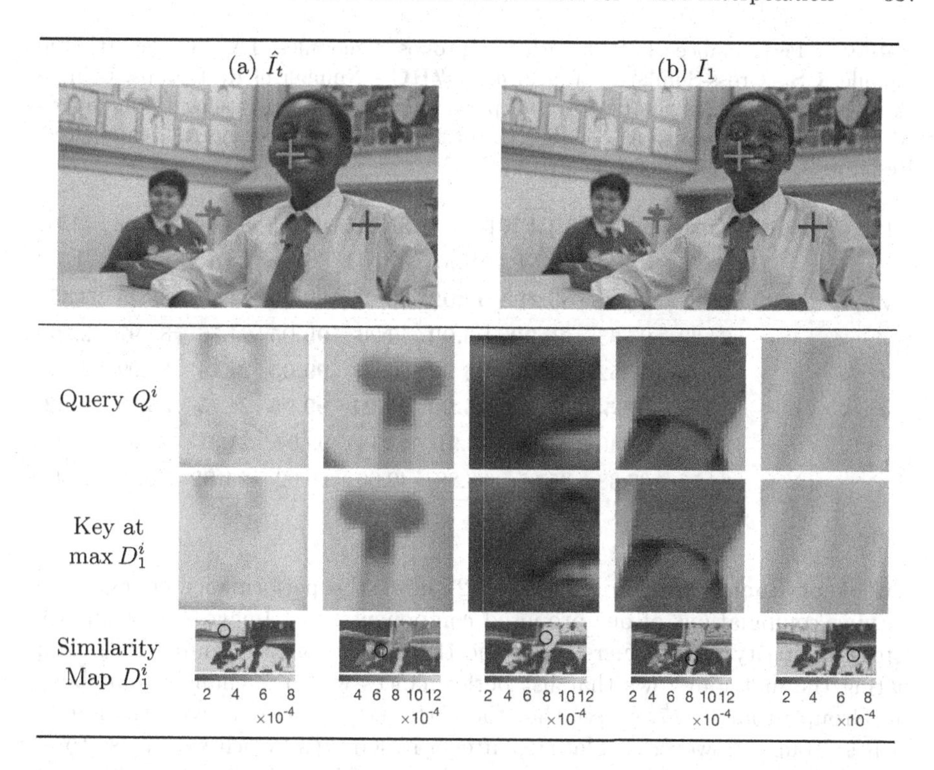

Fig. 6. Visualization of patches with the highest similarity scores from D in our proposed CS transformer. Top panel shows an example of (a) $\hat{I}_t$ and (b) I_1 from the test set of Vimeo90k [7] dataset. Example query features in $\hat{I}_t$ are shown with '+' mark, and their key feature with the highest similarity score are shown in I_1 in corresponding colors. Bottom panel shows 31×31 patches extracted from the query points ('+' in $\hat{I}_t$), their corresponding key patches with the highest similarity score ('+' in I_1), and their corresponding similarity map D_1^i with the highest score circled. Our CS module successfully aggregates similar appearance features when refining the interpolation prediction $\hat{I}_t$.

6 Ablation Study

Visualization of the Components of CS Module. We visualize the components of our proposed CS module using an example from the test set of Vimeo90k [7] dataset in Fig. 6. The top panel of this figure shows (a) $\hat{I}_t$ and (b) I_1, where example query features Q^i are shown with '+' mark in (a) and their corresponding key features with the highest similarity score are shown with '+' mark in (b) with the corresponding colors. Bottom panel shows 31×31 patches extracted from the location '+' of each query Q^i and key K_1^i features from (a) $\hat{I}_t$ and (b) I_1, respectively. We also include visualization of the corresponding similarity maps D_1^i and highlight their maximum score. As shown, our CS module successfully extracts similar appearance features when refining $\hat{I}_t$.

Table 2. Performance changes with our proposed modules: **IA** - Image Attention module; **CS** - Cross-Similarity transformer; **#RG** - Number of ResGroups. Using all the components yields the best overall performance. While increasing the number of ResGroups yields consistently higher performance, the performance plateaus after 5 ResGroups, which we use for TAIN (top row).

IA	CS	#RG	Vimeo90k		UCF101		SNU-easy		SNU-extreme		M.B
			PSNR	SSIM	PSNR	SSIM	PSNR	SSIM	PSNR	SSIM	IE
✓	✓	5	35.02	97.51	35.21	**96.92**	**40.21**	**99.05**	24.80	85.25	2.35
✓	✓	4	34.82	97.38	**35.22**	**96.92**	40.20	**99.05**	24.76	84.97	2.39
✓	✓	6	**35.06**	97.52	**35.22**	**96.92**	40.20	**99.05**	24.74	85.09	2.33
✓	✓	7	**35.06**	**97.53**	**35.22**	**96.92**	**40.21**	**99.05**	24.75	85.09	**2.32**
	✓	5	34.81	97.40	35.17	96.91	40.14	99.04	24.81	85.21	2.49
		5	34.76	97.38	35.05	96.88	40.00	99.02	**24.82**	**85.28**	2.66

Effect of Each Component. Table 2 shows the performance changes with varying combinations of our proposed components, *i.e.*, Image Attention (IA), Cross-Similarity transformers (CS), and the number of ResGroups (RG). Comparing the first four rows that list performances with the changing number of ResGroups from 4 to 7, we see that the performance increases with the number of ResGroups. However, it plateaus after 5 ResGroups, which we choose to use for TAIN (top row) for computational efficiency. In addition, each of the two main components, IA and CS module, contributes to the success of our method.

7 Conclusion

We propose TAIN, an extension of the CAIN network, for video interpolation. We utilize a novel vision transformer we call Cross Similarity module to aggregate input image features that are similar in appearance to those in the predicted frame to further refine the prediction. To account for occlusions in these aggregated features, we propose a spatial attention module we call Image Attention to suppress any features from occlusions. Combining both these components, TAIN outperforms existing methods that do not require flow estimation on multiple benchmarks. Compared to methods that utilize flow, TAIN performs comparably while taking a fraction of their time for inference.

Acknowledgments. This research is based upon work supported in part by the National Science Foundation under Grant No. 1909821 and by an Amazon AWS cloud computing award. Any opinions, findings, and conclusions or recommendations expressed in this material are those of the authors and do not necessarily reflect the views of the National Science Foundation.

References

1. Choi, M., Kim, H., Han, B., Xu, N., Lee, K.M.: Channel attention is all you need for video frame interpolation. In: AAAI (2020)
2. Niklaus, S., Mai, L., Liu, F.: Video frame interpolation via adaptive separable convolution. In: IEEE International Conference on Computer Vision (2017)
3. Niklaus, S., Liu, F.: Softmax splatting for video frame interpolation. In: IEEE Conference on Computer Vision and Pattern Recognition (2020)
4. Park, J., Ko, K., Lee, C., Kim, C.S.: Bmbc: bilateral motion estimation with bilateral cost volume for video interpolation. In: European Conference on Computer Vision (2020)
5. Bao, W., Lai, W.S., Ma, C., Zhang, X., Gao, Z., Yang, M.H.: Depth-aware video frame interpolation. In: IEEE Conference on Computer Vision and Pattern Recognition (2019)
6. Niklaus, S., Mai, L., Wang, O.: Revisiting adaptive convolutions for video frame interpolation. In: IEEE Winter Conference on Applications of Computer Vision (2021)
7. Xue, T., Chen, B., Wu, J., Wei, D., Freeman, W.T.: Video enhancement with task-oriented flow. Int. J. Comput. Vis. (IJCV) **127**, 1106–1125 (2019)
8. Bao, W., Zhang, X., Chen, L., Ding, L., Gao, Z.: High order model and dynamic filtering for frame rate up conversion. IEEE Trans. Image Process. **27**(8), 3813–3826 (2018)
9. Kuroki, Y., Nishi, T., Kobayashi, S., Oyaizu, H., Yoshimura, S.: A psychophysical study of improvements in motion-image quality by using high frame rate. J. Soc. Inf. Display **15**(1), 1–68 (2007)
10. Meyer, S., Cornillère, V., Djelouah, A., Schroers, C., Gross, M.H.: Deep video color propagation. In: BMVC (2018)
11. Jiang, H., Sun, D., Jampani, V., Yang, M.H., Learned-Miller, E., Kautz, J.: Super slomo: high quality estimation of multiple intermediate frames for video interpolation. In: 2018 IEEE/CVF Conference on Computer Vision and Pattern Recognition, pp. 9000–9008 (2018)
12. Wu, C., Singhal, N., Krähenbühl, P.: Video compression through image interpolation. In: European Conference on Computer Vision (ECCV) (2018)
13. Niklaus, S., Liu, F.: Context-aware synthesis for video frame interpolation. In: Proceedings of the IEEE Conference on Computer Vision and Pattern Recognition (CVPR) (2018)
14. Bao, W., Lai, W.S., Zhang, X., Gao, Z., Yang, M.H.: Memc-net: motion estimation and motion compensation driven neural network for video interpolation and enhancement. IEEE Trans. Pattern Anal. Mach. Intell. **43**(3), 933–948 (2018)
15. Hu, P., Niklaus, S., Sclaroff, S., Saenko, K.: Many-to-many splatting for efficient video frame interpolation. In: Proceedings of the IEEE/CVF Conference on Computer Vision and Pattern Recognition (CVPR), pp. 3553–3562 (2022)
16. Dosovitskiy, A., et al.: Flownet: learning optical flow with convolutional networks. In: Proceedings of the IEEE International Conference on Computer Vision (ICCV) (2015)
17. Sun, D., Yang, X., Liu, M.Y., Kautz, J.: Pwc-net: cnns for optical flow using pyramid, warping, and cost volume. In: Conference on Computer Vision and Pattern Recognition (2018)
18. Gui, S., Wang, C., Chen, Q., Tao, D.: Featureflow: robust video interpolation via structure-to-texture generation. In: Proceedings of the IEEE/CVF Conference on Computer Vision and Pattern Recognition (CVPR) (2020)

19. Liu, Z., Yeh, R., Tang, X., Liu, Y., Agarwala, A.: Video frame synthesis using deep voxel flow. In: Proceedings of International Conference on Computer Vision (ICCV) (2017)
20. Xiang, X., Tian, Y., Zhang, Y., Fu, Y., Allebach, J.P., Xu, C.: Zooming slow-mo: fast and accurate one-stage space-time video super-resolution. In: Proceedings of the IEEE/CVF Conference on Computer Vision and Pattern Recognition (CVPR) (2020)
21. Danier, D., Zhang, F., Bull, D.: St-mfnet: a spatio-temporal multi-flow network for frame interpolation. In: Proceedings of the IEEE/CVF Conference on Computer Vision and Pattern Recognition (CVPR), pp. 3521–3531 (2022)
22. Lu, L., Wu, R., Lin, H., Lu, J., Jia, J.: Video frame interpolation with transformer. In: Proceedings of the IEEE/CVF Conference on Computer Vision and Pattern Recognition (CVPR), pp. 3532–3542 (2022)
23. Park, J., Lee, C., Kim, C.S.: Asymmetric bilateral motion estimation for video frame interpolation. In: International Conference on Computer Vision (2021)
24. Choi, M., Lee, S., Kim, H., Lee, K.M.: Motion-aware dynamic architecture for efficient frame interpolation. In: Proceedings of the IEEE/CVF International Conference on Computer Vision (ICCV), pp. 13839–13848 (2021)
25. Teed, Z., Deng, J.: RAFT: recurrent all-pairs field transforms for optical flow. In: Vedaldi, A., Bischof, H., Brox, T., Frahm, J.-M. (eds.) ECCV 2020. LNCS, vol. 12347, pp. 402–419. Springer, Cham (2020). https://doi.org/10.1007/978-3-030-58536-5_24
26. Kim, H.H., Yu, S., Tomasi, C.: Joint detection of motion boundaries and occlusions. In: British Machine Vision Conference (BMVC) (2021)
27. Yu, S., Kim, H.H., Yuan, S., Tomasi, C.: Unsupervised flow refinement near motion boundaries. In: British Machine Vision Conference (BMVC) (2022)
28. Butler, D.J., Wulff, J., Stanley, G.B., Black, M.J.: A naturalistic open source movie for optical flow evaluation. In: Fitzgibbon, A., Lazebnik, S., Perona, P., Sato, Y., Schmid, C. (eds.) ECCV 2012. LNCS, vol. 7577, pp. 611–625. Springer, Heidelberg (2012). https://doi.org/10.1007/978-3-642-33783-3_44
29. Mayer, N., et al.: A large dataset to train convolutional networks for disparity, optical flow, and scene flow estimation. In: IEEE International Conference on Computer Vision and Pattern Recognition (2016) arXiv:1512.02134
30. Yuan, S., Sun, X., Kim, H., Yu, S., Tomasi, C.: Optical flow training under limited label budget via active learning. In: European Conference on Computer Vision (ECCV) (2022)
31. Lee, H., Kim, T., Chung, T.Y., Pak, D., Ban, Y., Lee, S.: Adacof: adaptive collaboration of flows for video frame interpolation. In: Proceedings of the IEEE/CVF Conference on Computer Vision and Pattern Recognition (CVPR) (2020)
32. Shi, Z., Xu, X., Liu, X., Chen, J., Yang, M.H.: Video frame interpolation transformer. In: Proceedings of the IEEE/CVF Conference on Computer Vision and Pattern Recognition (CVPR), pp. 17482–17491 (2022)
33. Simonyan, K., Zisserman, A.: Very deep convolutional networks for large-scale image recognition (2014)
34. He, K., Zhang, X., Ren, S., Sun, J.: Deep residual learning for image recognition. In: 2016 IEEE Conference on Computer Vision and Pattern Recognition (CVPR), pp. 770–778 (2016)
35. Deng, J., Dong, W., Socher, R., Li, L.J., Li, K., Fei-Fei, L.: ImageNet: a large-scale hierarchical image database. In: 2009 IEEE Conference on Computer Vision and Pattern Recognition, pp. 248–255 (2009)

36. Yang, M., Liu, S.C., Delbruck, T.: A dynamic vision sensor with 1% temporal contrast sensitivity and in-pixel asynchronous delta modulator for event encoding. IEEE J. Solid-State Circuits **50**, 2149–2160 (2015)
37. Tulyakov, S., et al.: Time lens: event-based video frame interpolation. In: Proceedings of the IEEE/CVF Conference on Computer Vision and Pattern Recognition, pp. 16155–16164 (2021)
38. Zhang, X., Yu, L.: Unifying motion deblurring and frame interpolation with events. In: Proceedings of the IEEE/CVF Conference on Computer Vision and Pattern Recognition (CVPR), pp. 17765–17774 (2022)
39. Tulyakov, S., Bochicchio, A., Gehrig, D., Georgoulis, S., Li, Y., Scaramuzza, D.: Time lens++: event-based frame interpolation with parametric non-linear flow and multi-scale fusion. In: Proceedings of the IEEE/CVF Conference on Computer Vision and Pattern Recognition (CVPR), pp. 17755–17764 (2022)
40. Dosovitskiy, A., et al.: An image is worth 16×16 words: transformers for image recognition at scale (2021)
41. Huang, Z., et al.: Ccnet: criss-cross attention for semantic segmentation (2020)
42. Vaswani, A., et al.: Attention is all you need (2017)
43. Jiang, S., Campbell, D., Lu, Y., Li, H., Hartley, R.: Learning to estimate hidden motions with global motion aggregation (2021)
44. Zhang, X., Wang, T., Qi, J., Lu, H., Wang, G.: Progressive attention guided recurrent network for salient object detection. In: Proceedings of the IEEE Conference on Computer Vision and Pattern Recognition (CVPR) (2018)
45. Long, G., Kneip, L., Alvarez, J.M., Li, H., Zhang, X., Yu, Q.: Learning image matching by simply watching video. In: Leibe, B., Matas, J., Sebe, N., Welling, M. (eds.) ECCV 2016. LNCS, vol. 9910, pp. 434–450. Springer, Cham (2016). https://doi.org/10.1007/978-3-319-46466-4_26
46. Niklaus, S., Mai, L., Liu, F.: Video frame interpolation via adaptive convolution. In: IEEE Conference on Computer Vision and Pattern Recognition (2017)
47. Ramachandran, P., Parmar, N., Vaswani, A., Bello, I., Levskaya, A., Shlens, J.: Stand-alone self-attention in vision models (2019)
48. Galassi, A., Lippi, M., Torroni, P.: Attention in natural language processing. IEEE Trans. Neural Netw. Learn. Syst. **32**, 4291–4308 (2021)
49. Liu, Z., et al.: Swin transformer: hierarchical vision transformer using shifted windows. In: Proceedings of the IEEE/CVF International Conference on Computer Vision, pp. 10012–10022 (2021)
50. Shi, W., et al.: Real-time single image and video super-resolution using an efficient sub-pixel convolutional neural network (2016)
51. Baker, S., Roth, S., Scharstein, D., Black, M.J., Lewis, J., Szeliski, R.: A database and evaluation methodology for optical flow. In: 2007 IEEE 11th International Conference on Computer Vision, pp. 1–8 (2007)
52. Brox, T., Bregler, C., Malik, J.: Large displacement optical flow. In: 2009 IEEE Conference on Computer Vision and Pattern Recognition, pp. 41–48. IEEE (2009)
53. Janai, J., Guney, F., Ranjan, A., Black, M., Geiger, A.: Unsupervised learning of multi-frame optical flow with occlusions. In: Proceedings of the European Conference on Computer Vision (ECCV) (2018)
54. Paszke, A., et al.: Pytorch: an imperative style, high-performance deep learning library. In: Advances in Neural Information Processing Systems, vol. 32, pp. 8024–8035. Curran Associates, Inc. (2019)
55. Kingma, D.P., Ba, J.: Adam: a method for stochastic optimization. CoRR abs/1412.6980 (2014)

56. Soomro, K., Zamir, A.R., Shah, M.: UCF101: a dataset of 101 human actions classes from videos in the wild. CoRR abs/1212.0402 (2012)
57. Wang, Z., Bovik, A., Sheikh, H., Simoncelli, E.: Image quality assessment: from error visibility to structural similarity. IEEE Trans. Image Process. **13**, 600–612 (2004)
58. Liu, Y., Liao, Y., Lin, Y.Y., Chuang, Y.Y.: Deep video frame interpolation using cyclic frame generation. In: AAAI (2019)

Deep Learning-Based Small Object Detection from Images and Videos

Evaluating and Bench-Marking Object Detection Models for Traffic Sign and Traffic Light Datasets

Ashutosh Mishra(✉), Aman Kumar, Shubham Mandloi, Khushboo Anand, John Zakkam, Seeram Sowmya, and Avinash Thakur

OPPO Research and Development Center, Hyderabad, India
{ashutosh.mishra1,aman.kumar1,shubham.mandloi, khushboo.anand,avinash.thakur}@oppo.com

Abstract. Object detection is an important sub-problem for many computer vision applications. There has been substantial research in improving and evaluating object detection models for generic objects but it is still not known how latest deep learning models perform on small road scene objects such as traffic lights and traffic signs. In fact, locating small object of interest such as traffic light and traffic sign is a priority task for an autonomous vehicle to maneuver in complex scenarios. Although some researchers have tried to investigate the performance of deep learning based object detection models on various public datasets, however there exists no comprehensive benchmark. We present a more detailed evaluation by providing in-depth analysis of state-of-the-art deep learning based anchor and anchor-less object detection models such as Faster-RCNN, Single Shot Detector (SSD), Yolov3, RetinaNet, CenterNet and Cascade-RCNN. We compare the performance of these models on popular and publicly available traffic light datasets and traffic sign datasets from varied geographies. For traffic light datasets, we consider LISA Traffic Light (TL), Bosch, WPI and recently introduced S2TLD dataset for traffic light detection. For traffic sign benchmarking, we use LISA Traffic Sign (TS), GTSD, TT100K and recently published Mapillary Traffic Sign Dataset (MTSD). We compare the quantitative and qualitative performance of all the models on the aforementioned datasets and find that CenterNet outperforms all other baselines on almost all the datasets. We also compare inference time on specific CPU and GPU versions, flops and parameters for comparison. Understanding such behavior of the models on these datasets can help in solving a variety of practical difficulties and assists in the development of real-world applications. The source code and the models are available at https://github.com/OppoResearchIndia/DLSOD-ACCVW.

Keywords: Traffic sign detection · Traffic light detection · CascadeRCNN · CenterNet · RetinaNet · Yolov3 · SSD · Autonomous driving

---

A. Kumar, S. Mandloi, K. Anand, J. Zakkam, S. Sowmya—Equal contribution.

Y. Zheng et al. (Eds.): ACCV 2022, LNCS 13848, pp. 345–359, 2023.
https://doi.org/10.1007/978-3-031-27066-6_24

1 Introduction

Detecting traffic light/traffic sign is one of the most difficult problem for the perception module of an autonomous agent due to factors such as size, imaging resolution, weather etc. Fundamentally, designing systems that can help in achieving automatic TLD/TLR/TSD/TSR (traffic light detection/traffic light recognition/traffic sign detection/traffic sign recognition) would be extremely helpful in substantially reducing the number of fatalities around the world. However, there are multiple hindrances in performing TLD/TLR or TSD/TSR. Some of these hindrances could be size of traffic lights/traffic sign, view distances, weather conditions and objects of confusion such as street lights, house bulbs or other light sources. Pre-deep learning era for TLD/TLR or TSD/TSR uses a combination of techniques such as color thresholding, template matching, pixel clustering etc [13,14,19,34,39]. But they have proven to be robust enough under certain specific conditions to give successful results.

Advancement in deep learning has led to many improvements in the generic object detection performance. The main objective of this paper is to evaluate the performance of recent deep learning based object detection models in detecting small scale objects related to scene understanding task for autonomous driving viz. traffic lights and traffic signs. This evaluation is of utmost importance in order to check the performance metrics that can lead to decrease in false detection rate in real time scenario. For this, we compare all the publicly available datasets on a common benchmark. We club the fine classes to their respective meta class and then compare the metrics of different object detection models. The process of conversion from fine class to meta class is explained in Sect. 3. Therefore, the main contributions of our work are:

- Evaluation and analysis of various state-of-the-art deep learning object detection models on traffic light datasets and traffic sign datasets - LISA TL [20], WPI [5], BOSCH [2], S2TLD [37], LISA TS [26], TT100K [41], GTSD [18] and MTSD [9] (See Fig. 1).
- We evaluate the performances of these datasets on common meta classes with six object detectors: Faster R-CNN [30], SSD [23], RetinaNet [22], Yolov3 [29], CenterNet [7] and Cascade-RCNN [4].

To our knowledge, this is the first comprehensive work to compare publicly available traffic sign and traffic light datasets from various geographies on the common meta classes. We choose these models for evaluation because of two reasons. First, they are widely accepted and publicly available across all the deep learning frameworks among industry and academia. Second, the inference time of these models make them deployable ready for real time applications. This evaluation necessitates the need for understanding how state-of-the-art object detection algorithms perform on small road scene objects such as traffic sign and traffic light which is critical for developing practical applications such as self driving vehicles.

Fig. 1. Sample images with ground truth from all traffic light and traffic sign datasets used for bench-marking. The first row (from L to R) contains images from all traffic light datasets: LISA-TL, BOSCH, WPI, S2TLD. The second row (from L to R) contains images from all traffic sign datasets: LISA-TS, GTSD, TT100K and MTSD *(Best viewed when zoomed)*.

2 Related Work

We are interested in bench-marking the performance of various object detection models on publicly available traffic sign and traffic light datasets. The performance obtained can be represented as the state-of-the-art on meta classes of these datasets.

Table 1. Statistics of different traffic light and traffic sign publicly available datasets. **Original Classes**: The classes originally present in the dataset folder. **Meta Classes**: The original classes converted to base class. For LISA TS [26] and S2TLD [37] datasets, the split of training and testing has been created by the authors. The original dataset has more images but we only consider the ones with proper annotations. The test annotations for MTSD [9] and GTSD [18] are not publicly available so we consider val set as test set for reporting results.

Dataset	TS/TL	Geography	Average resolution	Training images	Test images	Original classes	Meta classes
LISA TL [20]	TL	US	1280 × 960	20535	22481	7	3
WPI [5]	TL	US	1024 × 2048	1314	2142	21	2
BOSCH [2]	TL	US	1280 × 720	5093	8334	15	4
S2TLD [37]	TL	China	1920 × 1080	744	244	5	5
LISA TS [26]	TS	US	880 × 504	5027	1571	47	15
TT100K [41]	TS	China	2048 × 2048	6107	3073	128	3
GTSD [18]	TS	Germany	1360 × 800	600	300	43	4
MTSD [9]	TS	Diverse	3407 × 2375	36589	10544	313	4

2.1 Traffic Sign Detection

The objective of performing traffic sign detection(TSD) is to get the exact locations and sizes of traffic signs. The well-defined colors and shapes are two main cues for traffic sign detection. There have been various works on detecting traffic signs using traditional methods employing histogram of oriented gradients

(HOG) [14,19,39], SIFT (scale invariant feature transform) [15], local binary patterns (LBP) [8]. The main working concepts in TSD using traditional methods revolves around using a sliding window based or a region of interest (ROI) based approach. HOG and Viola-Jones-like detector [25] are examples of sliding based methods. Wang et al. [35] uses a hierarchical sliding window method to detect traffic signs.

With the advent of deep learning and the rise in the use of convolutional neural networks, there have been many works, such as Zhu et al. [40] developed a strategy to detect and recognize traffic signs based on proposals by the guidance of fully convolutional network. R-CNN using a proposal strategy gave good results on a small scale dataset [21]. R-CNN along with an object proposal method [42] was used to further improve the performance on the same dataset. In 2016, [1] proposed a method that implements the multi-scale sliding window technique within a CNN using dilated convolutions. In 2019, the multiscale region-based convolutional neural network (MR-CNN) [24] was proposed for small traffic sign recognition, where a multiscale deconvolution operation was used to upsample the features of deeper convolution layers that were concatenated with those of the shallow layer directly to construct fused feature map. Thus, the fused feature map could generate fewer region proposals and achieve a higher recall rate. multiresolution feature fusion network exploiting deconvolution layers with skip connecting and a vertical spatial sequence attention module was designed 501 for traffic signs detection.

2.2 Traffic Light Detection

Conventional methods of traffic light detection include selecting confident proposals from a probable candidate set of traffic light's generated using color and shape information.

In conventional traffic light detection(TLD), a candidate set of TL is, typically, generated using the colour and shape information [13,34]. Once the candidate TL's are identified, [13] and [6] employ Adaboost algorithm and other morphological operations to segment out the TL regions. But the disadvantage of using such algorithms using hand-crafted features is the lack of generalization ability of methods. Subsequently, methods such as HOG or SIFT tend to lose information which might help in the required task. This can lead to a very lower detection performance.

There have been many attempts to perform traffic light detection using modern convolutional networks as well. For instance, Weber et al. [36] used a convolutional network for traffic light detection modifying Alexnet network. The output of the network is a segmented image which is then given to a bounding box regressor for detection of traffic lights. In [31], the authors use a combination of SVM and CNN for the combined task of traffic light detection and recognition. Similarly, Behrendt et al. [3] uses a combination of detection, tracking, and classification using a convolutional neural network. Yudin et al. [38] propose another a fully convolutional network for traffic light detection. Heat-map is obtained highlighting areas of interest followed by a clustering algorithm to obtain final traffic light bounding boxes. Besides being a transfer learning approach, it has a very low precision of detection as compared to SSD-based models [27].

Table 2. **Tr**: Train, **Te**: Test. Publicly available traffic light datasets, their classes along with and it's corresponding number of instances per class. The counts per class for S2TLD [37] is based on the dataset split generated by authors.

Dataset	Meta classes	Tr. Instance Count	Te. Instance Count
LISA TL [20]	Go	24182	25222
	Stop	26089	30963
	Warning	1555	1464
WPI [5]	Green	1323	2763
	Red	1878	802
BOSCH [2]	Green	5422	7569
	Yellow	444	154
	Red	4164	5321
	Off	726	442
S2TLD [37]	Green	478	164
	Yellow	43	16
	Red	761	239
	Off	2	1
	Wait-on	178	64

3 Deep Learning for Object Detection

With the success of convolutional neural networks to outperform traditional methods on classification task, there have been similar trends for other computer vision tasks such as object detection. OverFeat [32] is an example of such a network which outputs bounding boxes along with the scores using a deep network in a sliding-window fashion. Later, R-CNN [12] was proposed which helped in increasing the detection accuracy and was faster than the previous counterparts. The main disadvantage of using R-CNN is that it is expensive both in time and memory because it executes a CNN forward-pass for each object proposal without sharing computation. Spatial Pyramid Pooling Network (SPPNet) [16] was proposed to improve R-CNN efficiency by sharing computation. Due to the multi-stage pipeline in SPPNet, the whole process of detection becomes quite slow. Moreover, the parameters below the spatial pyramid pooling layer cannot be updated while training. After SPPNet, Fast R-CNN [11] was introduced, which proposes a new training algorithm that provides solutions to fix the disadvantages of R-CNN and SPPNet by training in a single-stage using a multi-task approach. But the main bottleneck in this approach is the candidate proposal strategy which is still different from the network training process. To overcome such bottleneck, Faster R-CNN [30], replaced the use of Selective Search with a

Region Proposal Network (RPN) that shares convolutional feature maps with the detection network, thus enabling nearly cost-free region proposals. To improve the performance of Faster-RCNN even further, Cascade-RCNN [4] was introduced. It consists of a series of detectors sequentially feeding in to the output of previous detector trained with increasing threshold values making it very selective for false positives. All these approaches discussed so far are multi-stage methods where there are two or multiple pipelines involved. There are other family of networks known as single stage networks including Single Shot Multi-Box Detector (SSD) [23], YOLOv3 [29] and RetinaNet [22] which detects objects using a fully convolutional network rather than having separate tracks for detection and classification. This ability leads to a much faster object detection. Duan et al. have also proposed using anchor less technique for object detection which detects each object as a triplet of keypoints [7].

Standard object detector approaches can be broadly classified in two categories: (i) two-stage object detectors, (ii) one-stage object detectors. Two-stage object detectors combine a region-proposal step, region classification and regression step. On the contrary, one-stage detectors output boxes without a region proposal step. Two-stage and one-stage object detectors can also be called as anchor based since these models employ anchors to perform the detection. For bench-marking different traffic light and traffic sign datasets, we select publicly available and widely used Faster RCNN (2-stage), Yolov3 (one-stage), RetinaNet (one-stage) and SSD (one-stage) and Cascade-RCNN (mutli-stage) networks. Apart from the anchor based models, we also use CenterNet, anchor-less approach to bench-mark the datasets and analyse the quantitative results obtained. As mentioned earlier, we select these models for evaluation because of two reasons. Firstly, these models are known for real time performance. Secondly, these models are widely accepted in academia and industry. Almost all these models have been designed in such a way that they infer in near real-time and also validated through our experiments.

4 Experiments

The aim of the experimentation is to bench-mark different open source traffic light and traffic sign datasets against various state-of-the-art deep learning based object detection models. We consider datasets pertaining to different geographies in order to understand how these deep neural networks perform in different conditions.

4.1 TL and TS Datasets

We consider four publicly available datasets for traffic light performance evaluation namely: LISA Traffic Light Dataset, BOSCH Traffic Light Dataset, WPI Traffic Light Dataset and the recently introduced SJTU Small Traffic Light Dataset (S2TLD). For traffic sign detection performance, we consider the following publicly available datasets: TT100K Dataset, Mapillary Traffic Sign Dataset,

LISA Traffic Sign Dataset and German Traffic Sign Dataset. Figure 1 has representative images from all the datasets used for the bench-marking in this paper.

All the details for the respective datasets can be obtained from Table 1. We have merged all the fine classes to their respective meta class for all the datasets that we consider. For instance, LISA Traffic Light dataset contains seven classes namely, "*go*", "*goForward*", "*goLeft*", "*warning*", "*warningLeft*", "*stop*" which were converted namely to "*go*", "*warning*", "*stop*" for fair comparison amongst the common classes across various datasets. More details regarding the meta classes for various datasets is given in Table 2 and Table 3.

4.2 Training Setup

For training these models, we deploy frameworks such as Detectron2 [31] and MMDetection [5] which are modular and easy to train, validation and testing on custom datasets. The frameworks have been well documented and implemented in Pytorch [28] deep learning framework. The benchmarking was carried on a Linux machine having 2 Tesla V100 GPU's. Each of the detector model have been trained with some fixed parameters for fair experimentation and trained until convergence. The batch size is set to 4 with a learning rate of 0.00025, momentum of 0.9, and weight decay factor of 0.0001. For FasterRCNN and CascadeRCNN, the backbones considered are Resnet101 and Resnet50. The backbones used for Yolov3 and SSD are Darknet-53 [29] and VGG16 [33] with input resolution of 608×608 and 512×512 respectively.

4.3 Performance Evaluation

The evaluation of all the object detection performance models is done in terms of precision, recall and mean average precision using intersection-over-union (IoU). The mAP calculation is done based on the definition for the Pascal VOC 2007 competition with IoU threshold of 0.5 [10]. Equation (1) describes the formula of the calculation of mean average precision metric at 0.5 threshold. AP_i is the average precision per class in the dataset.

$$mAP@0.5 = \frac{1}{N}\sum_{i=1}^{N} AP_i \tag{1}$$

5 Results and Discussion

For the bench-marking task, we use YoloV3, Faster-RCNN, RetinaNet, SSD, CenterNet and Cascade-RCNN object detection model to serve as baselines.

Table 3. **Tr**: Train, **Te**: Test. Publicly available traffic sign datasets, their classes along with and it's corresponding number of instances per class. For MTSD [9] and GTSD [18], we consider validation set as test set since test set annotations are not publicly available.

Dataset	Meta classes	Tr. Instance Count	Te. Instance Count
LISA TS [26]	Warning	2611	651
	Prohibition	1459	360
	Speed-Limit	107	24
	Stop	1493	362
	Yield	187	49
	School	108	26
	School-Speed-Limit25	81	24
	Zone Ahead	55	14
	Ramp-speed-advisory	41	12
	Round-about	35	12
	Curve-left	27	12
	No-Left-Turn	25	12
	Thru-Traffic-Merge-Left	22	5
	Do-Not-Enter	19	4
	No-Right-Turn	14	4
GTSD [18]	Prohibitory	299	97
	Mandatory	84	30
	Danger	116	40
	Other	143	43
TT100K [41]	Warning	912	456
	Prohibitory	12393	6179
	Mandatory	3444	1626
MTSD [9]	Complementary	9082	1323
	Information	6507	948
	Regulatory	31574	4593
	Warning	14328	2073
	Others	118749	17209

5.1 Traffic Light Results

Table 4 contains quantitative results on the test set using Faster-RCNN, Yolov3, CenterNet, SSD, RetinaNet and Cascade-RCNN. We observe that almost on all datasets, there is a competition of scores between CenterNet and CascadeRCNN. Individually, WPI dataset has ground truth annotations lying in the range of medium and large box area. Hence, SSD performs comparatively better among all the competitive baselines because SSD focuses better on large sized objects.

Table 4. Results of various object detection models on all traffic light datasets using Faster R-CNN (FRCNN), SSD, Yolov3 (Y3), RetinaNet (RN) and CenterNet (CN) and Cascade-RCNN (CRNN). Mean Average Precision (mAP) at 0.5 threshold is indicated for different classes of the respective datasets. CPU and GPU inference time per image (in seconds) is also indicated in this table. The values in bold font represent the best results in each category across all methods on the respective datasets. Empty row values indicate that meta class is absent in the respective dataset.

	WPI						LISA TL					
Class	FRCNN	SSD	Y3	RN	CN	CRNN	FRCNN	SSD	Y3	RN	CN	CRNN
go	68.13	**91.80**	84.0	43.14	87.50	77.07	54.85	55.70	46.50	54.42	59.50	**62.27**
stop	49.83	**85.40**	75.50	13.09	81.90	81.23	35.40	53.50	8.80	40.61	**58.60**	45.36
warning	-	-	-	-	-	-	12.00	36.30	31.60	16.66	32.30	**39.63**
off	-	-	-	-	-	-	-	-	-	-	-	-
wait-on	-	-	-	-	-	-	-	-	-	-	-	-
mAP@0.5	58.98	**88.60**	79.80	28.12	84.70	79.15	34.09	48.50	29.00	37.23	**50.10**	49.09
GPU Inf Time (sec)	0.70	0.03	0.13	0.08	0.25	0.08	0.69	0.02	0.11	0.07	0.20	0.06
CPU Inf Time (sec)	14.32	3.24	7.27	2.24	5.77	5.79	14.14	3.00	5.15	4.05	5.84	6.65

	BOSCH						S2TLD					
Class	FRCNN	SSD	Y3	RN	CN	CRNN	FRCNN	SSD	Y3	RN	CN	CRNN
go	61.09	73.20	50.10	66.76	**83.50**	67.91	85.89	87.50	91.10	89.80	91.00	**94.15**
stop	60.15	64.40	71.70	46.39	79.00	**79.87**	85.73	87.5	92.3	90.54	**91.00**	86.08
warning	27.66	8.50	64.50	11.98	48.30	**62.51**	58.20	18.60	81.20	53.52	65.80	**92.26**
off	0.01	0.00	0.10	0.00	**0.10**	0.00	0.00	0.00	0.00	0.00	**0.40**	0.00
wait-on	-	-	-	-	-	-	93.76	95.90	90.40	98.20	**98.40**	92.56
mAP@0.5	37.23	36.50	46.60	31.28	**52.70**	52.57	64.72	57.90	71.00	66.38	69.30	**73.01**
GPU Inf Time (sec)	0.69	0.03	0.12	0.08	0.27	0.07	0.60	0.03	0.14	0.13	0.25	0.08
CPU Inf Time (sec)	14.04	3.11	8.75	4.62	5.74	5.69	17.91	2.96	5.76	5.37	5.83	7.42

Table 5. Number of trainable parameters(in million) and number of floating point operations for all the object detection models.

OD model	Backbone	Parameters (in million)	GFlops
FasterRCNN [30]	ResNet101	104.38	423.16
SSD [23]	VGG16	36.04	355.12
Yolov3 [29]	Darknet53	61.95	171.91
RetinaNet [22]	ResNet50	37.91	215.49
CenterNet [7]	ResNet18	14.44	44.78
CascadeRCNN [4]	ResNet50-FPN	69.10	214.44

SSD's performance is followed by CenterNet which is an anchor-less approach that outperforms other anchor based approaches.

For LISA TL and BOSCH and S2TLD, majority of the ground truth TL instances are located very far away from the camera view, thus pertaining to a small annotation box area. In such scenario, anchors have to adjust a lot to cater the needs of the algorithm in order to match the location of the object of interest. However, CenterNet and CascadeRCNN, both perform better. The reason is that Centernet is an anchor-less approach, eliminating the anchor dependency while CascadeRCNN applies repeated RPN blocks for better detection at uniform thresholds. An observation to note is that class "off" has almost zero mAP even though the class instances are present in BOSCH and S2TLD datasets. The plausible reason is that the class "off" indicates the no traffic light is activated but the network is trained ideally to detect colors since individually in both the datasets, class "off" has very few instances compared to other classes.

Figure 2 shows some qualitative results on few selected frames of the respective traffic light datasets. Visually we infer that all the models are able to detect traffic light at different angles of rotation. On BOSCH dataset, Faster-RCNN and RetinaNet are able to detect the traffic light present on the left in the presence of occlusion and lighting in the image. On the other hand, Yolov3 and CenterNet are able to detect the traffic light on the left. All the models are able to detect horizontal lights for S2TLD dataset and vertical lights for LISA TL and WPI.

5.2 Traffic Sign Results

Table 6 contains quantitative results obtained on the test set using Faster-RCNN, YoloV3, CenterNet, SSD, RetinaNet and CascadeRCNN. From the results, it can be inferred that CenterNet outperforms almost all the models on all the datasets. For LISA TS, CascadeRCNN achieved best mAP because of the presence of evenly distributed traffic signs in medium and small annotation box areas.

Figure 2 shows qualitative results on the frames of respective traffic sign datasets. Visual results indicate that models are able to predict correctly on all the classes that they have been trained on. SSD and RetinNet output some false detections on LISA TS and GTSD datasets respectively but CenterNet performs detection on all the datasets with higher confidence, even in slightly darker imaging environment as for the predictions of MTSD dataset.

Table 6. Results of various object detection models on traffic sign datasets using Faster R-CNN (FRCNN), SSD, Yolov3 (Y3), RetinaNet (RN) and CenterNet (CN). Mean Average Precision (mAP) at 0.5 threshold is indicated in the last row for different classes of the respective datasets. CPU and GPU inference time per image (in seconds) is also indicated in this table. The values in bold font represent the best results in each category across all methods on the respective datasets. Empty row values indicate that meta class is absent in the respective dataset.

	TT100K						GTSD					
Class	FRCNN	SSD	Y3	RN	CN	CRNN	FRCNN	SSD	Y3	RN	CN	CRNN
warning	85.19	79.90	92.3	88.72	**97.90**	95.48	-	-	-	-	-	-
prohibitory	78.65	79.90	92.50	92.40	97.20	**94.46**	66.40	91.60	98.20	81.98	**99.20**	75.13
mandatory	83.71	82.80	88.90	87.04	**97.20**	76.82	65.65	80.10	85.60	77.28	**97.60**	68.42
danger	-	-	-	-	-	-	78.50	96.70	100.00	96.37	**100.00**	86.71
regulatory	-	-	-	-	-	-	-	-	-	-	-	-
complementary	-	-	-	-	-	-	-	-	-	-	-	-
information	-	-	-	-	-	-	-	-	-	-	-	-
speed-limit	-	-	-	-	-	-	-	-	-	-	-	-
stop	-	-	-	-	-	-	-	-	-	-	-	-
yield	-	-	-	-	-	-	-	-	-	-	-	-
school	-	-	-	-	-	-	-	-	-	-	-	-
school-speed-limit25	-	-	-	-	-	-	-	-	-	-	-	-
zone-ahead	-	-	-	-	-	-	-	-	-	-	-	-
ramp-speed-advisory	-	-	-	-	-	-	-	-	-	-	-	-
round-about	-	-	-	-	-	-	-	-	-	-	-	-
curve-left	-	-	-	-	-	-	-	-	-	-	-	-
no-left-turn	-	-	-	-	-	-	-	-	-	-	-	-
thru-traffic-merge-left	-	-	-	-	-	-	-	-	-	-	-	-
do-not-enter	-	-	-	-	-	-	-	-	-	-	-	-
no-right-turn	-	-	-	-	-	-	-	-	-	-	-	-
other	-	-	-	-	-	-	60.06	73.70	90.00	86.81	**92.20**	65.03
mAP@0.5	82.51	80.90	91.60	89.39	**97.40**	88.92	67.65	85.50	93.40	85.68	**97.20**	73.85
GPU Inf Time. (sec)	0.69	0.03	0.10	0.07	0.21	0.07	0.67	0.002	0.13	0.08	0.21	0.07
CPU Inf Time. (sec)	15.86	2.53	9.23	2.09	5.83	6.69	14.90	3.04	9.81	2.14	6.13	8.15

	MTSD						LISA TS					
Class	FRCNN	SSD	Y3	RN	CN	CRNN	FRCNN	SSD	Y3	RN	CN	CRNN
warning	66.78	59.20	60.00	66.68	**81.00**	75.98	84.82	90.20	95.00	88.74	88.90	**96.38**
prohibitory	-	-	-	-	-	-	82.07	90.40	93.90	84.94	92.80	**95.77**
mandatory	83.71	82.80	88.90	87.04	**97.20**	76.82	-	-	-	-	-	-
danger	-	-	-	-	-	-	-	-	-	-	-	-
regulatory	57.70	44.40	53.30	54.02	**81.50**	73.94	-	-	-	-	-	-
complementary	50.74	38.90	48.50	41.49	73.00	**74.27**	-	-	-	-	-	-
information	-	-	-	-	-	-	36.89	28.70	38.10	35.13	**67.50**	30.61
speed-limit	-	-	-	-	-	-	25.82	48.90	94.60	**96.70**	80.90	96.78
stop	-	-	-	-	-	-	74.18	84.30	**92.10**	84.20	85.10	94.27
yield	-	-	-	-	-	-	30.92	55.30	**88.70**	44.65	66.80	87.50
school	-	-	-	-	-	-	62.79	91.5	**99.30**	25.74	93.50	87.56
school-speed-limit25	-	-	-	-	-	-	84.73	74.50	95.40	**100.00**	91.10	95.60
zone-ahead	-	-	-	-	-	-	46.37	11.20	80.00	88.6	59.80	**100.00**
ramp-speed-advisory	-	-	-	-	-	-	73.62	74.30	**100.00**	79.86	**100.00**	89.25
round-about	-	-	-	-	-	-	72.70	51.50	71.40	48.67	89.10	**97.69**
curve-left	-	-	-	-	-	-	67.06	27.50	88.30	40.77	88.30	**94.54**
no-left-turn	-	-	-	-	-	-	25.49	46.60	**100.00**	67.85	95.30	80.01
thru-traffic-merge-left	-	-	-	-	-	-	80.19	33.50	**100.00**	60.13	**100.00**	63.51
do-not-enter	-	-	-	-	-	-	88.61	5.00	93.30	83.31	88.60	**96.70**
no-right-turn	-	-	-	-	-	-	0.00	24.00	20.00	75.24	69.50	**100.00**
other	44.24	27.60	30.30	34.00	**55.50**	30.01	-	-	-	-	-	-
mAP@0.5	82.51	80.9	91.6	89.39	**97.40**	88.92	59.96	53.92	87.50	71.27	86.0	**91.71**
GPU Inf Time. (sec)	0.69	0.03	0.10	0.07	0.21	0.07	0.68	0.03	0.11	0.077	0.23	0.06
CPU Inf Time. (sec)	15.86	2.53	9.23	2.09	5.83	6.69	18.03	2.95	8.28	3.72	5.20	2.62

5.3 Object Detection Model's Statistics

Table 5 shows the number of trainable parameters and the floating point operations (GFlops) of the respective object detection models. From the figures, it is evident that FasterRCNN has the highest number of parameters since it has Resnet101 as the backbone architecture while CascadeRCNN and RetinaNet uses Resnet50 as the backbone architecture. This is also evident in Table 4 and Table 6 for GPU and CPU inference times. SSD operates on lower image resolution compared to Yolov3, hence the run-time for SSD is much lower. CenterNet uses Resnet18 [17] as the backbone having the lowest number of trainable parameters and the number of floating point operations surpassing all anchor based models in terms of detection and also real-time performance.

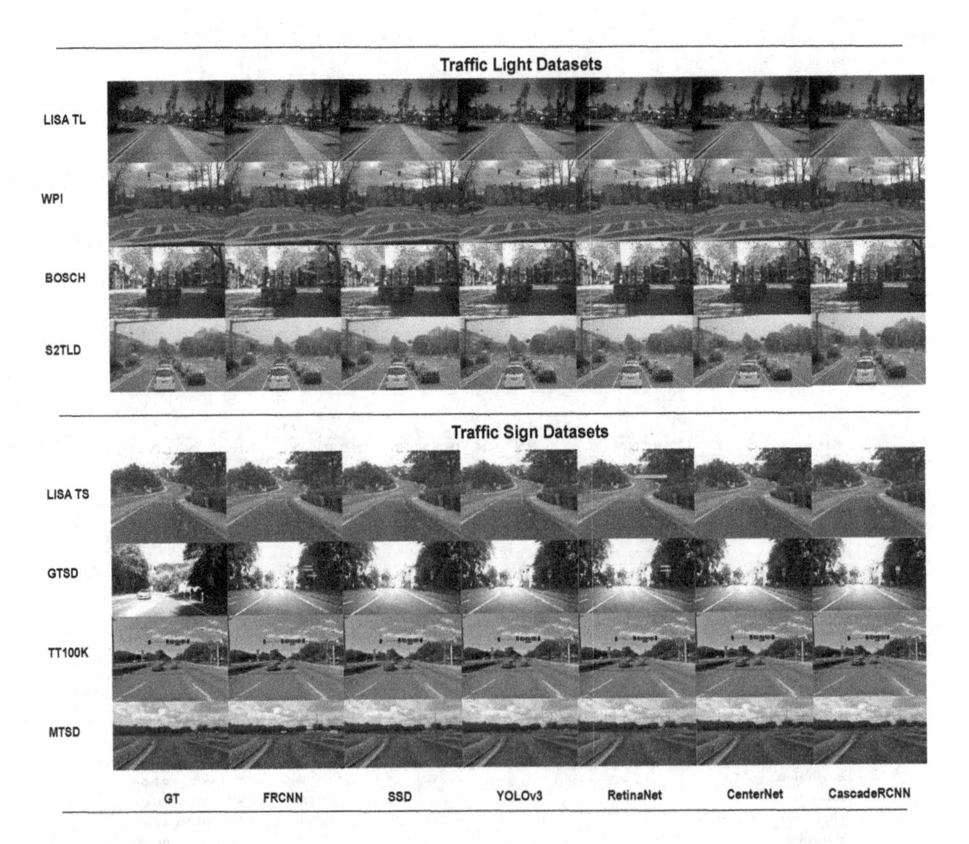

Fig. 2. Qualitative results of various object detection models on all traffic sign and traffic light datasets. The first four rows have visual results from traffic light datasets. The next four rows have visual results from traffic sign datasets. The first column is the ground truth image along with annotations in green color *(Best viewed when zoomed)*.

6 Conclusion

In this work, we used various state-of-the-art models for object detection and evaluate their performance on publicly available traffic light datasets: LISA TL, WPI, BOSCH and S2TLD and traffic sign datasets: LISA TS, GTSD, TT100K and MTSD. To the best of our knowledge, this work is the first detailed analysis of different object detection models (anchor based and anchor-less) on common meta classes of various traffic light and traffic sign datasets. From the results, it is evident that anchor less methods outperform anchor-based methods on almost all traffic light and traffic sign datasets irrespective of the instance location. For traffic light instances located nearby, even SSD performs better compared to any other anchor based model for WPI dataset. Specifically for LISA TS, Yolov3 marginally outperforms CenterNet due to a balance of traffic sign instances in medium and large annotation box areas from the camera view. In future work, we would like to explore the direction of the effect of weather patterns and the role of domain adaptation techniques to increase the detection accuracies of small objects such as traffic sign and traffic light.

References

1. Aghdam, H.H., Heravi, E.J., Puig, D.: A practical approach for detection and classification of traffic signs using convolutional neural networks. Robot. Auton. Syst. **84**, 97–112 (2016)
2. Behrendt, K., Novak, L.: A deep learning approach to traffic lights: detection, tracking, and classification. In: 2017 IEEE International Conference on Robotics and Automation (ICRA). IEEE (2017)
3. Behrendt, K., Novak, L., Botros, R.: A deep learning approach to traffic lights: detection, tracking, and classification. In: 2017 IEEE International Conference on Robotics and Automation (ICRA), pp. 1370–1377. IEEE (2017)
4. Cai, Z., Vasconcelos, N.: Cascade R-CNN: delving into high quality object detection. In: Proceedings of the IEEE Conference on Computer Vision and Pattern Recognition, pp. 6154–6162 (2018)
5. Chen, Z., Huang, X.: Accurate and reliable detection of traffic lights using multiclass learning and multiobject tracking. IEEE Intell. Transp. Syst. Mag. **8**(4), 28–42 (2016). https://doi.org/10.1109/MITS.2016.2605381
6. De Charette, R., Nashashibi, F.: Traffic light recognition using image processing compared to learning processes. In: 2009 IEEE/RSJ International Conference on Intelligent Robots and Systems, pp. 333–338. IEEE (2009)
7. Duan, K., Bai, S., Xie, L., Qi, H., Huang, Q., Tian, Q.: Centernet: keypoint triplets for object detection. In: Proceedings of the IEEE/CVF International Conference on Computer Vision, pp. 6569–6578 (2019)
8. Ellahyani, A., Ansari, M., Jaafari, I., Charfi, S.: Traffic sign detection and recognition using features combination and random forests. Int. J. Adv. Comput. Sci. Appl. **7**(1), 686–693 (2016)
9. Ertler, C., Mislej, J., Ollmann, T., Porzi, L., Neuhold, G., Kuang, Y.: The mapillary traffic sign dataset for detection and classification on a global scale. In: Vedaldi, A., Bischof, H., Brox, T., Frahm, J.-M. (eds.) ECCV 2020. LNCS, vol. 12368, pp. 68–84. Springer, Cham (2020). https://doi.org/10.1007/978-3-030-58592-1_5

10. Everingham, M., Van Gool, L., Williams, C.K.I., Winn, J., Zisserman, A.: The PASCAL Visual Object Classes Challenge 2007 (VOC2007) Results. http://www.pascal-network.org/challenges/VOC/voc2007/workshop/index.html
11. Girshick, R.: Fast R-CNN. In: Proceedings of the IEEE International Conference on Computer Vision, pp. 1440–1448 (2015)
12. Girshick, R., Donahue, J., Darrell, T., Malik, J.: Rich feature hierarchies for accurate object detection and semantic segmentation. In: Proceedings of the IEEE Conference on Computer Vision and Pattern Recognition, pp. 580–587 (2014)
13. Gong, J., Jiang, Y., Xiong, G., Guan, C., Tao, G., Chen, H.: The recognition and tracking of traffic lights based on color segmentation and camshift for intelligent vehicles. In: 2010 IEEE Intelligent Vehicles Symposium, pp. 431–435 (2010). https://doi.org/10.1109/IVS.2010.5548083
14. Greenhalgh, J., Mirmehdi, M.: Real-time detection and recognition of road traffic signs. IEEE Trans. Intell. Transp. Syst. **13**(4), 1498–1506 (2012). https://doi.org/10.1109/TITS.2012.2208909
15. Haloi, M.: A novel PLSA based traffic signs classification system (2015)
16. He, K., Zhang, X., Ren, S., Sun, J.: Spatial pyramid pooling in deep convolutional networks for visual recognition. IEEE Trans. Pattern Anal. Mach. Intell. **37**(9), 1904–1916 (2015)
17. He, K., Zhang, X., Ren, S., Sun, J.: Deep residual learning for image recognition. In: Proceedings of the IEEE Conference on Computer Vision and Pattern Recognition, pp. 770–778 (2016)
18. Houben, S., Stallkamp, J., Salmen, J., Schlipsing, M., Igel, C.: Detection of traffic signs in real-world images: the German traffic sign detection benchmark. In: International Joint Conference on Neural Networks, no. 1288 (2013)
19. Huang, Z., Yu, Y., Gu, J., Liu, H.: An efficient method for traffic sign recognition based on extreme learning machine. IEEE Trans. Cybern. **47**(4), 920–933 (2016)
20. Jensen, M.B., Philipsen, M.P., Møgelmose, A., Moeslund, T.B., Trivedi, M.M.: Vision for looking at traffic lights: issues, survey, and perspectives. IEEE Trans. Intell. Transp. Syst. **17**(7), 1800–1815 (2016). https://doi.org/10.1109/TITS.2015.2509509
21. Larsson, F., Felsberg, M.: Using fourier descriptors and spatial models for traffic sign recognition. In: Heyden, A., Kahl, F. (eds.) SCIA 2011. LNCS, vol. 6688, pp. 238–249. Springer, Heidelberg (2011). https://doi.org/10.1007/978-3-642-21227-7_23
22. Lin, T.Y., Goyal, P., Girshick, R., He, K., Dollár, P.: Focal loss for dense object detection. In: Proceedings of the IEEE International Conference on Computer Vision, pp. 2980–2988 (2017)
23. Liu, W., et al.: SSD: single shot MultiBox detector. In: Leibe, B., Matas, J., Sebe, N., Welling, M. (eds.) ECCV 2016. LNCS, vol. 9905, pp. 21–37. Springer, Cham (2016). https://doi.org/10.1007/978-3-319-46448-0_2
24. Liu, Z., Du, J., Tian, F., Wen, J.: MR-CNN: a multi-scale region-based convolutional neural network for small traffic sign recognition. IEEE Access **7**, 57120–57128 (2019)
25. Møgelmose, A.: Visual analysis in traffic & re-identification (2015)
26. Mogelmose, A., Trivedi, M.M., Moeslund, T.B.: Vision-based traffic sign detection and analysis for intelligent driver assistance systems: Perspectives and survey. IEEE Trans. Intell. Transp. Syst. **13**(4), 1484–1497 (2012). https://doi.org/10.1109/TITS.2012.2209421

27. Müller, J., Dietmayer, K.: Detecting traffic lights by single shot detection. In: 2018 21st International Conference on Intelligent Transportation Systems (ITSC), pp. 266–273 (2018). https://doi.org/10.1109/ITSC.2018.8569683
28. Paszke, A., et al.: Pytorch: an imperative style, high-performance deep learning library. Adv. Neural. Inf. Process. Syst. **32**, 8026–8037 (2019)
29. Redmon, J., Farhadi, A.: Yolov3: an incremental improvement. arXiv preprint arXiv:1804.02767 (2018)
30. Ren, S., He, K., Girshick, R., Sun, J.: Faster R-CNN: towards real-time object detection with region proposal networks. Adv. Neural. Inf. Process. Syst. **28**, 91–99 (2015)
31. Saini, S., Nikhil, S., Konda, K.R., Bharadwaj, H.S., Ganeshan, N.: An efficient vision-based traffic light detection and state recognition for autonomous vehicles. In: 2017 IEEE Intelligent Vehicles Symposium (IV), pp. 606–611. IEEE (2017)
32. Sermanet, P., Eigen, D., Zhang, X., Mathieu, M., Fergus, R., LeCun, Y.: Overfeat: integrated recognition, localization and detection using convolutional networks. arXiv preprint arXiv:1312.6229 (2013)
33. Simonyan, K., Zisserman, A.: Very deep convolutional networks for large-scale image recognition. arXiv preprint arXiv:1409.1556 (2014)
34. Siogkas, G., Skodras, E.: Traffic lights detection in adverse conditions using color, symmetry and spatiotemporal information, vol. 1 (2012)
35. Wang, G., Ren, G., Wu, Z., Zhao, Y., Jiang, L.: A robust, coarse-to-fine traffic sign detection method. In: The 2013 International Joint Conference on Neural Networks (IJCNN), pp. 1–5. IEEE (2013)
36. Weber, M., Wolf, P., Zöllner, J.M.: DeepTLR: a single deep convolutional network for detection and classification of traffic lights. In: 2016 IEEE Intelligent Vehicles Symposium (IV), pp. 342–348 (2016). https://doi.org/10.1109/IVS.2016.7535408
37. Yang, X., Yan, J., Yang, X., Tang, J., Liao, W., He, T.: SCRDet++: detecting small, cluttered and rotated objects via instance-level feature denoising and rotation loss smoothing. arXiv preprint arXiv:2004.13316 (2020)
38. Yudin, D., Slavioglo, D.: Usage of fully convolutional network with clustering for traffic light detection. In: 2018 7th Mediterranean Conference on Embedded Computing (MECO), pp. 1–6. IEEE (2018)
39. Zaklouta, F., Stanciulescu, B.: Real-time traffic-sign recognition using tree classifiers. IEEE Trans. Intell. Transp. Syst. **13**(4), 1507–1514 (2012)
40. Zhu, Y., Zhang, C., Zhou, D., Wang, X., Bai, X., Liu, W.: Traffic sign detection and recognition using fully convolutional network guided proposals. Neurocomputing **214**, 758–766 (2016)
41. Zhu, Z., Liang, D., Zhang, S., Huang, X., Li, B., Hu, S.: Traffic-sign detection and classification in the wild. In: The IEEE Conference on Computer Vision and Pattern Recognition (CVPR) (2016)
42. Zitnick, C.L., Dollár, P.: Edge boxes: locating object proposals from edges. In: Fleet, D., Pajdla, T., Schiele, B., Tuytelaars, T. (eds.) ECCV 2014. LNCS, vol. 8693, pp. 391–405. Springer, Cham (2014). https://doi.org/10.1007/978-3-319-10602-1_26

Exploring Spatial-Temporal Instance Relationships in an Intermediate Domain for Image-to-Video Object Detection

Zihan Wen, Jin Chen, and Xinxiao Wu(✉)

Beijing Laboratory of Intelligent Information Technology, School of Computer Science, Beijing Institute of Technology, Beijing, China
{wenzihan,chen_jin,wuxinxiao}@bit.edu.cn

Abstract. Image-to-video object detection leverages annotated images to help detect objects in unannotated videos, so as to break the heavy dependency on the expensive annotation of large-scale video frames. This task is extremely challenging due to the serious domain discrepancy between images and video frames caused by appearance variance and motion blur. Previous methods perform both image-level and instance-level alignments to reduce the domain discrepancy, but the existing false instance alignments may limit their performance in real scenarios. We propose a novel spatial-temporal graph to model the contextual relationships between instances to alleviate the false alignments. Through message propagation over the graph, the visual information from the spatial and temporal neighboring object proposals are adaptively aggregated to enhance the current instance representation. Moreover, to adapt the source-biased decision boundary to the target data, we generate an intermediate domain between images and frames. It is worth mentioning that our method can be easily applied as a plug-and-play component to other image-to-video object detection models based on the instance alignment. Experiments on several datasets demonstrate the effectiveness of our method. Code will be available at: https://github.com/wenzihan/STMP.

Keywords: Deep learning · Object detection · Domain adaptation

1 Introduction

Tremendous progress has been achieved on object detection in videos [3,9,10,19,23] thanks to the great success of deep neural networks. However, training deep object detectors requires annotating large-scale video frames, which is often time-consuming and labor-intensive. On the other hand, images are much easier and cheaper to be annotated, and there are also many existing labeled image datasets that can be readily utilized.

Therefore, image-to-video object detection has been proposed to leverage annotated images for detecting objects in unannotated videos, as to break the

Y. Zheng et al. (Eds.): ACCV 2022, LNCS 13848, pp. 360–375, 2023.
https://doi.org/10.1007/978-3-031-27066-6_25

Fig. 1. Illustration of the domain discrepancy between images from the PASCAL-VOC dataset and video frames from the Youtube-Objects dataset. Although both datasets have the samples of "horse" and "car", the domain discrepancy caused by appearance variance (a) and motion blur (b) still make it challenging to apply an object detector learned from source images to target video frames. In order to reduce the domain discrepancy between images and video frames, we propose to generate an intermediate domain and some intermediate domain images are shown in (c).

heavy dependency on the expensive annotation of frames. However, directly applying the detector trained on source images may significantly hurt its performance on detection in target videos, since there exists serious domain discrepancy between images and frames caused by appearance variance and motion blur, as shown in Fig. 1. To address this problem, several prior works [1,4,17,22] mainly focus on doing image-level and instance-level alignments of source images and target video frames to minimize the domain discrepancy. Although these methods have achieved promising results, the bounding box deviation, occlusion and out of focus may lead to the false alignments between cross-domain instances, which degrades their performance and limits their applications in real scenarios.

Taking into account the spatial-temporal context of instances as a potentially valuable information source for alleviating the false instance alignments, in this paper, we propose a novel spatial-temporal message propagation method to model the contextual relationships between instances for image-to-video object detection. By propagating message over the graph, the visual information of neighboring object proposals both within the same frame and from the adjacent frames are adaptively aggregated to enhance the current object instance representation. This feature aggregation strategy can mitigate the influence of bounding box deviation, occlusion and out of focus in images or frames. For example, spatially neighboring proposal features that are aggregated according to the intersection over union are helpful for relieving the deviation of bounding boxes, and temporally adjacent proposal features that are aggregated under the guidance of optical flow are beneficial to the alleviation of effects from the occlusion and motion blur. To be more specific, we first build an undirected graph where the nodes are represented by the region proposals and the edges are represented by the spatial-temporal instance relationships between the nodes. Then we introduce a single-layer graph convolution network with normal propagation rule [12] for message propagation over the graph and update the node representation by aggregating the propagated features from the spatial-temporal neighboring nodes.

Moreover, to adapt the source-biased decision boundary to the target domain, we generate an intermediate domain between the source images and target video frames via generative adversarial learning. It serves as a bridge for connecting the source and target domains, and the cross-domain alignments on both the image level and the instance level are performed on this domain, which further boosts the positive transfer of the object detector between different domains.

The main contributions of this work are three-fold:

- We propose a novel spatial-temporal graph to model the contextual relationships between object instances for alleviating the false instance alignments in image-to-video object detection. It can be easily and readily applied as a plug-and-play component for other detection models based on the instance alignment.
- We propose an intermediate domain between the source images and the target video frames to reduce the domain discrepancy, thereby facilitating the cross-domain alignment.
- Extensive experiments on several datasets demonstrate that our method achieves better performance than existing methods, validating the effectiveness of modeling the instance relationships via a spatial-temporal graph.

2 Related Work

Video object detection is vulnerable to motion blur, illumination variation, occlusion and scale changes. To address this problem, many methods have been proposed to utilize temporal context for detection, roughly falling into two categories: box-level propagation and feature-level propagation. The box-level propagation methods [9,10] explore bounding box relations and apply temporal post-processing to suppress false positives and recover false negatives. T-CNN [10] incorporates temporal and contextual information from tubelets obtained in videos which dramatically improves the baseline performance of existing still image detection frameworks. Seq-NMS [9] uses high-scoring object detections from nearby frames to boost scores of weaker detections within the same clip. The feature-level propagation methods [3,19,23] use the temporal coherence on features to solve the problem. FGFA [23] improves the per-frame features by the aggregation of nearby features along the motion paths, and thus improves the video detection accuracy. Based on FGFA, MANet [19] jointly calibrates the features of objects on both pixel-level and instance-level in a unified framework. MEGA [3] augments proposal features of the key frame by effectively aggregating global and local information. All these methods of video object detection heavily depend on the manually per-frame annotations of bounding box coordinates and categories that is usually time-consuming and labor-expensive.

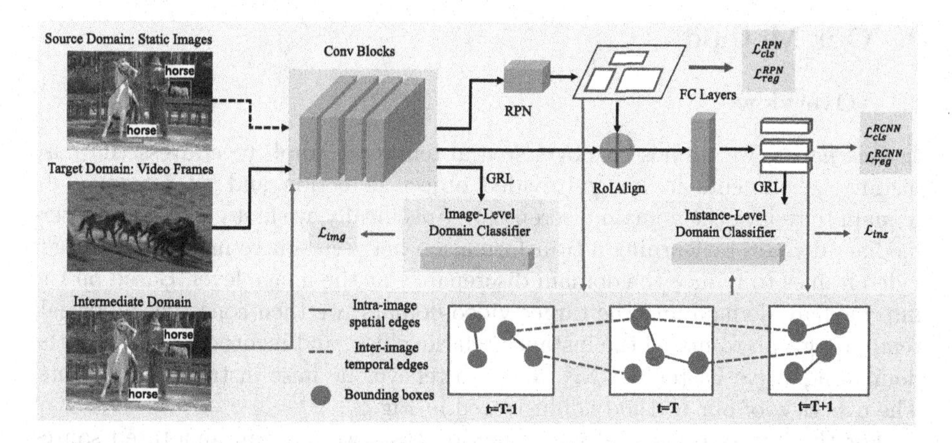

Fig. 2. Overview of our framework. The top of the framework is a basic image-to-video detector based on [4]. The bottom of the framework is our proposed spatial-temporal message propagation (STMP) and intermediate domain, which serves as a plug-and-play component of basic image-to-video detector for positive alignments on both image and instance levels. The source images linked with a dotted line are utilized to generate the intermediate domain and not used for the training of the image-to-video detector.

To break the dependency on the large-scale annotated video frames, image-to-video object detection is proposed to leverage existing annotated images to help object detection in unannotated videos. Several image-image object detection methods have been proposed in recent years. DA-Faster [4] is a prominent and effective approach for domain adaptive object detection, which introduces two domain classifiers on both image and instance levels to alleviate the performance drop caused by domain shift. SW-Faster [17] is an improvement of DA-Faster, which proposes a novel approach based on strong local alignment and weak global alignment. SCDA [22] improves DA-Faster by replacing the plain image-level alignment model with a region-level alignment model. HTCN [1] extends the ability of previous adversarial-based adaptive detection methods by harmonizing the potential contradiction between transferability and discriminability. These cross-domain object detection methods can also be used to reduce the domain gap between images and video frames in image-to-video object detection.

In the aforementioned methods, there often exists false alignments between instances across the domains due to appearance variations and motion blur. In this paper, we attempt to handle the false instance alignments by taking full advantage of spatial and temporal information within and across frames in videos to enhance primal instance representations, thus boosting the positive instance alignments.

3 Our Method

3.1 Overview

In this paper, we propose a novel spatial-temporal graph to address the false instance alignments in image-to-video object detection and an intermediate domain to reduce the domain discrepancy. Specifically, we first generate an intermediate domain by learning a transformation between source images and target video frames to reduce the domain discrepancy at the image level. Based on the intermediate domain and the target video domain, we then construct a spatial-temporal graph to model the instance relationships, and incorporates it into the domain adaptive Faster R-CNN [4,17] to relieve the false instance alignments. The overview of our method is illustrated in Fig. 2.

For the image-to-video object detection task, we have an annotated source image domain of N_s images, denoted as $\mathcal{D}_s = \{\boldsymbol{x}_i^s, y_i^s\}_{i=1}^{N_s}$, and an unannotated target video domain that consists of N_t video frames, denoted as $\mathcal{D}_t = \{\boldsymbol{x}_i^t\}_{i=1}^{N_t}$. The region proposals are generated via a region proposal network. Let $B_i^s = \{b_{i,j}^s|_{j=1}^{N_s^b}\}$ represent a set of region proposals in $\boldsymbol{x}_i^s$, where N_s^b is the number of region proposals in $\boldsymbol{x}_i^s$. Let $B_i^t = \{b_{i,j}^t|_{j=1}^{N_t^b}\}$ represent a set of region proposals in $\boldsymbol{x}_i^t$, where N_t^b is the number of region proposals in $\boldsymbol{x}_i^t$.

3.2 Intermediate Domain

Due to the large discrepancy between the source and target domains, the performance of detector degrades substantially. In this paper, we propose to generate an intermediate domain $\mathcal{D}_f$ to bridge the source and target domains by learning to translate the source images into the target video frames. A typical and powerful image-to-image translation network, CycleGAN [21], is employed to learn a transformation between the source image domain $\mathcal{D}_s$ and the target video domain $\mathcal{D}_t$. Since ground truth labels are only accessed for source domain, we merely consider the transformation from source images to target frames after training CycleGAN and then translate the source domain $\mathcal{D}_s$ into an intermediate domain $\mathcal{D}_f = \{\boldsymbol{x}_i^f, y_i^s\}_{i=1}^{N_s}$. $\mathcal{D}_s$ and $\mathcal{D}_f$ are similar in image content, but diverges in visual appearances, while $\mathcal{D}_f$ and $\mathcal{D}_t$ differ in image content, but have similar distributions on the pixel-level. Therefore, our intermediate domain constitutes an intermediate feature space distributed neutrally between the source and target domains. As shown in Fig. 1, we give some visualization results of the generated intermediate-domain images.

3.3 Domain Adaptive Faster R-CNN

Based on the generated intermediate domain $\mathcal{D}_f$ and the target domain $\mathcal{D}_t$, We utilize DA-Faster [4] to enable a basic image-to-video object detection model, which consists of an object detector (Faster R-CNN [16]) and an adaptation module.

Faster R-CNN is a two-stage detector mainly consisting of three main components: a deep convolutional neural network (*i.e.* "Conv Blocks" in Fig. 2) to extract features, a region proposal network (*i.e.* "RPN" in Fig. 2) to generate region proposals, and a full connected network (*i.e.* "FC Layers" in Fig. 2) to focus on bounding box detection and regression. The loss function of Faster R-CNN is summarized as

$$\mathcal{L}_{det} = \mathcal{L}_{cls}^{RPN} + \mathcal{L}_{reg}^{RPN} + \mathcal{L}_{cls}^{RCNN} + \mathcal{L}_{reg}^{RCNN}. \quad (1)$$

The adaptation module aims at aligning the distribution between the intermediate domain and the target domain at both image and instance levels, which consists of an image-level domain classifier D_{img} and an instance-level domain classifier D_{ins}. Specifically, let d_i denote the domain label of the i-th training image either in the intermediate domain or the target domain, with $d_i = 0$ for the intermediate domain and $d_i = 1$ for the target domain. By denoting the output of D_{img} located at (u, v) as $p_i^{(u,v)}$, the image-level adaptation loss can be written as

$$\mathcal{L}_{img} = \sum_{i,u,v} \left[d_i \log p_i^{(u,v)} + (1 - d_i) \log \left(1 - p_i^{(u,v)} \right) \right]. \quad (2)$$

Let $p_{i,j}$ denote the output of the instance-level domain classifier D_{ins} for the j-th region proposal in the i-th image. The instance-level adaptation loss can be written as

$$\mathcal{L}_{ins} = \sum_{i,j} [d_i \log p_{i,j} + (1 - d_i) \log (1 - p_{i,j})]. \quad (3)$$

To align the domain distributions, the parameters of image-level and instance-level domain classifiers should be optimized to minimize the above corresponding domain classification loss, while the base network should be optimized to maximize the training loss. For the implementation, the gradient is reversed by Gradient Reverse Layer (GRL) [7] to conduct the adversarial training between (D_{img},D_{ins}) and the base network of Faster R-CNN.

3.4 Spatial-Temporal Instance Relationships Construction

To avoid the false instance alignments during the learning of domain adaptive Faster R-CNN, we propose a spatial-temporal graph to model the contextual relationships between instances on both the spatial and temporal dimensions. As shown in Fig. 2, an intra-image spatial sparse graph and an inter-image temporal sparse graph are successively constructed for spatial-temporal message propagation. After message propagation, we obtain more accurate instance representations, which are fed into the instance-level domain classifier D_{ins} for positive instance alignment. Note that the inter-image temporal sparse graph is an extension of the intra-image spatial sparse on the temporal dimension.

Intra-image Spatial Sparse Graph Construction. We construct an intra-image spatial sparse graph for spatial message propagation. Specifically, for an

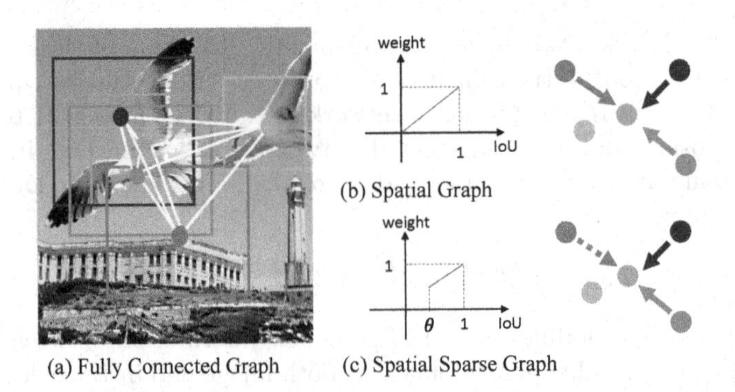

(a) Fully Connected Graph (b) Spatial Graph (c) Spatial Sparse Graph

Fig. 3. Different choices of constructing a graph to encode spatial relationships: (a) Fully connected graph: implicitly learning a fully connected graph between region proposals. The learned graph is redundant and ignores spatial information between region proposals. (b) Spatial graph: using Intersection Over Union (IoU) between region proposals to learn a spatial graph. However, there still exist redundant edges, which are useless to get a better instance representation, and even damage the primal features. (c) Spatial sparse graph: a spatial sparse graph is learned via adding constraints on the sparsity of a spatial graph.

image $\boldsymbol{x}_i^f$ in the intermediate domain $\mathcal{D}_f$ or a frame $\boldsymbol{x}_i^t$ in the target domain $\mathcal{D}_t$, we structure its region proposals as an undirected graph $\mathcal{G} = (\mathcal{V}, \mathcal{E})$. $\mathcal{V}$ is the set of nodes corresponding to the proposal set B_i^f of $\boldsymbol{x}_i^f$ or B_i^t of $\boldsymbol{x}_i^t$ and each node is corresponding to one proposal in the proposal set. $\mathcal{E} \in \mathcal{V} \times \mathcal{V}$ is the set of edges and represents the relationships between proposals within an image or a video frame. Intuitively, two spatially neighboring proposals are likely to represent the same object and should be highly correlated. Intersection Over Union (IoU) is a broadly used metric that measures the spatial correlation of two proposals by their bounding box coordinates. Hence, we employ IoU to construct a spatial-aware adjacency matrix $\mathbf{A}$, formulated as

$$A_{j,k} = \mathrm{IoU}\left(b_{i,j}, b_{i,k}\right) = \frac{b_{i,j} \cap b_{i,k}}{b_{i,j} \cup b_{i,k}}, \tag{4}$$

where $A_{j,k}$ is the element in the j-th row and the k-th column of $\mathbf{A}$. $b_{i,j}$ and $b_{i,k}$ denote the j-th and k-th region proposals in $\boldsymbol{x}_i^f$ or $\boldsymbol{x}_i^t$, respectively. Although some irrelevant edges have been removed by using spatial information, the subsequent graph still over-propagates information and leads to confused proposal features. To solve this problem, we impose constraints on graph sparsity by only retaining the edges between region proposal $b_{i,j}$ and $b_{i,k}$ only if $\mathrm{IoU}\left(b_{i,j}, b_{i,k}\right)$ is greater than a threshold θ_{spt}. In other words, for a node $b_{i,j}$, we select some relevant nodes as its neighborhoods, formulated as

$$\mathrm{Neighbour}(\mathrm{Node}\ b_{i,j}) = \{b_{i,k} | \mathrm{IoU}(b_{i,j}, b_{i,k}) > \theta_{spt}\}. \tag{5}$$

As shown in Fig. 3, compared to a fully connected graph between region proposals, our spatial sparse graph picks up relevant neighborhoods for each node, which greatly reduces the noise of irrelevant nodes and leads to low computation cost.

Inter-image Temporal Sparse Graph Construction. An inter-image temporal sparse graph is constructed to model the instance relationships on the temporal dimension, where the optical flow between consecutive video frames is utilized to improve the coordinates of region proposals from neighboring frames. Given two adjacent video frames $\boldsymbol{x}_{i-1}^t$ and $\boldsymbol{x}_i^t$ with their region proposals B_{i-1}^t and B_i^t, the temporal region proposal propagation is formulated by

$$B_{i-1\to i}^t = \text{flowbox}(B_{i-1}^t, \boldsymbol{\mathcal{F}}_{i-1\to i}), \tag{6}$$

$$\overline{B}_i^t = [B_{i-1\to i}^t,\ B_i^t], \tag{7}$$

where $\boldsymbol{\mathcal{F}}_{i\to i+1}$ is the optical flow map between $\boldsymbol{x}_{i-1}^t$ and $\boldsymbol{x}_i^t$, and flowbox() is a propagation function to generate pseudo proposals for $\boldsymbol{x}_i^t$ by adding the mean flow vectors to the region proposals in $\boldsymbol{x}_{i-1}^t$. After propagation, we construct an inter-image temporal sparse graph for the current frame $\boldsymbol{x}_i^t$ with region proposals $\boldsymbol{x}_i^t$. Different from the intra-image sparse spatial graph introduced above, the set of proposals $\mathcal{V}$ contains not only the proposals from its self frame, but also the pseudo proposals from adjacent frames. Accordingly, $\mathcal{E}$ denotes the set of edges between these proposals. We constantly use the IoU metric to form an adjacency matrix $\mathbf{A}$ and put constraints on sparsity by a threshold θ_{tmp}.

Spatial-Temporal Message Propagation. By utilizing the intra-image spatial sparse graph and inter-image temporal sparse graph constructed above, we conduct spatial-temporal message propagation (STMP) to achieve accurate instance representation. With the assistance of adjacency matrix $\mathbf{A} \in \mathbb{R}^{N^b \times N^b}$ ($N^b = 256$ for spatial graph, while $N^b = 512$ for temporal graph), region proposals either from the intermediate or target domain of proposal features $\mathbf{F} \in \mathbb{R}^{N^b \times d}$ (d is the dimension of proposal feature) are aggregated by

$$\widetilde{\mathbf{F}} = \mathbf{D}^{-\frac{1}{2}}\mathbf{A}\mathbf{D}^{-\frac{1}{2}}\mathbf{F}, \tag{8}$$

where $\mathbf{D} = \sum_k A_{j,k}$ is the diagonal degree matrix of $\mathbf{A}$. After aggregating the intra-image and inter-image adjacent proposals, proposal feature $\widetilde{\mathbf{F}} \in \mathbb{R}^{N^b \times d}$ is discriminative enough and expresses more accurate instance-level information especially for low image quantity cases. Compared to the conventional graph convolution [12], we leave the trainable weight matrix $\mathbf{W}$ out. After STMP, we use $\widetilde{p}_{i,j}$ to denote the output of the instance-level domain classifier D_{ins} for the j-th region proposal in the i-th image in the intermediate domain or target domain. The instance-level alignment loss (*i.e.*, Eq. (3)) is rewritten as

$$\mathcal{L}_{ins}^{STMP} = \sum_{i,j} [d_i \log \widetilde{p}_{i,j} + (1 - d_i) \log (1 - \widetilde{p}_{i,j})]. \tag{9}$$

Objective Function. The overall objective function for jointly learning the domain adaptive Faster R-CNN and spatial-temporal message propagation is formulated by

$$\mathcal{L}_{DA-STMP} = \mathcal{L}_{det} + \lambda(\mathcal{L}_{img} + \mathcal{L}_{ins}^{STMP}), \tag{10}$$

where λ is a balance hyper-parameter to control the relative importance of detection and adaptation, and is set to 0.1 in our experiments.

4 Experiment

4.1 Datasets

To evaluate the effectiveness of our method, we conduct experiments on three public datasets, including PASCAL-VOC (VOC) [6], COCO [14], and Youtube-Objects (YTO) [15]. With these three datasets, we construct two image to video transfer tasks: VOC→YTO and COCO→YTO.

VOC→YTO. The VOC has 20 categories and consists of about 5,000 training images with bounding box annotations. The YTO is a sparsely annotated video frame dataset for video object detection, which has about 4,300 frames for training and 1,800 frames for test with 10 categories. There are 10 common categories between VOC and YTO, including "aeroplane", "bird", "boat", "car", "cat", "cow", "dog", "horse", "bike", and "train". The images of the common 10 categories on the VOC dataset are used as the source image domain. We use unannotated sparse training frames associated with their adjacent six frames as the target video domain.

COCO→YTO. The COCO is a large-scale real-world image dataset with 80 object categories. We randomly select 4,000 images of the common 10 categories between COCO and YTO from the training set as the source domain. For the target domain, we use the same setting as the VOC→YTO task.

4.2 Experiment Settings

Baselines and Comparison Methods. In our experiment, DA-Faster [4] and SW-Faster [17] are adopted as our baseline detectors. The modified overall objectives of DA-Faster and SW-Faster are both equipped with Eq. (9). We also compare our method with Faster R-CNN [16] that directly adapts the model trained on images to videos and several other image-to-video object detection methods [2,13,20].

Implementation Details. The instance-level domain classifier D_{ins} for "SW-Faster" is constructed by three full-connected layers ($4096 \rightarrow 100 \rightarrow 100 \rightarrow 2$) and the first two layers are activated by the ReLU [8] function. For the domain classifiers in "SW-Faster" and "DA-Faster", we follow the settings in original papers [4,17]. The learning ratio of each domain classifier to the backbone network of Faster R-CNN is set as $1:1$, *i.e.*, setting the parameter of GRL layer as 1. We adopt the VGG-16 [18] pretrained on ImageNet [5] as the backbone of Faster R-CNN, and finetune the overall network with a learning rate of 1×10^{-3} for 50k iterations and then reduce the learning rate to 1×10^{-4} for another 30k iterations. Each batch consists of one image from the intermediate domain and one video frame from the target domain. We employ RoIAlign (*i.e.*"RoIAlign" in Fig. 2) for RoI feature extraction. As for the training of CycleGAN, we set the batch size to 1, and adopt the Adam optimizer [11] with a momentum of 0.5 and an initial learning rate of 0.0002. For evaluation, both the Average Precision (AP) of each category and the mean Average Precision (mAP) of all categories are computed with an IoU threshold 0.5 for both two transfer tasks.

Table 1. Experimental results (%) on the VOC→YTO task.

Methods	Aero	Bird	Boat	Car	Cat	Cow	Dog	Horse	Bike	Train	mAP
Faster R-CNN [16]	75.0	90.4	37.3	71.4	58.0	52.8	49.1	42.2	62.8	39.2	57.8
CycleGAN [13]	78.5	**97.2**	31.5	72.3	66.3	59.7	45.9	43.6	66.3	49.4	61.1
SW-ICR-CCR [20]	79.3	95.2	36.9	**75.6**	58.7	61.4	38.7	45.4	66.6	42.6	60.0
SIR [2]	78.9	95.3	31.1	66.1	61.3	56.3	48.1	42.7	64.6	34.4	57.9
DA-Faster [4]	77.1	96.8	31.8	72.5	60.3	59.4	39.6	43.0	63.7	40.3	58.4
DA-Faster-inter-STMP	81.2	95.4	**42.6**	71.4	**66.5**	**72.0**	48.2	49.3	67.1	47.1	64.1
SW-Faster [17]	81.2	95.2	29.4	74.4	56.4	55.3	41.2	44.3	66.5	40.9	58.5
SW-Faster-inter-STMP	**84.4**	95.8	39.2	75.1	64.9	61.7	**53.8**	**50.6**	**68.4**	**52.9**	**64.7**

Table 2. Experimental results (%) on the COCO→YTO task.

Methods	Aero	Bird	Boat	Car	Cat	Cow	Dog	Horse	Bike	Train	mAP
Faster R-CNN [16]	57.9	91.4	29.6	68.9	51.9	51.4	**64.2**	55.2	63.2	52.3	58.6
CycleGAN [13]	75.5	87.8	37.6	69.4	52.3	62.8	61.0	57.8	64.4	58.0	62.7
SW-ICR-CCR [20]	69.8	90.6	34.6	**72.4**	54.9	61.1	58.0	58.3	**66.4**	47.5	61.4
SIR [2]	65.6	91.6	25.8	67.6	47.6	60.7	54.8	54.3	60.9	50.1	57.9
DA-Faster [4]	**84.9**	**93.3**	32.2	71.6	61.7	**66.9**	47.8	47.9	64.8	43.6	61.5
DA-Faster-inter-STMP	78.3	92.7	**41.8**	69.7	60.1	64.5	54.7	55.4	63.8	63.3	**64.4**
SW-Faster [17]	71.7	90.7	29.9	71.6	53.0	60.8	59.3	**58.8**	60.7	56.8	61.3
SW-Faster-inter-STMP	81.8	91.3	30.1	70.2	**64.5**	60.5	58.2	55.6	64.3	**67.4**	**64.4**

4.3 Results

VOC→YTO. Table 1 shows the comparison results on YTO. First, our method outperforms all the compared methods on mAP, clearly demonstrating the effectiveness of our proposed method. Second, our proposed intermediate domain and spatial-temporal message propagation consistently boosts the performance of "DA-Faster" and "SW-Faster" detectors with gains of 5.7% and 6.2% on mAP, respectively. Third, it is noteworthy that for some difficult categories, "DA-Faster" and "SW-Faster" perform worse than "Faster R-CNN", probably due to that there exist false alignments across domains and the performance of domain adaptive detectors is limited. As for the false alignment categories in "SW-Faster" such as "boat", "cat" and "dog", "SW-Faster-inter-STMP" improves them by 9.8%, 8.5% and 12.6%, respectively. The performance drops a lot on the false alignment categories such as "boat" and "dog" in "DA-Faster", and

Table 3. Ablation studies (%) on VOC→YTO and COCO→YTO tasks.

Methods	mAP	
	VOC→YTO	COCO→YTO
DA-Faster [4]	58.4	61.5
DA-Faster-inter	61.9	61.9
DA-Faster-inter-SMP	63.4	62.8
DA-Faster-inter-STMP	64.1	64.4
SW-Faster [17]	58.5	61.3
SW-Faster-inter	60.6	62.2
SW-Faster-inter-SMP	64.1	63.5
SW-Faster-inter-STMP	64.7	64.4

"DA-Faster-inter-STMP" can greatly improve these difficult categories by 10.8% and 8.6%, respectively.

COCO→YTO. As shown in Table 2, our method outperforms all the compared methods on mAP. Moreover, we observe that our proposed framework improves "DA-Faster" and "SW-Faster" by 2.9% and 3.1%, respectively. Similar to the observation on the VOC→YTO task, we significantly improve the detection result of some false alignment categories such as "bike" by 3.6% for "SW-Faster" and "dog" by 6.9%, "horse" by 7.5%, "train" by 19.7% for "DA-Faster". In addition to these difficult categories, we further promote positive alignments in other simple categories, which validates the effectiveness of our method.

4.4 Ablation Study

To evaluate the effectiveness of each component, we conduct ablation studies on both the VOC→YTO and COCO→YTO tasks. The results are shown in Table 3, where "inter", "SMP", and "STMP" denote the intermediate domain, spatial message propagation, and spatial-temporal message propagation, respectively.

Effect of the Intermediate Domain: To evaluate the effect of the intermediate domain, we compare "DA-Faster-inter" with "DA-Faster" and "SW-Faster-inter" with "SW-Faster". For the VOC→YTO task, we observe that "DA-Faster-inter" and "SW-Faster-inter" achieve 3.5% and 2.1% improvements over "DA-Faster" and "SW-Faster", respectively. Similar improvements can be found for the COCO→YTO task, clearly demonstrating the effectiveness of the intermediate domain on promoting positive alignments between the source and target domains.

Effect of the Spatial Message Propagation: To evaluate the effectiveness of the spatial message propagation, we compare "DA-Faster-inter-SMP" with "DA-Faster-inter" and "SW-Faster-inter-SMP" with "SW-Faster-inter". Their difference is whether handling false instance alignments by propagating spatial message within an image or video frame. From the results, "DA-Faster-inter-SMP" achieves better results compared to 'DA-Faster-inter" by spatial-temporal

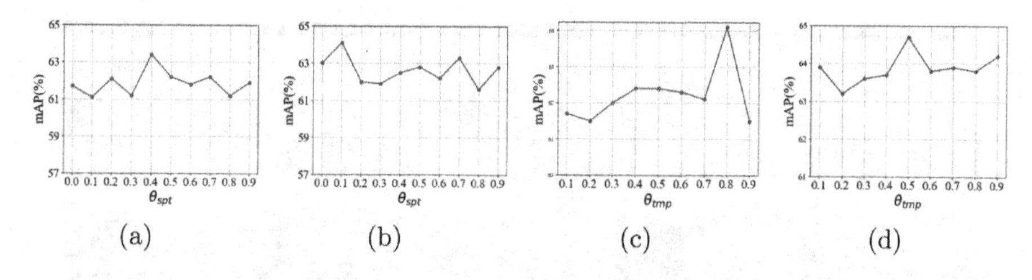

Fig. 4. Analysis on spatial graph sparsity threshold θ_{spt} and temporal graph sparsity threshold θ_{tmp} on the VOC→YTO task. (a) DA-Faster-inter-SMP, (b) SW-Faster-inter-SMP, (c) DA-Faster-inter-STMP, (d) SW-Faster-inter-STMP.

message propagation on two transfer tasks. Similar improvements are achieved with "SW-Faster" as the base detector. These improved results strongly validate the effectiveness of spatial message propagation in relieving the false instance alignments.

Effect of the Temporal Message Propagation: To evaluate the effectiveness of the temporal message propagation, we compare "DA-Faster-inter-STMP" with "DA-Faster-inter-SMP" and "SW-Faster-inter-STMP" with "SW-Faster-inter-SMP". From the results shown in Table 3, "SW-Faster-inter-STMP" works better than "SW-Faster-inter-SMP" for VOC→YTO and COCO→YTO, respectively. Also, "DA-Faster-inter-STMP" outperforms "DA-Faster-inter-SMP". It validates that temporal message propagation can contribute to better instance representations for improving instance-level alignment.

4.5 Parameter Analysis

To analyze the influence of the spatial graph sparsity threshold θ_{spt} and the temporal graph sparsity threshold θ_{tmp} on spatial-temporal message propagation, we conduct experiments using "DA-Faster" and "SW-Faster" as the baseline detectors for the VOC→YTO task. We select θ_{spt} and θ_{tmp} in the range of {0.1, 0.2, 0.3, 0.4, 0.5, 0.6, 0.7, 0.8, 0.9}, and show the mAP-threshold curve in Fig. 4, where the horizontal axis represents the value of θ_{spt} or θ_{tmp} and the vertical axis represents the mAP.

We conduct experiments on "DA-Faster-inter-SMP" and "SW-Faster-inter-SMP" to analyze the performance of θ_{spt}. From the results in Fig. 4(a) and (b), we can find that small and large spatial graph sparsity thresholds θ_{spt} both lead to the decreasing of mAP. This is probably because that the smaller the spatial graph sparsity θ_{spt}, the more noise will be introduced and the larger the spatial

Table 4. Setting of spatial-temporal graph sparsity thresholds for both VOC→YTO and COCO →YTO tasks.

Base detector	θ_{spt}	θ_{tmp}
DA-Faster	0.4	0.8
SW-Faster	0.1	0.5

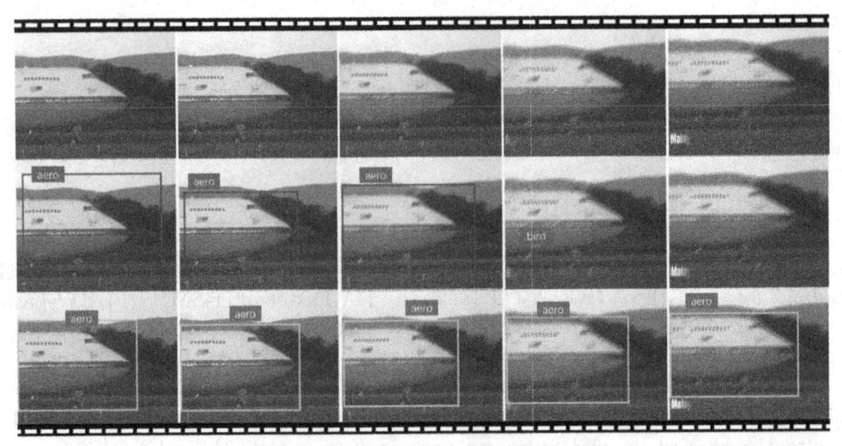

(a) Missing detections

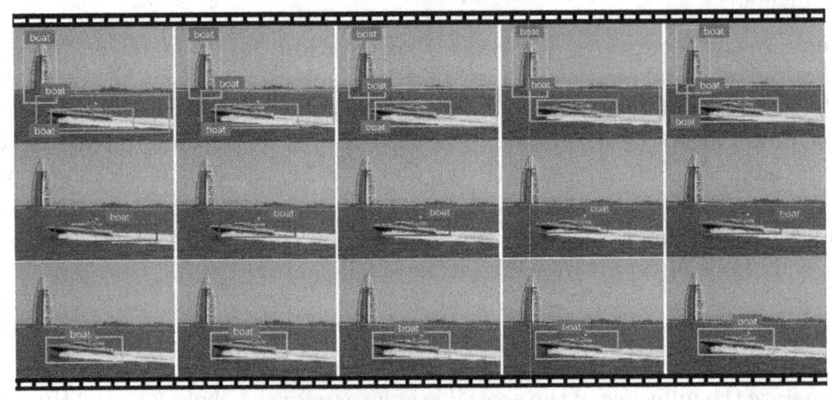

(b) Redundant detections

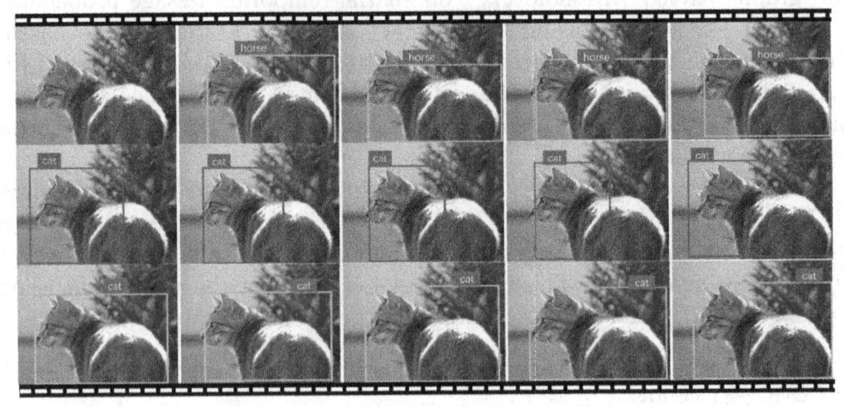

(c) Wrong detections

Fig. 5. Detection examples on the COCO→YTO task. Bounding boxes in blue, red, and green denote the detection results of "SW-Faster", "SW-Faster-inter-STMP" and ground truth, respectively. (**Best viewed in color.**) (Color figure online)

graph sparsity θ_{spt}, the less message will be propagated between instances to form better instance representations. Based on the experimental results, we set $\theta_{spt} = 0.4$ using the base detector of "DA-Faster" and $\theta_{spt} = 0.1$ using the base detector of "SW-Faster" to balance message propagation and noise filtering. To analyze the performance of θ_{tmp}, we use the θ_{spt} selected before to conduct experiments on "DA-Faster-inter-STMP" and "SW-Faster-inter-STMP". From the results in Fig. 4(c) and (d), we select $\theta_{tmp} = 0.8$ and $\theta_{tmp} = 0.4$ as the sweet spot between message propagation and noise filtering for "DA-Faster" and "SW-Faster", respectively. For the COCO→YTO task, we use the same θ_{spt} and θ_{tmp}. Table 4 gives a detailed summary of the graph sparsity thresholds using two base detectors.

4.6 Qualitative Analysis

Figure 5 shows some detection examples of the COCO→YTO task by "SW-Faster" and "SW-Faster-inter-STMP (Ours)". There are three examples of five consecutive frames from YTO. As shown in Fig. 5(a), the base object detector fails to detect the blurred "aeroplane", while our method could partially recover the false negatives. It probably benefits from our proposed intermediate domain module that reduces the domain discrepancy. As shown in Fig. 5(b) and (c), the base object detector misclassifies the background tall building into a "boat", and misclassifies the foreground "cat" as a "horse". However, our method perfectly solves the redundant and wrong detections. This is probably because our proposed spatial-temporal message propagation module can successfully relieve the false instance alignments. In general, our method achieves more accurate detection results under the domain shift with poor image quality.

5 Conclusion

In this paper, we have presented a novel spatial-temporal graph to exploit spatial-temporal contextual relationships between object instances for alleviating the false instance alignments in image-to-video object detection. The generated intermediate domain can bridge the source image domain and the target video domain. With this intermediate domain and the target video domain, an intra-image spatial sparse graph and an inter-image temporal sparse graph have been constructed to enable the spatial-temporal message propagation, which can enrich the instance representation according to the guidance of spatial-temporal contextual. Extensive experiments on several datasets have demonstrated the effectiveness of our method.

Acknowledgments. This work was supported in part by the Natural Science Foundation of China(NSFC) under Grant No 62072041.

References

1. Chen, C., Zheng, Z., Ding, X., Huang, Y., Dou, Q.: Harmonizing transferability and discriminability for adapting object detectors. In: Proceedings of the IEEE/CVF Conference on Computer Vision and Pattern Recognition (CVPR), pp. 8869–8878 (2020)
2. Chen, J., Wu, X., Duan, L., Chen, L.: Sequential instance refinement for cross-domain object detection in images. IEEE Trans. Image Process. (TIP) **30**, 3970–3984 (2021)
3. Chen, Y., Cao, Y., Hu, H., Wang, L.: Memory enhanced global-local aggregation for video object detection. In: Proceedings of the IEEE/CVF Conference on Computer Vision and Pattern Recognition (CVPR), pp. 10337–10346 (2020)
4. Chen, Y., Li, W., Sakaridis, C., Dai, D., Van Gool, L.: Domain adaptive faster R-CNN for object detection in the wild. In: Proceedings of the IEEE Conference on Computer Vision and Pattern Recognition (CVPR), pp. 3339–3348 (2018)
5. Deng, J., Dong, W., Socher, R., Li, L.J., Li, K., Fei-Fei, L.: Imagenet: a large-scale hierarchical image database. In: 2009 IEEE Conference on Computer Vision and Pattern Recognition (CVPR), pp. 248–255. IEEE (2009)
6. Everingham, M., Van Gool, L., Williams, C.K., Winn, J., Zisserman, A.: The pascal visual object classes (VoC) challenge. Int. J. Comput. Vis. (IJCV) **88**(2), 303–338 (2010)
7. Ganin, Y., Lempitsky, V.: Unsupervised domain adaptation by backpropagation. In: International Conference on Machine Learning (ICML), pp. 1180–1189. PMLR (2015)
8. Glorot, X., Bordes, A., Bengio, Y.: Deep sparse rectifier neural networks. In: Proceedings of the Fourteenth International Conference on Artificial Intelligence and Statistics (AISTATS), pp. 315–323. JMLR Workshop and Conference Proceedings (2011)
9. Han, W., et al.: Seq-NMS for video object detection. arXiv preprint arXiv:1602.08465 (2016)
10. Kang, K., et al.: T-CNN: tubelets with convolutional neural networks for object detection from videos. IEEE Trans. Circ. Syst. Video Technol. (TCSVT) **28**(10), 2896–2907 (2017)
11. Kingma, D.P., Ba, J.: Adam: a method for stochastic optimization. arXiv preprint arXiv:1412.6980 (2014)
12. Kipf, T.N., Welling, M.: Semi-supervised classification with graph convolutional networks. arXiv preprint arXiv:1609.02907 (2016)
13. Lahiri, A., Ragireddy, S.C., Biswas, P., Mitra, P.: Unsupervised adversarial visual level domain adaptation for learning video object detectors from images. In: 2019 IEEE Winter Conference on Applications of Computer Vision (WACV), pp. 1807–1815. IEEE (2019)
14. Lin, T.-Y., et al.: Microsoft COCO: common objects in context. In: Fleet, D., Pajdla, T., Schiele, B., Tuytelaars, T. (eds.) ECCV 2014. LNCS, vol. 8693, pp. 740–755. Springer, Cham (2014). https://doi.org/10.1007/978-3-319-10602-1_48
15. Prest, A., Leistner, C., Civera, J., Schmid, C., Ferrari, V.: Learning object class detectors from weakly annotated video. In: 2012 IEEE Conference on Computer Vision and Pattern Recognition (CVPR), pp. 3282–3289. IEEE (2012)
16. Ren, S., He, K., Girshick, R., Sun, J.: Faster R-CNN: towards real-time object detection with region proposal networks. In: Advances in Neural Information Processing Systems (NeurIPS), vol. 28, pp. 91–99 (2015)

17. Saito, K., Ushiku, Y., Harada, T., Saenko, K.: Strong-weak distribution alignment for adaptive object detection. In: Proceedings of the IEEE/CVF Conference on Computer Vision and Pattern Recognition (CVPR), pp. 6956–6965 (2019)
18. Simonyan, K., Zisserman, A.: Very deep convolutional networks for large-scale image recognition. arXiv preprint arXiv:1409.1556 (2014)
19. Wang, S., Zhou, Y., Yan, J., Deng, Z.: Fully motion-aware network for video object detection. In: Proceedings of the European Conference on Computer Vision (ECCV), pp. 542–557 (2018)
20. Xu, C.D., Zhao, X.R., Jin, X., Wei, X.S.: Exploring categorical regularization for domain adaptive object detection. In: Proceedings of the IEEE/CVF Conference on Computer Vision and Pattern Recognition (CVPR), pp. 11724–11733 (2020)
21. Zhu, J.Y., Park, T., Isola, P., Efros, A.A.: Unpaired image-to-image translation using cycle-consistent adversarial networks. In: Proceedings of the IEEE International Conference on Computer Vision (ICCV), pp. 2223–2232 (2017)
22. Zhu, X., Pang, J., Yang, C., Shi, J., Lin, D.: Adapting object detectors via selective cross-domain alignment. In: Proceedings of the IEEE/CVF Conference on Computer Vision and Pattern Recognition (CVPR), pp. 687–696 (2019)
23. Zhu, X., Wang, Y., Dai, J., Yuan, L., Wei, Y.: Flow-guided feature aggregation for video object detection. In: Proceedings of the IEEE International Conference on Computer Vision (ICCV), pp. 408–417 (2017)

Author Index

Y. Zheng et al. (Eds.): ACCV 2022, LNCS 13848, pp. 377–378, 2023.
https://doi.org/10.1007/978-3-031-27066-6

S

T

V

W

Y

Z

Printed in the United States
by Baker & Taylor Publisher Services